THE
NUCLEAR
AXIS

Other books by Zdenek Červenka

The Organization of African Unity and Its Charter
The Nigerian War 1967-70
Landlocked Countries of Africa
The Unfinished Quest for Unity: Africa and the OAU

Other books by Barbara Rogers

White Wealth and Black Poverty: American Investments in Southern
  Africa
Divide and Rule: South Africa's Bantustans

THE NUCLEAR AXIS

Secret collaboration between
West Germany and South Africa

Zdenek Červenka and Barbara Rogers

NYT
Times
BOOKS

Manufactured in the United States. Published simultaneously in
Canada by Fitzhenry & Whiteside, Ltd., Toronto.

First published in 1978 in Great Britain by Julian Friedmann Books Ltd.

Library of Congress Cataloging in Publication Data
  Červenka, Zdenek, 1928-
  The nuclear axis.
  Includes index.
  1. Atomic power industry – Germany, West.   2.   Atomic power
industry – Africa, South.   3.   Germany, West – Foreign relations –
Africa, South.   4.   Africa, South – Foreign relations – Germany,
West.   5.   Atomic power – International control.   I.   Rogers,
Barbara, 1945-   joint author.   II.   Title.
HD9698.G42C43   1978   327.43'068   77-93055
ISBN 0-8129-0760-4

Contents

Contents

Illustrations

Photographs (follow page 208)
The nuclear-industrial complex in West Germany *(chart)*
South African Embassy, Bonn
SA Foreign Minister H. Muller and Ambassador D. B. Sole
A.J.A. Roux
Georg Leber
Pelindaba/Valindaba
Pelindaba
Tank containing separation nozzle tubes
Separation tube containing nozzles being inserted into tank
Inside Pelindaba

Illustrations

Acknowledgements

I would like to thank Dieter Habicht-Benthin, Director of the German-Arab-African Bureau and of the publishing branch of the Bureau, Progress Dritte Welt (PDW), for allowing us to reproduce the documents and photographs published in 3 WELT MAGAZIN and in 'Israel-South Africa: Co-operation of Imperialistic Outposts'. My thanks also go to Ingeborg Wick of the Anti-Apartheid Movement in Bonn, Peter Ripken of the Information Center on Southern Africa (ISSA) and Gesa Liethschmidt of Bonn University. A number of German friends whose help and advice on the first part of the book was indispensable prefer not to be named.

I am also greatly indebted to the collective of researchers at the Stockholm International Peace Research Institute (SIPRI), namely Dr Frank Barnaby (the Director), Dr Bhupendra Jasani, Dr Jozef Goldblad, Signe Landgren-Bäckström and Jean-Louis Sainz for their comments and advice on the manuscript and for allowing me to quote extensively from SIPRI publications and to use SIPRI files.

I also would like to thank Colin Legum of *The Observer*, Frene Ginwala of the African National Congress (ANC),

Acknowledgements

Professor Larry Bowman of the University of Connecticut, researchers at the Swedish Ministry of Defence and members of the Swedish diplomatic service for their advice and comments on the various political, scientific and military aspects raised in the book.

Finally, my thanks go to the Scandinavian Institute of African Studies for providing me with an excellent research base for my writing and to my colleagues both for their help with my work and their patience with my race against the deadline.

Z.C.

I am indebted to many friends and colleagues for their contributions, suggestions and comments. I was generously given access to the files of the United Nations in New York, the Natural Resources Defence Council (NRDC) in Washington DC, the American Committee on Africa (ACOA) in New York, and the Wiener Library in London.

Invaluable material was contributed by Friends of the Earth in Australia and Britain; Wolff Geisler and the Anti-Apartheid Bewegung in Bonn; Stephen Salaff in Toronto; Jean-Bernard Curial in Paris; Alexandre Kum'a N'dumbe in Lyon; Jack Spence of the University of Leicester; Anthony Wilkinson, Patrick Keatley, Martin Walker and Ann Fullerton in London; Ruurd Huisman in Groningen, the Netherlands; Sean Gervasi in New York; and in Washington DC, Jacob Scherr of the NRDC, Robert Alvarez of the Environmental Policy Center, Dan Matthews of the African Bibliographic Center, James Morrell of the Center for International Policy, Helen Hopps of the Institute for Policy Studies, Chris Root of the Washington Office on Africa, and Harleigh Ewell. Officials of the United States Government, who prefer to remain anonymous, have also been very helpful.

Documents and related information were supplied by the Department of Information and Publicity of the South West African People's Organisation (SWAPO) of Namibia. I am

x

most grateful to them and to all those, named and unnamed, who helped in the investigation of this issue.

Finally, and most important of all, this book would never have appeared without the understanding and personal commitment of Julian Friedmann, Alison Burns and Jean Beith of Julian Friedmann Books.

B.R.

Publisher's Note

The Nuclear Axis is a joint venture by the two authors, but, for the record, each bears major responsibility for their own section. Zdenek Červenka contributed Part One, together with Chapters 10 and 11, the Postscript and Appendix 1. Barbara Rogers contributed Part Two, except for the sections in Chapter 4 on Namibia, the Rössing mine and West German involvement there, which were written by Roger Murray. She also contributed Chapters 8 and 9.

The Publisher would like to thank Al, Jean, Marion, Paul, Dick, Roger, John, Bob and particularly Barbara, for their unstinting help in getting the book out so quickly under pressure.

Thanks are also due to Peter Davis, whose television documentary for WGBH Boston and Swedish TV, entitled *South Africa: The Nuclear File*, provides excellent background to the subject of this book, and to the *Sunday Times*, London, who shared with us the results of their world-wide investigations into this issue.

Finally, special thanks are due to David, who was instrumental in foiling the attempted theft of the manuscript at the 1977 Frankfurt International Book Fair.

Foreword

*by Frank Barnaby, Director of the Stockholm
International Peace Research Institute.*

The *Nuclear Axis* has been written at a time when preparations
are under way for one of the most important international
conferences of our time: the Special Session of the UN
General Assembly on Disarmament, due to take place in May
1978 in New York.

There is growing apprehension about the possibility of a
nuclear war. The severe shortcomings of nuclear deterrent
doctrines are becoming increasingly clear. In both the US and
the USSR, groups which think in terms of a nuclear war being
feasible may be gaining political influence. Certain qualitative
developments in offensive and defensive strategic weapons
and in tactical nuclear weapons may encourage the notion
that a nuclear war is both fightable and 'winnable'. That such
a notion, if ever put to the test, is likely to be proved wrong is
cold comfort.

Some South African politicians seem to regard nuclear
weapons as a source of power and prestige, locally and in the
world at large. In this respect they belong to the category of
politicians interested in acquiring a nuclear military capability,
who claim that world security would be better served if

countries other than the United States, USSR, Great Britain, France and China had their own nuclear forces. It is argued, for example, that nuclear weapons may enhance a country's independence by preventing more powerful countries, even the Great Powers, from interfering in its affairs. Are not these weapons, it is asked, the great equalizers?

But of what real use would nuclear weapons be to the South African régime? The Reverend Dr C. F. Beyers Naudé, Director of the Christian Institute of South Africa, and Mr Horst Kleinschmidt, an official of the Institute now in exile in the Netherlands, stated in 1976, one year before it became public, that South Africa was about to test a nuclear device:

> 'There is in our opinion no doubt that the development of a nuclear industry in South Africa is undesirable at this point of time. We make this statement because we believe that access to nuclear power by the South African Government would strengthen their hand not only against the voteless majority within South Africa but also against economically captive communities around South Africa. The development of nuclear energy in South Africa would consequently constitute an increase in the existing threat to the peace of the sub-continent. A white minority government in South Africa with nuclear technology would be even more disdainful of the aspirations of both the black majority and Organisation of African Unity demands.'

Professor Ronald W. Walters, Director of the Social Science Research Center of Howard University, discussed this issue in testimony before the Sub-committee on Africa of the House Committee on International Relations. He emphasised the volatile nature of South Africa's domestic and international situation and reminded us that nuclear weapons are designed for use against human targets:

> 'The growing sophistication of nuclear weapons suggests their eventual use, especially where it can be determined that maximum control of destruction might be attained

xiii

without self-inflicted, mutual damage. In this case, the military targets would be outside the territory of South Africa but significant enough to warrant the use of such force. Significant targets would be transport systems, large troop bases, food sources or agricultural works, small villages and towns, and industrial installations.'

It is a sad discovery made by the authors of *The Nuclear Axis* that it is the Federal Republic of Germany, the first country to have embarked on a purely peaceful nuclear-power programme, which has provided South Africa with a key to the manufacture of fissionable material by helping it develop uranium-enrichment technology. Despite South African denials, it would appear that the uranium-enrichment method used by the South Africans is based on technology developed at the Karlsruhe Nuclear Centre in West Germany.

According to *The Nuclear Axis* the study conducted by Steag and Ucor on the comparison of the South African and West German uranium-enrichment processes implied that both processes were closely scrutinized and whatever each party did not know about the other's secrets it learned in the course of the comparative study.

The claim repeatedly made by the West German Government, that the 'jet-nozzle' process is not suitable for the manufacture of enriched uranium for nuclear weapons, is not correct. For use in nuclear weapons, the concentration of uranium-235 in uranium has to be increased to 40% or more. In reactor fuel, enrichment of up to 4% is normal. Weapons-grade uranium could be produced in a plant built for the production of reactor fuel by recycling the uranium gas many times.

The Nuclear Axis shows how easily international efforts aimed at stopping the proliferation of nuclear weapons can be thwarted and how commercial considerations are often given priority over international security. The case of South Africa is of particular concern because of the mounting conflict between independent Africa and the white minority régimes in Southern Africa.

A South African nuclear force would drastically alter the situation on the African continent, which until now has been a 'nuclear-free' zone, with most of its countries wishing to keep it that way. But if South Africa acquires a national nuclear force then other African countries are likely to wish to counter the threat by acquiring nuclear weapons.

Some African countries already have experience in nuclear technology, others plan to get it. Zaire has been operating a research reactor at Kinshasa since 1959. Egypt has had a research reactor at Inshas, near Cairo, since 1961. The Egyptian reactor, supplied by the USSR, is effectively not subject to international or bilateral safeguards. Libya is constructing a research reactor imported from the USSR; it has a bilateral agreement with Argentina for nuclear co-operation — a country which has not signed the NPT. And in 1977 Nigeria announced plans to build a nuclear-power reactor.

A number of African countries have significant resources of cheap uranium. Apart from South Africa, with one of the world's largest deposits, the Central African Empire, Gabon, Niger and Zaire also have substantial amounts of cheap uranium; and other African countries almost certainly have exploitable uranium reserves yet to be discovered.

Unless effective measures are adopted not only to prevent the spread of weapons but also to prevent their use by the states which already have them, the possibility of a nuclear war remains dangerously near.

Many people believe that civilisation as we know it is most likely to survive if nuclear proliferation is controlled.

The Nuclear Axis shows that this is no longer enough. Its conclusions convincingly support the SIPRI argument that unless all existing nuclear weapons are destroyed our very survival is at stake.

Stockholm, February 1978

PART ONE   WEST GERMANY

1

The Ambassador loses a file

Wednesday morning 24 September 1975 held a promise of another fine day in Europe's warmest summer for a century. By 9 o'clock the sun had already won its daily battle with the smog and was climbing into the blue skies above the tops of the Siebengebirge hills, still wrapped in the grey veil of haze. In Bonn the rush-hour had already passed and the city was settling down to the leisurely pace of provincial life. Across the river the narrow roads of Bad Godesberg, once a quiet spa suburb of Bonn, today the world's largest diplomatic village with a community of over 10,000, entitled to tax-free drinks, cigarettes and petrol, were still jammed with CD number-plated cars.

On the corner of Auf der Hostert, an olive-green armoured-car of the Federal Border Guard covering the area drove up the pavement to allow a black Mercedes to enter the drive in No.3, the South African Embassy. It is a large dark brown building of steel and concrete, with tinted bullet-proof windows. Situated next to the run-down garden of the Portuguese Embassy, its front commands a magnificent view over the Rhine valley, while its back faces the small row of buildings which include the Uruguayan and Bulgarian embassies. The entrance from Auf der Hostert is protected by a thick stone wall and its ground floor is shielded by an iron grid pulled over the windows and glass doors. Its functional, austere architecture is reminiscent of a mental hospital or of a police headquarters built to resist a siege.

On the second floor, in a spacious office, Ambassador Donald Bell Sole was sitting behind his desk. He was reading through the draft reports on the visit of his Minister, who had left for Pretoria the previous night after three weeks in the Federal Republic of Germany. Sole had every reason to be pleased with himself. Dr Hilgard Muller's visit was a great diplomatic success and it was Sole who had prepared the ground. The Minister had seen almost everyone who had any say in West Germany policy towards South Africa. Relations between the two countries were good in all respects. In 1974 West Germany overtook all other Western powers to become South Africa's biggest trading partner with a volume of trade totalling over £620 million. Half of all South African credits, over £400 million, were financed by German banks; three-quarters of this amount was provided by three private banks – Deutsche Bank, Dresdner Bank and Commerzbank. West Germany was catching up with Britain as the third biggest investor in South African industry, which in turn was rewarding its investors with an 18% profit. South Africa's détente policy towards Black Africa enjoyed full West German political support.

Pretoria was now expecting Sole's suggestions for a follow-up to Dr Muller's visit. Sole's most immediate task was to persuade Foreign Minister Hans Dietrich Genscher to accept Muller's invitation to South Africa. Sole was going through a memo dealing with a dinner he had given on 11 September at the Hotel Steigenberger in Bonn. It was an important 'working dinner' for the Minister. The small party of German guests was made up of ex-Minister of Defence and Foreign Affairs, Dr G. Schröder; General Lemm; the Quartermaster General of the *Bundeswehr*, Herr Lichtenberg; the Chairman of the Commerzbank, Herr Hansen; the Chairman of Bayer AG; and Secretaries of State Haunschild and Rohwedder — all good friends of Sole and of his country.

At 11 a.m. his secretary, Miss Prinsloo, informed him that Herr Heine from *Der Stern* was on the telephone. Heine wanted to know if it was true that General Günther Rall (of the *Luftwaffe*) had been to South Africa in 1974 as a guest of the South African Ministry of Defence. Sole apparently told him that it was a personal trip and that the Embassy had not had anything to do with it.

However, Heine had in his possession a copy of a secret cable sent on 13 August 1974 by Sole to the South African Embassy in London, which he described to Sole, and the telephone conversation ended. Sole was not particularly disturbed about Heine's allegation. A number of people did know about Rall's trip, too many perhaps: the travel agent in Cologne, Embassy people in Brussels and London, and somebody could have seen Rall in South Africa. He asked his secretary to bring him Rall's file so that he could check on his own copies of all the cables. But the file was missing.

Ten minutes later Sole's deputy, John Becker, the Military Attaché, General Pieter Bosman, and his deputy, Captain F. S. Bellingan, were summoned to Sole's office and told about the disappearance of the file. Their first task was an immediate enquiry without alerting any of the staff. A call was made to Inspector General Zimmerman of the *Bundeswehr* informing him of a possible leak and asking him to notify Minister Georg Leber (Minister of Defence). An attempt was also made to get in touch with General Rall to ensure that he stuck to the original story that it was a strictly private visit he had made to South Africa. The *Verfassungschutz* (Federal Security Bureau) headquarters were informed and were asked to send someone to help with the enquiry.

What was discovered was that 9 files were missing – in effect a car-load of classified documents. An urgent coded cable was sent to the Foreign Ministry in Pretoria with a copy to the Bureau of State Security (BOSS). Within an hour Prime

Minister John Vorster and General van den Bergh of BOSS learned about one of the most serious breaches in South African security.

On the following morning, 25 September, a brown registered envelope addressed to Ambassador Sole arrived at the Embassy. It contained a glossy brochure. On the cover was a photograph of the mushroom cloud of a nuclear explosion and it was entitled *The Nuclear Conspiracy*. The African National Congress (ANC) of South Africa was the publisher.

Anti-apartheid booklets, leaflets and posters are not a novelty in the incoming mail of the South African Embassy. No matter to whom at the Embassy they are addressed, they are all passed on to the information department headed by Counsellor Nicolaas Willem du Bois and rarely reappear. *The Nuclear Conspiracy* would possibly never have reached Ambassador Sole had not Miss Prinsloo, who signed for the mail, perused the brochure out of curiosity. She was stunned by what she saw. Reproduced were photocopies of six letters and documents from the Embassy's missing files. They were quoted as evidence for the ANC charge against the Government of the Federal Republic of Germany that it was helping South Africa to produce a nuclear bomb.

The South African Embassy was not the only recipient of the booklet. The Anti-Apartheid Movement distributing the brochure on behalf of the ANC sent it to every Embassy in Bonn, to all Ministries, members of Parliament and to all major West German newspapers and institutions – a total of 3,000 copies.

From 8.30 a.m., the opening hour of the Embassy, the telephones at the switchboard began to ring. Most of the callers asked for the Ambassador, but he was not available. He was reading the ANC brochure and was worried not only about what was in it but also what was left out. There was no mention of the trip by General Rall that *Der Stern*'s reporter had asked him about. This suggested that there was more to come.

The Ambassador loses a file

One week later, on 2 October, a fresh copy of *Der Stern*, a weekly magazine with a circulation of over 1 million, appeared on the news-stands in Germany. Inside was a full-page picture of *Luftwaffe* General Günther Rall in uniform. The story was about how a *Bundeswehr* General had travelled incognito to South Africa to discuss questions relating to NATO and nuclear research with the South African Government.

On 6 October, General van den Bergh of BOSS arrived in Bonn to supervise the investigations personally. The whole network of the South African Embassy's contacts was suspect. It would be necessary to comb all the files of every institution with which the Embassy corresponded. These files contained not only letters received and copies of letters sent, but also handwritten memos and notes as well as personal correspondence that could not be checked. The intelligence officers sent by the *Verfassungschutz* were eager to help but suspected that the South Africans were not telling them the whole truth. The four BOSS men who arrived from Pretoria to join the task-force of the *Verfassungschutz* found that the four people who had access to the files at the Embassy (Ambassador Sole, his deputy Becker, and their secretaries Miss Prinsloo and Miss van den Bergh) were all above suspicion. Their interrogation yielded no results.

'How did the Bonn bungle happen?', asked the Johannesburg *Rand Daily Mail* in its editorial on 7 October:

> '... Amid the accusations, statements, denials and counter-denials, what is dismayingly evident is that there has been a major breach of South African security.
>
> It is not only one piece of paper that has gone missing, but a series of important documents – with, it seems, even more to come. Documents marked "secret", documents dealing with intimate dealings between South Africa's Ambassador and highly-placed Germans ... all are seeing the embarrassing light of day.
>
> It is unforgiveable. No country can afford to have its secret papers filched in this way – and particularly not South Africa, with its already delicate international

5

position.

For it must be realised that, apart from the damage being done to South Africa's relations with Germany – which is now our major trading partner and therefore of especial significance to us – much wider harm can ensue. Who wants to enter into quiet dealings and negotiation with a country that cannot ensure the integrity of its confidential State papers?'

One vague trace led to the Cologne-based removal firm Roggendorf and to the frontier police at St Augustin. The BOSS men were looking into the possibility of a link between the disappearance of the files and the moving of the South African Embassy from Cologne to Bonn-Bad Godesberg in April 1975.[1]

The investigation at Röggendorf did not yield any results and at St Augustin the inquiries were met with indignation from *Oberstleutnant* Ulrich Wegener, who produced a letter from South Africa's acting Military Attaché, Captain F. S. Bellingan, dated 17 April 1973. Captain Bellingan, who personally supervised the move, had conveyed his and Ambassador Sole's thanks for the help of *Oberstleutnant* Wegener for what he described as 'an impressive operation carried out without any problems', and expressed a hope that 'should I need your help again I am sure that with these soldiers one can always get the best possible help and protection'. (A section of this letter, showing the good relations between the South African Embassy and the *Bundesgrenzschutz*, the elite corps of the West German armed forces, is in Document 22.) In order to appease disturbed white opinion in South Africa, in 1976 BOSS arrested Breyten Breytenbach, a South African writer, who had left his Paris exile to pay a visit to his home country. Among the 'confessions' alleged to have been made by Breytenbach was that the theft of the documents had been organized by members of the African National Congress.[2]

Ambassador Donald Bell Sole himself survived the storm and stayed in Bonn until May 1977. On Friday 6 May, the

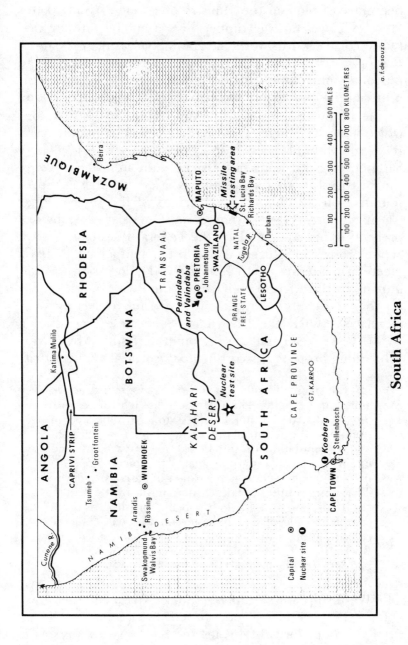

South Africa

a. f. de souza

7

Secretary of State at the Ministry of Foreign Affairs, Peter Hermes, gave a farewell dinner for Sole and his wife at which many 'cordial words were said about Sole', as the *Frankfurter Rundschau* of 7 May 1977 put it. Hermes praised Sole's personal contribution to the 'constructive development in mutual relations despite the existing differences in opinion'.

Sole is now South African Ambassador in Washington, which shows how highly Pretoria values his nuclear diplomacy. It is not without interest that his successor Kurt Robert Samuel von Schirnding has a background in nuclear energy similar to Sole's. He had also served as South Africa's representative on the Board of Governors of the International Atomic Energy Agency in Vienna. Only six weeks before presenting his credentials to the Federal President, Walter Scheel, Herr von Schirnding had lost his temper at the Board meeting in Vienna on 16 June 1977. The occasion was the moving by the African members of the Board of a proposal not to re-elect South Africa to the Board and to elect Egypt in its place.[3] When the Chairman, Senegalese Ambassador A. M. Cissé, read out the list omitting South Africa, von Schirnding protested, according to the official minutes, that:

> '. . . the group of States which had resolved to take action aimed at depriving South Africa of its rightful position on the Board was not, needless to say, interested in upholding the Statute of the Agency; it was not interested in the reputation of the Board for impartiality and integrity; it had only one goal – the pursuance of a narrow political vendetta, regardless of the consequences. That was a ruthless political action in which all considerations of law and fact were being brushed aside, a completely illegal action based on internal domestic considerations which were of no more concern to the Board than were the domestic policies of the States which had initiated the action.'[4]

The Western powers supported South Africa, but their total votes – 13 – were not enough. The African proposal was carried by a majority of 4 votes and South Africa was struck from the list of the candidates for the Board which was to be put to the General Conference in September 1977.

The troubles continued in Bonn. When von Schirnding was presenting his credentials on 25 July 1977, *Der Spiegel* marked the occasion by publishing a story headlined 'What is the truth?', which began as follows:

> 'The hot stuff goes over the border in car boots. For several months now official cars of the South African Embassy in Bonn with CD plates cross the German-French custom check-points quite unmolested. Our source knows what they carry: precision jet-nozzles manufactured by a big Munich-based engineering company. They are destined for the *apartheid* Republic. Their purpose: by means of these jet-nozzles, used in the uranium enrichment process, the natural uranium can be converted into reactor fuel or into fissionable material for bombs. Export of such equipment without special permission is illegal.'[5]

The same article revealed other illicit traffic:

> 'On 20 December 1976 crates weighing 8.5 tons were loaded on a freighter at the port of Bremen. Consignment papers stated the contents to be "Engines with transmissions and accessories". Destination: PROKURA DIESEL SERVICE LTD, DURBAN, South Africa. Enquiries about PROKURA brought negative results. No company of such a name is listed in the directory of Durban firms. And yet the crates with engines have not been returned with a note "addressee unknown", hence they must have reached the South African recipient.
>
> The clue to the mystery has been provided by the engines themselves: engines of this type are used for high-speed Navy launches. Export of such machinery is strictly forbidden.'

And this is not all. *Der Spiegel* also recalled the case of the delivery to South Africa of four Airbus aircraft jointly produced by West Germany and France. The delivery of these planes was defended by the Foreign Ministry at the time

9

on the grounds that they were 'purely civilian aircraft'. 'What the Foreign Ministry omitted', wrote *Der Spiegel*, 'was that all four Airbuses were equipped with special fuel tanks used for Air Force jet-fighter refuelling in the air.'

Thus the same day that he was officially installed, Ambassador Kurt von Schirnding was obliged to issue his first denial: that the entire article in *Der Spiegel* rested on hearsay and not a word of it was true. The story was based on the same documents the South Africans were so desperately trying to trace, the documents which also led to the writing of this book.

2

The rise of West Germany as a nuclear power

West Germany's first five years: from an occupied territory to ally of the Occupying Powers

> 'There is no substitute for the German alliance. German manpower, industrial capacity, location, and will are all of paramount importance to the United States. The capabilities of the Federal Republic are unparalleled among the other allies of America and are thus of central importance to the continued operation of the Atlantic Alliance. From the point of view of both conventional and nuclear defense the Ruhr, the Rhineland, the central territories of Germany, and the stamina of a nation, strong to counter aggression, have an importance sufficiently obvious to render explanation superfluous.'[1]
>
> Eleanor Lansing Dulles

The establishment of the Federal Republic of Germany on 28 May 1949 resulted from the failure of the United States, the Soviet Union, Great Britain and France to agree on a common postwar policy towards Germany, which they had jointly defeated four years ealier. The disagreements between the Western powers and the Soviet Union, as well as discord among the Western powers themselves, extended to nearly every aspect of the occupation: the crucial issue of Germany's unity, the reconstruction of the economy, claims for reparations

11

and the political structure of a future German state.[2]

The decision to fuse American, British and French zones of occupation, and to establish the Federal Republic of Germany, was made at a conference of the US, Britain, France and the Benelux countries (Belgium, the Netherlands and Luxembourg), convened on 23 February at India House in London. General Lucius D. Clay, the Supreme Commander of the Allied Forces in Europe, led the American delegation and made it clear that the US was determined to go ahead with the establishment of West Germany even if France continued to oppose it.[3]

But the French delegation gave way to the combined American-British pressure and agreed to the plan for a federation of German *Länder* occupied by the three Western powers. The implementation of the agreement made in London took place as the rapidly worsening relationship between the Western powers and the Soviet Union reached its nadir with the Soviet blockade of Berlin in June 1948, to which the West responded by mounting an air-bridge to the Western sectors of Berlin.

On 1 September 1948 the Parliamentary Council, a political body set up by the agreement in London and consisting of the representatives of all the *Länder* of Germany occupied by the Western powers, met to elect its Executive. The Presidency went to Konrad Adenauer, the 72-year-old leader of the Christian Democratic Union (CDU), a political party which he had founded in the British zone of occupation. He had started his political career as a City Council official in Cologne, becoming Mayor of the city in 1917. His administration took Cologne to the brink of bankruptcy, and the opposition party succeeded in removing him in 1933 for abuse of power. The Nazis, who are held responsible in some of Adenauer's official biographies for his being deposed, actually had nothing to do with his departure from office.

In March 1945 the Americans reinstated him as Mayor of Cologne but on 6 October of the same year a British brigadier in charge of military and civil affairs in North Rhine-

Westphalia fired him 'for obstruction and non-cooperation'. In November 1945 the ban on his political activity was lifted and from then on his political star rose. On 23 May 1949 the Parliamentary Council adopted the Constitution of the Federal Republic of Germany, which came into force five days later. Dr Adenauer was elected first Chancellor of the Federal Republic, a post he held until 1963.

Concurrently with the transformation of the Western occupational zones into the Federal Republic, American military strategists were re-appraising the situation in Europe. The first indication of the hopes the United States were entertaining about Germany came from the American Secretary of State, James F. Byrnes. On 6 September 1946 he addressed an assembly of the American and German administrators of the US zone of occupation in Stuttgart. He told his audience that '. . . the American people wanted to help the German people to win their way back to an honorable place among the free and peace-loving nations of the world.'[4] Those Americans most concerned to help were wearing generals' uniforms. They were considering US policy towards Europe in the light of a new strategic concept that included the possibility of a Third World War. While never doubting American military supremacy over the Soviet Union they were worried about the massive Soviet ground forces which they feared were capable of over-running much of Europe, at least in the first phase of a war. This could provide the Soviets with the industrial resources of the Ruhr and Saar and the excellent naval facilities of the French ports, as well as large quantities of Allied military supplies.

In order to attack the Soviet Union the US would have to invade Eastern Europe. To the American generals this appeared to be a risky and costly venture. They concluded that it would be better to hold Western Europe. 'Once this was seriously considered,' wrote one of the top RAND specialists on European military strategy, Dr Hans Speier, consultant to the Scientific Advisory Board to the Chief of Staff of the US Air Force, 'the United States had to work toward mobilizing,

13

equipping and organizing ground forces that would be available in Europe at the outbreak of the war.'[5]

West Germany was part of the American plan. In Speier's words

> 'West Germany had a population of fifty million people, while France for example had forty-two and one-half million. German military manpower resources were untapped. Unlike France, Germany had no colonies and no territory to defend in Asia or Africa. Germany's geographical location would place German reserves where they might be needed in the event of a Soviet attack; they would not first have to be moved into the combat area. Former German officers of all ranks, and millions of German men, had had experience in fighting Soviet forces. Millions of Germans had had close contact with Soviet communism, and with the Red army as an occupation force; they were not, therefore, ready victims of communist propaganda, whereas the communist parties of France and Italy had been able to muster considerable popular support in those countries. Finally, the participation of Germany in Western defence could ease the burden of armament which the other European powers had to bear.'[6]

The policy of the disarmament and demilitarization of Germany, agreed upon by the Allied Summit at Potsdam in May 1945, was unilaterally reversed by the United States in the summer of 1950 without consultation with her allies. The Pentagon concept of Western defence in Europe based on a German contribution was advocated by the Secretary of Defence, General George C. Marshall, and by General Omar Bradley for the Joint Chiefs of Staff, emphasizing the need for full exploitation of West German military manpower. This view was first made public on 23 August 1950 by John McCloy, the US High Commissioner for Germany, who announced that 'the defense arrangements for Western Europe must include German participation'.[7]

The details of the plan were revealed to the NATO partners at the meeting of the NATO Council in New York in

September 1950, where the US Secretary of State, Dean Acheson, indicated that unless the Germans were brought into the Western military alliance the United States might not take part in it.

Two approaches towards German re-armament evolved from the American initiative. The first was the Pleven Plan.[8] This constituted a compromise between France's strong opposition to the idea of re-arming its recent enemy, and the need to meet the American demand for integrating West Germany into NATO. France was urged by Britain and the Benelux countries not to oppose this. The Pleven Plan called for the establishment of a 'European Defence Community' based on the creation of a European army into which national contingents would be integrated at the level of the smallest possible units. This meant that there would be no German army, but rather German battalions integrated into European brigades. Provisions for the political integration of Europe were proposed as a pre-requisite for consolidation of Europe's military resources. The force was to be under a European Defence Minister responsible to the European Defence Council (EDC) of Ministers, which in turn was responsible to the European Assembly which was also to approve a common defence budget. The Pleven Plan was silent on the question of the political equality of West Germany with other partners of the EDC. It permitted French forces stationed outside Europe to be exempt from the European command.

A conference to discuss the Pleven Plan was convened in Paris in February 1951. Another meeting was held in Bonn to consider an alternative plan, favoured by the United States and Britain, providing for German membership of NATO. But the French would not hear of it and the EDC approach seemed to be the one likely to succeed. Its failure was later attributed solely to France, but in fact Britain also helped to sink it, although for different reasons. The then British Foreign Secretary, Anthony Eden, while publicly favouring the Pleven Plan, had never seriously considered Britain's

15

supporting it.[9]

The third, then unknown, alternative was the proposal put forward by the German generals who were following political developments in Europe attentively from their enforced retirement. Most of them shared the belief that '... the current period of the cold war is but a phase of a continuing total and global war'.[10] Hans Speier, of the RAND Corporation, interviewed 120 former generals of Hitler's *Wehrmacht* and *Luftwaffe*, and recorded, amongst others, a statement by an unnamed general who told him that he had startled his American interrogators immediately after the war by remarking that 'Germany can now wash her hands of it; it will be up to the Western Allies to cope with the Soviet Union.'[11]

In early October 1950, at the invitation of Adenauer, General von Schwerin convened a meeting of a group of his former colleagues. It included Generals Adolf Heusinger, Hans Speidel and Count Baudissin. These three were the chief draftsmen of a secret memorandum submitted to Adenauer. It bore the title *Memorandum on the build-up of a German contingent within the framework of an international force for the defence of Western Europe*.[12] The military strategy put forward in the memorandum was based on the Western deterrent: '... the greater the provisions made by the West for its own protection the less would the Soviet Union be tempted to adopt an aggressive policy.' For that purpose they proposed to raise twelve German divisions to provide the protective shield for the 12-14 Allied divisions in the rear. The memorandum put forward two basic conditions for German co-operation: (1) the abolition of the Occupation Statute, and the transfer of sovereignty from the Allied powers to the Germans; (2) the review of judgements on war criminals, among whom were many of their fellow officers found guilty of war crimes as defined by the Nürnberg Military Tribunal.

The memorandum also reflected the views expressed as early as 1948 by General Hans Speidel on rearmament. According to General Speidel:

'Troops whose armament was unequal to that of the forces surrounding them not only would suffer from feelings of inferiority but also would prove a weak link, an attractive target for enemy attack. Given Germany's strategic position, these theoretical principles took on a special significance. Troop morale had to be developed among a population educated to demilitarization as well as to denazification; destruction of the German battle-field had to be minimized. Moreover, commanders of second-class troops from a second-class member-state could not expect to exercise more than limited influence over the formulation of the strategic plans they were to execute.'[13]

This view eventually prevailed over the ideas in the Pleven Plan. The influence of General Speidel, former Chief-of-Staff under General Rommel, on NATO military thinking had been remarkable. In 1957 he became the first German commander to hold a major NATO post – that of Commander, Allied Ground Forces, Central Europe, with headquarters at Fontainebleau.

The memorandum was endorsed by the Federal Chancellor who took it up with General Eisenhower, Supreme Commander of NATO Forces in Europe. The precision of the strategy outlined in the memorandum impressed the Pentagon and was subsequently incorporated into the American proposals on the structure of the EDC made by General Eisenhower in June 1951. Adenauer was attracted by the European framework of the EDC, because he regarded European integration as the only way that the Federal Republic could achieve full rehabilitation.

After the collapse of the Third *Reich* in 1945 the German people were led to build up hope in a united Europe. Adenauer had made this hope, later transformed into a quest for a united Germany and an integrated Western Europe, a corner-stone of his foreign policy. He was, however, dissatisfied by some 'discriminatory' features of the EDC and by the fact that Britain wished to remain outside the EDC. When, for example, the Germans demanded that the EDC

17

treaty should contain a clause which would commit EDC forces automatically to resist an attack on any member of the Community, the Netherlands flatly rejected this on the grounds that they did not wish to commit themselves to fight for a German interest without Britain. It took two years to hammer out the final draft of the EDC Treaty.

The signing ceremony was held on 27 May 1952 in the Salon de l'Horloge of the French Foreign Office, the Quai d'Orsay. Under the EDC, the European forces were to consist of several types of national units of divisional strength, integrated from Army Corps level up; that is, each Corps was to consist of divisions of different nationals. The plans in respect of the German contingent included twelve divisions with the appropriate operational headquarters, supporting troops, service units and schools, as well as a tactical air force, naval units and territorial troops. Altogether the Federal Republic was to provide about 500,000 men.

The French insistence on some sort of US-British guarantees of the EDC was met, at the last minute, by the Anglo-American undertaking which read as follows:

> 'Accordingly, if any action from whatever quarter threatens the integrity or unity of the Community, the two Governments will regard this as a threat to their own security. They will act in accordance with Article 4 of the North Atlantic Treaty.'[14]

The 'from whatever quarter' clause was supposed to cover not only the Soviet threat, which was the sole purpose of the whole exercise as far as the Americans were concerned, but also a possibility that '... Germany might at a later stage secede from the EDC and set up her own national army.'[15] But this did not allay French fears about German militarism, and the EDC was never realized. After two years of bitter debates on 30 August 1954, the French National Assembly refused to ratify it by 319 votes to 264, with twelve abstentions. On the last day of the debate, Eduard Herriot, the veteran radical statesman and 'Président d'Honneur' of the Assembly made a moving speech in which he called the EDC 'the end of France' and said:

> 'Who does not see the danger of a provision which already permits Germany to raise a supplementary army of 100,000 or 200,000 men? It was a problem which arose in the time of the Doumergue Cabinet, when Hitler proposed an "entente" with France if she would agree that Germany should raise forces of this type. . . . It was from these forces that the *Wehrmacht* was raised – an army, which, though possessing certain professional traditions of honour, was accompanied during the occupation of France by butchers who massacred our brothers – a fact which I cannot forget.'[16]

Adenauer, who guided the EDC Treaty through the *Bundestag* with difficulty on 19 May 1953, by a majority of 58, was deeply disappointed by the French vote. Churchill and Eden were annoyed; Dulles, the American Secretary of State, was furious. In a statement issued by the State Department on 31 August 1954 Dulles called the decision of the French a 'tragedy' brought about by 'nationalism abetted by Communism' (a reference to the fact that the EDC treaty was rejected on the strength of 99 Communist votes). He issued a thinly veiled threat, saying that '. . . the French negative action . . . obviously imposes on the USA the obligation to re-appraise its foreign policies, particularly those in relation to Europe.' He reiterated the commitment made on 30 July 1954 by the US Senate, which unanimously adopted a resolution calling for the restoration of sovereignty to Germany, '. . . to enable her to contribute to the maintenance of international peace and security.'[17]

Adenauer, comforted by Dulles's speech and by a personal message from Sir Winston Churchill delivered to him by Sir Frederick Hayer Millar, the British High Commissioner in Germany, was more guarded in expressing his sorrow. In a broadcast on 4 September 1954 he called the situation created by the French rejection of the EDC Treaty 'grave' and said that improving the situation would demand '. . .great wisdom

19

and careful consideration'. He disclosed that new negotiations with the US and Britain on this issue had already begun.

The conflict over the European Defence Community between 1950 and 1954 was one example of the United States' clash with France over European policy. Washington wanted the rearmament of West Germany and an integrated Western defence (under American command), while Paris insisted on armament control (especially that of West Germany) and on the right of France to have a say in shaping European policy. In the postwar years there has been an important difference between the attitude of the United States, and that of the Europeans (both West and East) towards West Germany. The difference stemmed from the European experience with Nazi Germany's conquest of Europe. While to most Americans the reign of terror by the *Gestapo* and *SS*-Stormtroopers, and the devastation of the war, were dramas played out on film-screens and on the pages of their national papers, to the Europeans they were an unforgettable and horrible personal experience which did not fade easily. At the time of the debates about the EDC, the Nazi atrocities were overshadowed in the minds of Americans by their violent fear of communism, which went well with the McCarthy mood in the United States. It was 'the Reds' who were made to seem like America's worst enemies. But in Europe Nazi war crimes were still an open wound. As a result there was a sharp contradiction between the attitude towards the successors of the Third *Reich* adopted by the Americans and that held by its former devastated subjects. In short, the United States saw West Germany as a future ally while Europe still saw her as yesterday's enemy.

There were, of course, differences among the Western Europeans themselves. The British Government, if not the people, was certainly more ready than the others to 'forgive and forget'. British lenience had very little effect on France, which resented the haste with which the Americans had pushed through the rehabilitation of Germany.

However, these differences were minor in relation to the

unanimity in the Western Alliance about confronting the 'Iron Curtain' countries. It was the bitter anti-communism of the time which ultimately ensured West Germany's re-armament and rehabilitation in NATO without the programme of denazification being carried out, as originally intended at the end of the war. West Germany is now the most ardently anti-communist of all Western countries, a stance originating in its Nazi past. Its problem now is no longer one of joining the Western Allies, but one of ensuring its own survival in the event either of a withdrawal of other NATO troops from its soil, or of being destroyed in its position as a buffer state, a future battlefield for the East-West Conflict.

How Adenauer's non-nuclear weapons pledge secured West German membership of NATO

In most contemporary writings on the history of West Germany, the pledge made by Chancellor Konrad Adenauer – 'The Federal Republic of Germany undertakes not to manufacture in its territory . . . atomic, biological and chemical weapons'[18] — has been regarded as a gesture demonstrating the peaceful intentions of the State. In reality it was a shrewd manoeuvre on the part of the West German Chancellor and the US Secretary of State, John Foster Dulles, who jointly wrote the script for the ceremony. Its setting was as follows:

The place: Lancaster House, London
The date: 1 October 1954
The occasion: The nine-power conference on European Security and German Association in Western Defence (US, UK, France, Canada, Belgium, the Netherlands, Luxembourg, West Germany and Italy).

The purpose: To find a way out of the deadlock caused by the French Assembly's rejection of the proposal for the creation of the European Defence Community (EDC).

The issue: West Germany's accession to the Brussels Treaty of 17 March 1948 which provided for 'collaboration in economic, social and cultural matters and for legitimate self-defence', which was to be extended to become the Western European Union (WEU).

The main difficulty: How to combine French insistence on strict control of West Germany's arms production with American plans for the re-militarization of West Germany.

The deal: West Germany's non-nuclear weapons pledge in exchange for French consent to West German membership of the Western European Union (WEU) and NATO, and termination of the Occupation Statute. The Adenauer/Mendès-France agreement on the Saar on 23 October 1954 helped to clear the way for its approval by the French Parliament.

The London conference was a result of the initiative undertaken by the British Foreign Secretary, Anthony Eden. On Sunday 6 September, when he was contemplating world affairs in the bath at his country residence at Pewsey, Wiltshire, it occurred to him that the EDC could be resurrected through the Brussels Treaty. Signed on 17 March 1948 by Britain, France, Belgium, the Netherlands and Luxembourg, it pledged the signatories to give all military and other aid in their power to any member to the treaty who was subjected to an armed attack: it had been principally directed against a revival of German aggression.[19] By 1954, in other words, the Soviet Union had taken the place of Germany as the potential aggressor, while Germany was to be made an ally of its previous victims. This was a considerable *volte face* by Eden on the matter. Only one year earlier, on 19 March 1953, he had

22

been at pains to explain to the Dutch Foreign Minister, Dr Stikker, over a dinner at the British Embassy in Paris that to extend the Brussels Treaty, concluded with the 'closest European friends and allies', to Germany and Italy was 'inconceivable'. In September 1954 it was suddenly the way to build 'a new political framework for Europe without discrimination'.[20]

Eden set out on a lightning tour of Brussels, Bonn, Rome and Paris. In Bonn he exchanged with Adenauer misgivings about the French who, as both feared, might not realize the dangers inherent in the 'agonizing reappraisal' of American policy of which Dulles kept reminding them. Adenauer was pleased by Eden's plan for bringing Germany into NATO via the extended Brussels Treaty and showed more optimism about the possibility of winning French approval for the idea than Eden did.

Dulles, though kept fully informed by Eden about the progress of his mission, decided to make sure that this time the US plan for the rearmament of Germany would not fail. On 15 September 1954 when Eden was about to leave Rome for Paris, he received a cable from Dulles informing him that Dulles was flying that day to Bonn and asking him to see Dulles two days later in London. The move was typical of John Foster Dulles. He was, in President Eisenhower's own words, '. . . the first Secretary of State to adopt on a major scale the practice of flying personally to trouble spots of the world instead of depending on letters, cables and messages.'[21] Twenty years later Henry Kissinger institutionalised the practice as 'shuttle diplomacy'. Eden thought that Dulles's intervention was 'unhappily timed',[22] and was very disturbed that Dulles had not told him exactly why he wanted to talk to Adenauer. The description of the meeting between Adenauer and Dulles on 15 and 16 September 1954 occupies six pages in Adenauer's memoirs.[23]

Dulles and Adenauer were friends; they liked each other and were working for the furtherance of their respective positions and policies. These had a common foundation in both men's fear of communism, represented by the Soviet Union. They both regarded the Soviet Union as the potential aggressor against Western Europe. In Moscow and the other East European capitals the thinking was just the reverse: West Germany was the future aggressor, and indeed the spectre of West German *revanchism* rearmed by its imperialist accomplice, the United States, has dominated the foreign policies of Eastern Europe since the Second World War. Adenauer clung to the Americans not only because of the US nuclear umbrella, which was at that time regarded as the guarantee of West European security, but also because he realized that the full restoration of German sovereignty was possible only through rearmament and could only be attained through Washington. He had feared, and often warned those who opposed his foreign policy, that the United States might lose interest in West Germany if West Germany failed to meet American expectations of the German military contribution to the Western defence. The almost colonial attitude adopted by the United States, behaving towards Germany like a true 'metropolitan' power, had not come to an end with the establishment of the Federal Republic in 1949.

The replacement of the Military Governors, who had exercised supreme authority in their respective zones (and jointly in matters affecting West Germany as a whole), by Allied High Commissioners, did not have any effect on the severe limitations imposed on the new State even before it came into existence. The International Authority for the Ruhr, set up by the United States, Britain, France and the Benelux countries on 28 April 1949, imposed economic control over the industrial heart of West Germany.[24]

West Germany was also to remain totally disarmed. On 17 January 1949 the Military Governors signed a directive setting up a Military Security Board at Koblenz, '. . . to ensure the maintenance of disarmament and demilitarization in the interests of security'. Accordingly, no provision on defence had been made in the Constitution establishing the Federal Republic. The new Occupation Statute proclaimed by the three Military Governors on 10 April 1949 made it clear that the Western powers would retain the supreme authority over Germany which they had assumed by the Berlin Declaration of 5 June 1945. The Occupation Statute came into force simultaneously with the Constitution.[25]

It was not until 1951 that West Germany was permitted to have its own Foreign Office. Foreign Affairs were run by the *Dienstelle für Auswärtige Angelegenheiten* (Foreign Affairs Desk) set up within the Chancellor's office. Adenauer was his own Foreign Minister until 6 June 1955, when Heinrich von Brentano was given the post. Similarly, the Chancellor's office also contained the so-called *Amt Blank* ('amt' means 'office') which was in charge of defence matters. It was established in October 1950 as an 'Office for the question of increases of allied troops' and headed by Theodor Blank, a Member of Parliament who on 6 June 1955 became West Germany's first Defence Minister.

However, Adenauer's position *vis-à-vis* the United States was not as weak as it appeared on the surface; he knew that Germany had an important role in American policy towards Europe. Dulles's plan, to make Western Europe an impregnable fortress protecting the Western world from communism, was called 'the sword and the shield' in military circles. The 'sword' stood for the nuclear strike-force of Strategic Air Command (SAC), while the US Navy was the 'shield' for the NATO forces in Europe; it was to be guarded by a large conventional force, the core of which was to consist of a 500,000-strong German army capable of halting the Soviet advance until the Strategic Air Command could intervene by deploying its nuclear arsenal.

Dulles intensely disliked the French, as did most American politicians at that time. Adenauer, while on the one hand cultivating Franco-German relations, on the other had skilfully exploited Dulles's hostility towards France by his public display of loyalty to the United States in contrast to Paris's disregard of Washington. France was one of the topics discussed between Dulles and Adenauer on 15 September.

Adenauer listened to the angry tirade about French obstinacy, and approved of Dulles's snub in leaving Paris out of his itinerary on his way to Bonn. But when Dulles told him that he was considering economic sanctions against France he counselled caution: 'One should tighten the tap a little, but not turn it off completely.'[26]

The talks further included such details as Adenauer's proposal for building barracks, airports and training-grounds for the future German army. The 'reliability' of Western allies in the event of a conflict with the Soviet Union was also discussed and Adenauer told Dulles that in his view only Turkey, Greece, Spain (though not a member of NATO), the Benelux countries, and of course West Germany were fully reliable, while Britain could be trusted only to a certain extent. France and Italy were not even mentioned. The most important purpose of Dulles's visit to Bonn — to work out a strategy which would guarantee French approval of Germany's entry into NATO – was only vaguely mentioned in Adenauer's memoirs. And yet a firm undertaking was offered to Adenauer by Dulles, namely:

1. the United States would support the speedy restoration of German sovereignty and full equality with other members of the Western alliance;
2. it would also support the *Bundesrepublik*'s entry to NATO and would undertake to build up the German army by supplying the necessary military equipment.[27]

This suggests that Adenauer convinced Dulles that he would be able to persuade the French to accept it.

Adenauer's own explanation of his 'non-nuclear pledge', made at the London conference, is unconvincing. He claimed to have made the decision entirely on his own because of the difficulty that would have been involved in getting Cabinet agreement, and because the '. . . necessity arose during the course of negotiation'. It is more likely that the matter was discussed with Dulles when Adenauer went to see him off from Bonn at Wahn airport on 16 September. Dulles and Adenauer travelled in the same car with only the German

Secretary of State, Hallstein Platz, present as interpreter. After arriving at the airport they remained seated in the car and concluded the discussion with the following exchange: 'Herr Adenauer, will you come to London when you get the invitation?' asked Dulles. 'Only if you are coming too, Mr Dulles,' answered Adenauer. Dulles smiled and said 'I will go there only if you come too.'[28] The likelihood that the non-nuclear pledge came from Dulles rather than Adenauer was mentioned to the authors by a source close to Adenauer, who recalled a meeting between Generals Speidel and Heusinger with Adenauer in November 1954 at which the two men tried to explain to Adenauer what nuclear weapons, which Adenauer used to refer to as 'just another kind of artillery', were really about.

Adenauer described Dulles's reaction after he made the non-nuclear pledge in the following words:

> 'Dulles rose from his seat on the other side of the Conference table, walked to me and spoke to me in a loud voice so that everybody would hear it. "Mr Chancellor, you have just declared that the Federal Republic of Germany renounces the production of ABC weapons. You meant this declaration, I assume, to be valid only *rebus sic stantibus* as all declarations and obligations in international law are." I answered him equally loudly. "You have interpreted my declaration correctly." The others remained silent.'[29]

None of these loud exchanges was reported by the press at that time, nor were they mentioned by Eden who has described the particular session in detail in his memoirs. The reason behind Adenauer's recollection of this alleged exchange in his memoirs is obvious. *Rebus sic stantibus* is a clause which is not very popular with international lawyers. In effect it means that the State which is undertaking an obligation by signing an international treaty or agreement, or (as in this case) making a unilateral declaration, claims the right to abrogate its international commitment whenever the conditions under which that State agreed to it have changed. The arbitrator of whether or not the conditions have changed is the State itself, thus *rebus sic stantibus* is nothing but a back-door through which the State can evade its obligations the moment they cease to offer political advantages. As Sir Hersh Lauterpacht succinctly put it, '*rebus sic stantibus* has, on occasions, been abused by providing a cloak for lawless violation of treaties.'[30] *Rebus sic stantibus*, moreover, was not the only loophole in Adenauer's solemn pledge. His chief draftsmen, Ambassadors Wilhelm Grewe and Herbert Blankenhorn, had provided a bigger one.

The emphasis on banning the production of nuclear weapons *on German soil* was a formula to dispel French fears that the German arms industry would rise from the ashes to make another bid for the *Grossdeutsches Reich* with another Hitler. What mattered in 1954 to France (and to the whole of Europe for that matter) was what was happening inside Germany. Indeed, who would have thought of a West German extraterritorial nuclear programme at that time? It is very unlikely that men with the experience of Blankenhorn and Grewe would overlook such a weakness in the self-imposed prohibition. While they may not have thought of the possibility of Germany's manufacturing nuclear weapons in another country, they were undoubtedly aware of the possibility that Germany could obtain a nuclear capability from another country. This was precisely the point made by the Gaullist opponents to the agreements during the discussions in the French National Assembly. They argued that firstly, under the accords, Germany was still free to receive nuclear weapons from abroad (the United States being named as the obvious supplier) and that secondly, uncontrolled German development of atomic energy for commercial purposes would, in effect, mean access to atomic weapons.[31] In order to meet the French concern, Chancellor Adenauer sent a letter to the British Foreign Secretary Anthony Eden (convenor and Chairman of the London conference) offering guarantees for the limitation of atomic energy development in the Federal Republic for a two-year period. During this period the Federal Republic would install only one nuclear reactor with a maximum capacity of 10,000 KW, and its stocks of enriched uranium would not exceed 3,500 grammes per year.[32]

The explanation of the draftsmen for the wording of the non-nuclear pledge was that 'We were under such enormous pressure during the negotiations that the matter was not taken into consideration'. This was accepted by Herr Theo Sommer of *Die Zeit*, who went on to describe the idea of purchasing nuclear weapons or producing them elsewhere as 'harebrained' and a formal renunciation of such a possibility as 'unnecessarily hilarious'.[33]

The West German non-nuclear pledge and Britain's commitment to participate in European defence finally won French agreement, with the condition imposed by Mendès-France (Prime Minister and Minister of Foreign Affairs) that an inspection agency be formed to control the armaments of WEU members. The nine-power London Conference ended on 3 October 1954 with full agreement that (a) West Germany and Italy should be admitted to the expanded Brussels Treaty Organization, and (b) West Germany 'freed of the occupation regime and with her sovereignty fully restored should enter NATO'.[34]

Three weeks later, on 23 October 1954, the Foreign Ministers of the nine powers met again in Paris to sign the agreements reached in London. The last obstacle to the signing – the problem of the Saar – was resolved by an almost all-night meeting between Adenauer and Mendès-France on 23 October. Adenauer offered France an agreement by which the Saar was given European status within the framework of the WEU. The *Bundestag* approved the treaties on 27 February 1955 by 473 votes to 314.

On 8 May, the tenth anniversary of the German Armed Forces' unconditional surrender, the Federal Republic became a member of NATO. On 9 May 1955 the black, red and gold flag of West Germany joined those flying on the poles outside NATO headquarters in Paris.

'One of the important gains made by the Federal Republic of Germany by joining the Western alliance,' Konrad Adenauer told the *Bundestag* on 15 December 1954, 'was that Germany became free to engage in civil atomic research and to exploit atomic energy for peaceful purposes.'[35]

West Germany goes nuclear

In Hitler's Germany a major research programme was set up for the development of nuclear weapons – although this never had the priority given to some of the other secret weapons, such as the V-1 and V-2, the Messerschmidt jet fighters and new submarines.[36]

The uranium enrichment project was led by the Hamburg University physical chemists Paul Harteck and Wilhelm Groth, whose joint letter to the Third *Reich*'s War Office of 24 April 1939 called attention to '. . . the newest development in nuclear physics, which, in our opinion, will probably make it possible to produce an explosive many orders of magnitude more powerful than conventional ones.'[37] Most of the nuclear research was concentrated at the Kaiser Wilhelm Institute for Physics in Berlin. It was staffed by top German scientists who, at the outbreak of war, had occupied many of the world's leading positions in atomic physics.

However, most of those who had belonged to the *Uran Verein* ('Uranium Society'), such as Walther Gerlach, Otto Hahn, Werner Heisenberger and Karl Friedrich von Weizsäcker, had fled from Nazi Germany in 1939. The project was also seriously crippled by Heinrich Himmler's frequent and arbitrary arrests of nuclear scientists for suspected disloyalty to the *Führer*, and by their transfer to Himmler's own scientific experiments.

Not much more progress was made during the war at the other laboratories manned by the scientists of the Ministry for

Armaments and Munitions, the Ministry of Education and other research centres of the Third *Reich*. Professor Samuel Goudsmit, who in 1944 led an American scientific mission to Europe which followed the Allied invasion forces, reported that German scientists had not succeeded in separating U_{235} nor come any closer to the construction of a chain-reaction uranium pile which at that time was already operating in the United States.[38] The German programme for nuclear weapons production had been based on the use of natural uranium and heavy water; once the heavy water plant in Norway had been sabotaged by resistance fighters, prospects for developing a nuclear weapon by the end of the war faded.

According to one of the German scientists, von Weizsäcker, the members of the *Uran Verein* regarded the difficulties inherent in the production of an atomic bomb as so formidable that it never occurred to them that the United States might be able to solve the problem so quickly.[39] The question of just how close the Nazis were to developing a nuclear weapon at the end of the war is the subject of some dispute. What is clear, however, is that the basis for subsequent nuclear development after the war had been firmly laid; it led to an unbroken line of further research which has given West Germany a strong lead among the countries of Western Europe in the development of nuclear power, backed by a corps of highly-trained scientists and a financial and industrial network with a history of promoting this novel form of power in spite of various attempts from outside to limit nuclear development in that country.

After the war, nuclear science was officially 'taboo' in West Germany and subject to sweeping restrictions by the occupying powers. Explicit authorisation from the tripartite Allied Military Security Board (MSB) was required for the construction of any facilities capable of separating isotopes of uranium, and even then the permission was to be granted only provided the result did not exceed 1 milligramme of U_{235} within 24 hours. Nevertheless, serious work on nuclear technology, including enrichment, was carried on by the same

31

scientific groups as had been active during the war, with backing from the same industrial conglomerates. Professor Hans Martin of Kiel University, who elaborated counter-current centrifuge theory in 1939 and whose subsequent researches received top-priority funding from the Nazis,[40] published a major study of the centrifuge gas flow profile in 1950.[41] In 1949 the wartime centrifugists under Professor Groth were funded to continue their experiments by the Emergency Association for German Science.[42] Groth is now the senior centrifugist in the Federal Republic.

Leading financial roles in the Emergency Association were assumed in 1949 by representatives of Metallgesellschaft AG and Gutehoffnungshütte AG – both companies heavily involved in West German nuclear industry and the centrifuge 'Troika' (see p. 300). Both, of course, have strong historical and cross-ownership ties with IG-Farben and the industrial complex which helped to ensure Hitler's access to power and then formed a vital part of the Nazi war machine (see below).[43]

After the war it was the industrialist Friedrich Flick, in prison for war crimes at the time,[44] who organised secret uranium prospecting by the Eisenwerk Maxmilianshütte AG in the Fichtengebirge, north-east of Nürnberg, in contravention of occupation regulations. This was done in collaboration with the American authorities,[45] who presumably placed the need for uranium for the American nuclear weapons stockpile above the need to eliminate the traces of Nazi aspirations for a nuclear capability.

When the Allied military administration was first established in the occupied zones of Germany they included a large number of people described by US Senator Kilgore in 1945 as people to whom '... Nazi industrial organization is not repugnant and who had shown every disposition to make their peace with it'.[46] For example, the director of the economic section of the British military government, Sir Percy Mills, had been a member of the Federation of British Industries delegation which had concluded an agreement with the *Reichsgruppe Industrie* at Düsseldorf in March 1939.

His American colleague, William H. Draper, Jr, was a former Treasurer of the American bank whose credit had fostered the prodigious growth of the Vereinigte Stahlwerke as the largest steel trust in Germany.[47]

The German industrial giants such as Siemens, Vereinigte Stahlwerke and IG Farben ended the war considerably less damaged than has been generally realized. The sight of German cities reduced by Allied bombers to heaps of ruins led people to believe that the German factories had met a similar fate. This was an illusion. 'After the defeat in 1945 German industry was at a standstill but not destroyed', wrote Professor Alfred Grosser in his book *Germany of our Time*. He also said:

'... the destruction of plant was only 10 per cent for metallurgy, 10-15 per cent for the chemical industry, 15-20 per cent for engineering and 20 per cent for textiles. There were a number of reasons for this comparative immunity: the efficient camouflaging of factories and the German system of dispersal, the Allies' concentration on the bombing of towns to sap the population's morale, which left the factories relatively intact, and, finally, the "selective" bombings that spared certain factories in which non-German firms had interests. The immense I.G. Farben head office, which remained intact at Frankfurt while three-quarters of the town was destroyed, stood as the symbol of a war that was more total for some people than for others.'[48]

The present-day atomic industry in West Germany consists of a tightly-interlocked group of companies (see illustrations) backed by certain major banks, which are directly descended from the corporate structure under Hitler which, as David Irving points out, was responsible for the early successes of the Nazis' nuclear project.[49] Many of the nuclear companies are 'successor firms' of IG Farben, the chemical concern founded in 1925, which by the outbreak of the Second World War was at the centre of the international cartel structure and the greatest conglomeration of industrial

33

and military strength ever assembled under a single management. Groth and others were conducting nuclear research on IG Farben contracts during the war.

IG Farben's war crimes were such that two large volumes of the Nürnberg Tribunal hearings are devoted to the case against the company and some of its senior officials. Among the allegations in the indictment it is said that:

'Hitler, with his programme of war, and Farben, which could make Germany...self-sufficient for war, found a basis for close collaboration as early as 1932. The Farben leaders and other industrialists saw the Nazi movement growing and saw in it the opportunity to extend their economic dominion.'[50]

'As a result of the basis for collaboration established between Hitler and Farben in 1932, Farben concentrated its vast resources on the creation and equipment of the German military machine for war, invented new production processes, and produced huge quantities of materials for war...Without them Germany could not have initiated and waged aggressive war...'[51]

'After the Nazi government came into power, Farben used the international cartel as an economic weapon in the preparation for aggressive war through trade penetration, political propaganda, collection of strategic information about foreign industries, and in weakening other countries by crippling production and stifling scientific research...From 1935 on, all cartel agreements and extensions and modifications thereof, were cleared by Farben with the Wehrwirtschaftsstab (Military Economics Staff) of the Wehrmacht.

The result was a tragic retardation of the development of strategic industries in countries which the Nazi government planned to invade and attack.'[52]

This record led the American military government after the war to the conclusion that:

'The war-making power represented by the industries
owned or controlled by IG Farbenindustrie AG constitutes
a major threat to the peace and security of the post-war
world so long as such industries remain within the control
of Germany.'[53]

The Allied Control Council for Germany passed a special law
on the dissolution of IG, and it was 'decartelised' along
geographic lines. Nevertheless, the individual turnover of
Hoechst, BASF and Bayer, all 'successor firms' of IG, puts
each of them in the top eight West German firms. The three
remain interconnected through joint participation in subsidiaries. Their representatives share the leadership of the
Chemical Industry Association and head its working committees. They also retain in their top management important
participants in IG's war-time operations.

Dr Carl Wurster, for example, served under Hitler as a
wehrwirtschaftsfüher (military economic leader), member of
the IG Farben *vorstand* (executive board of directors),
production chief for inorganic chemicals, and plant leader at
the BASF Ludwigshafen plant, which used slave labour from
concentration camps. Wurster was also involved in the
attempted takeover by IG Farben, in collaboration with the
Reich Government, of the Czech company Prager Verein
before the Munich Pact of 29 September 1938.[55] In 1952
Wurster became the first Chairman of the Managing Board
of BASF, and in 1965 Chairman of the Supervisory Board.
Several other directors of West Germany's nuclear industry
have similar backgrounds.

A prominent member of the network of nuclear-chemical
companies formed by the IG successors and related firms is
Nukem, whose main shareholder is Degussa AG (the other
shareholders of Nukem are Metallgesellschaft/RWE, a major
electricity producer, and Rio Tinto-Zinc of Britain). Degussa
gained its initial nuclear-chemical experience in the Second
World War, where it supplied the Nazis with all of the
uranium used in their enrichment efforts.[56] Conversion of
Degussa's uranium oxide to uranium hexafluoride, the first

Europe

Kiel •

Hamburg •

WEST GERMANY

EAST
GERMANY

Almelo ◑

NETHERLANDS

• Rotterdam

Eindhoven •

Göttingen •

Rhine R.

• Essen

RUHR

• Düsseldorf

Jülich
◑

• Cologne
• St. Augustin
BONN ◉ •Siebengebirge Hills

BELGIUM

• Koblenz
• Frankfurt

Fichtel-
gebirge

LUXEMBOURG

SAAR

Nürnberg •

◑ Karlsruhe

• Stuttgart

FRANCE

Capital   ◉
Nuclear site   ◑

Munich •

0   500 Mls

0   500 Kms.

SWITZERLAND

AUSTRIA

a. f. de souza

36

stage of enrichment, was undertaken by IG Farben.[57] It was Degussa which subsequently fabricated the new Federal Republic's first uranium fuel rods, using the ore illegally obtained by Flick's organization.[58] Degussa then took the leadership in the fabrication of fuel elements for all types of nuclear reactors. The decisive conceptual advance in centrifuge research and development, the Zippe centrifuge, was patented in 1957 by Degussa who thereafter developed it.[59] It was carried to the experimental stage in the 1960s at the Jülich Nuclear Research Centre under the scientific control of Degussa metallurgist and fuel element specialist Dr Alfred Böttcher.

German atomic industry has, almost since the beginning of the Third *Reich*, aimed at acquiring a complete and autonomous fuel cycle within West Germany, particularly the capacity for uranium enrichment. In this, the Federal Government itself has played a critical role, through the German Atomic Commission (DAtK), the Jülich and Karlsruhe Nuclear Research Centres, and with various forms of support including credits, subsidies and loan guarantees.

The year 1955, when official repression of nuclear research in West Germany came to an end, also saw the first breakthrough in the secrecy which had surrounded all atomic energy activities until then. The occasion was the First International Conference on the Peaceful Uses of Atomic Energy, held in Geneva from 8 to 20 August 1955. The conference resulted from an initiative by President Eisenhower, who put before the 9th General Assembly of the United Nations in 1953 his 'Atoms for Peace' proposals. However, at the Geneva conference the veil of atomic secrecy was lifted not by the Americans but, to the surprise of all, by the Soviets. They were the first to come out with information which until then had been guarded by the West as top secret. 'One of the reasons for witholding information or refusing to "declassify" nuclear data had always been that it could not be proved that the Soviets knew the facts,' explained R. E. Lapp, a US delegate to the conference: 'So once the Soviets published the data, we were smoked out and had to release much more information from the secret category.'[60] The conference was attended by 1,428 delegates from 73 states, and 1,350 observers representing the 'atomic energy business'. In order to show their superiority, the United States revealed considerably more than the Soviets.

West Germany sent a 70-member delegation, a surprisingly big team for a country which was still on probation. The West German delegation was led by Professor Otto Hahn, the President of the Kaiser Wilhelm Society for Promotion of Sciences (renamed the Max Planck Society) of Göttingen. The contribution of the German delegation to the conference was rather modest in terms of scientific papers, but its composition reflected the keen interest of German big business. Professor Dr Karl Winnaker, Director General of Farbwerke Hoechst – one of the successors of IG Farben – was accorded a high place in the German delegation, ranking third after Professor Hahn and Professor Getner of the University of Freiburg. Among the members of the German delegation who followed the three specialized sessions of the conference (physics and reactors, chemistry; metallurgy and technology; and medicine, biology and radioactive isotopes), where more than 1,000 papers were read, were Professor Dr Bartholome of Badische Anilin und Sodafabrik (BASF), Ludwigshafen; Dr Wilhelm Boveri, head of Brown Boveri, Mannheim; Dr Duhm of Bayer, Wuppertal; Professor Dr Finkelburg of Siemens, Erlangen; Professor Dr Houdremont of Krupp, Essen; Professor Ing. Löbl of Rheinish-Westfäliche Elektrizitätswerke AG; Herr Schimmelbusch, Director of Degussa, Frankfurt; and 24 other scientists and engineers who preferred not to reveal their connections and were listed by name only.

The Geneva conference was the first opportunity for West German scientists and businessmen to widen their international contacts, confined until then largely to the US and Britain. This was also the task of the delegation from the Republic of South Africa, which consisted of six members. It was led by

Dr S. M. Naudé, President of the Council for Scientific and Industrial Research, and also included T. K. Prentice, Uranium Consultant of the Central Mining Corporation, and Dr H. J. Van Eck of the Industrial Development Corporation.

It was at the Geneva conference that nuclear co-operation between West Germany and South Africa was initiated. The diplomatic and economic team of the West German delegation consisted of top envoys from the Ministry for Economic Affairs: Walter Hinsch and three Senior State Counsellors, Dr Karl Pretsch, Dr Karl Schmidt-Amelung and Dr Emil Pohland. Their task was to look for new partners in the promising field of atomic energy, partners who would be interested in the export of German technology and who might have something to offer in return. The grand vistas of the future of atomic energy outlined by Dr B. F. J. Schonland, a South African scientist attached to the British Atomic Energy Research Establishment at Harwell, included a considerable market in Africa for medium-sized nuclear power plants.[61] This did not fail to impress Hinsch and his colleagues. They also registered Schonland's warning, which was echoed by many other delegates to the conference: uranium was going to be scarce; amongst the 'nuclear' countries of the seventies it would become one of the most sought-after commodities.

Compared with the other countries in which the German delegation had also shown interest – Brazil and Argentina, the first because of its mineral potential and the other because of already existing economic ties – South Africa looked good. The Brazilian electrification plans and prospects for atomic power (presented to the conference by E. Motta Rezede,[62] as well as in the report by B. C. de Mattas and J. Costa Ribeiro, 'Economic significance of nuclear power for Brazil')[63] were not based on indigenous uranium: Brazilian uranium was still to be discovered. South Africa, by contrast, had both uranium and the industrial base to make use of it. A report on the Geneva conference signed by Walter Hinsch went to West German Economic Minister Dr Ludwig Erhard. A copy also went to Franz Josef Strauss, the Minister for Atomic Affairs. Dr Abraham Johannes Andries Roux, President of the South African Atomic Energy Board which was established in 1957, is believed to have met Strauss during his six-month information tour abroad in the same year. After 1966, Strauss became a regular visitor to South Africa.

Atomic energy for anything other than peaceful uses was a very delicate topic to raise publicly in Germany in 1955. The Germans were still recovering from the shock of a NATO military nuclear exercise, performed under the code name *Carte Blanche* from 20 to 28 June 1955. The Allied Tactical Air Forces and national air force units of France and Benelux (a total of 3,000 aircraft) took part in the manoeuvres which covered Holland, Belgium, north-eastern France and West Germany. During a simulated attack 400 atomic bombs were used against selected targets which included 100 targets between Hamburg and Munich. The West German 'casualties' were estimated to be 1.7 million dead and 3.5 million wounded. The reaction of the German population was one of horror. The Social Democratic Party led by Erich Ollenhauer attacked the Adenauer government's rearmament policy as 'suicidal', and a group of eighteen German Nobel Prize winners issued warnings of an apocalyptic future.[64] German military circles, however, had drawn the opposite conclusion from the *Carte Blanche* exercise, which is discussed below.

The year 1955 also saw a great leap towards the economic integration of Europe, within which nuclear co-operation occupied an important place. On 10 June 1955 the Council of Ministers of the Organisation for European Economic Co-operation (OEEC)[65] established a working group to explore the possibilities of the economic and financial co-operation of OEEC members in the peaceful uses of atomic energy. Two and a half years later, on 17 December 1957, the Council approved:

1. The convention establishing the European Nuclear Agency of the OEEC.
2. The establishment of the European company for the

40

Chemical Processing of Irradiated Fuels (Eurochemic).
3. The Convention on Security Control which was to ensure that nuclear co-operation between the OEEC members should 'not further any military purposes'.[66]

Among the first joint undertakings of the OEEC were the construction and operation of the experimental 5MW boiling water reactor at Halden, Norway, and a 10MW high temperature gas-cooled reactor at Winfreth Heath in Great Britain. The next step, leading to an even closer interlocking of the emergent European nuclear industries, was the establishment of the European Atomic Energy Committee (Euratom). This formed an integral part of the European Community, which was set up in Rome on 25 March 1957, with treaties coming into effect on 1 January 1958. The United States gave consistent support to the negotiations leading to the establishment of Euratom; but these negotiations were not without difficulties. The supranational character of Euratom and its monopoly on transfers of technology and nuclear fuel met with strong objections from France and Britain. France was assured that Euratom would not stand in the way of her own *force de frappe* which, after the Suez crisis in 1956, it regarded as the only way to end its military dependence on the United States, so in the end France agreed to join; but Britain stayed out.

The United States had tried to overcome the reservations of France and the Benelux countries by offers of large amounts of both technological equipment and enriched uranium for Euratom programmes. However, the fact that military and peaceful aspects of atomic energy are inseparable, and fear that the participation of West Germany would bring the country within reach of getting the atomic bomb, disturbed the Euratom partners. Strong protests were made by the Soviet government, which called Euratom a serious threat to European security, at a time when détente between Washington and Moscow had opened new opportunities for trade with Eastern Europe. 'One cannot fail to see that the creation of Euratom would, in fact, lead to the removal of any

41

restrictions in the production of atomic energy in relation to West Germany,' read the official statement by the Soviet government on 12 July 1956. It continued: 'This would permit revenge-seeking West German circles to organize in their country the production of atomic weapons, which would create a serious threat to the cause of peace in Europe.'[67]

In 1957 the International Atomic Energy Agency (IAEA) came into existence, with headquarters in Vienna, as a specialized United Nations agency. The IAEA was a result of President Eisenhower's 'Atoms for Peace' proposal put before the UN in 1953, and of subsequent negotiations in 1954 between the United States and the Soviet Union and since 1956 amongst eight Western nations, which included South Africa in its capacity as the Western world's major uranium producer. The group consisted of the United States, Britain, France, Canada, South Africa, Belgium, Australia and Portugal. Later on Brazil, Czechoslovakia, India and the Soviet Union were invited to participate in the preparatory work on the Statute of the IAEA which was to implement the 'Atoms for Peace' programme on a global scale. There were two factors behind this 'internationalising' of co-operation in the field of atomic energy:

1. The termination of the American atomic weapons monopoly on 9 September 1952, when the Soviet Union exploded its first bomb, was followed by the explosion of the first British bomb on 3 October 1952. The Soviet hydrogen bomb tested on 20 August 1953 removed any remaining doubts about how far the Soviet scientists had progressed in this respect.

2. There was also pressure from the American nuclear industry lobby, which made the point repeatedly to the Administration that the continuing secrecy surrounding the production of nuclear reactors and the relevant technological and chemical processes had become an expensive nuisance, hindering exports in a field in which the United States was without any serious competitors.

The establishment of the IAEA set the seal on an era of

euphoria about atomic energy, when it was seen as the ultimate solution to the world's energy problems — a view which is reflected in the Agency's Statute.[68] This enthusiasm subsequently gave way to concern about the crucial issue of the Agency's safeguards, which after twenty years of refinement are still far from adequate. The nuclear explosion by India in 1974, a country whose representative on the Board of Governors of the IAEA, the late Dr Bhaba, had been one of the most vociferous advocates of the need for improving agency safeguards, illustrates the point.

West Germany became an active participant in all forms of international co-operation, and German science and industry kept abreast of the rapid technological advances. It was not long before West Germany moved into a leading position. West German nuclear technology has created a new dimension in the national economy, and the sale of its pressurized water reactor, Biblis, constitutes one of its most significant breakthroughs into the world market.

Nuclear research in West Germany has enjoyed top priority, receiving generous government grants. According to the Minister for Research, Hans Hermann Matthöfer, the ratio between government funding of nuclear and non-nuclear research rose from 1:7.8 in 1972 to 1:4.3 in 1977.[69] One of the principal recipients has been the Gesellschaft für Kernforschung at Karlsruhe, founded in 1956. In 1959 Professor Erwin Willi Becker, Director of one of the GfK establishments, the Institut für Kernverfahrenstechnik des Kernforschungszentrums (the Institute for Nuclear Processing Techniques), began work on the development of a uranium enrichment process known as the *Trenndüsenssystem* (jet-nozzle system).[70] This process was given to South Africa (in circumstances described later in the book), and it was the jet-nozzle system which became the key to South Africa's own enrichment process. Another new method for the enrichment of uranium, gas centrifuge, is being developed at the nuclear centre at Jülich, which has received DM600 million for the construction of an enrichment plant. The present nuclear network of the Federal Republic of Germany can be outlined as follows:

— The two government-owned and -controlled institutions, the Society for Nuclear Research at Karlsruhe and the Nuclear Research Establishment at Jülich (Kernforschungsanlage Jülich GmbH), are linked both with companies owned by the government, such as the Corporation for Energy Affairs (Gesellschaft für Energiebeteilung), and private concerns such as Bayer AG, Hoechst AG, Ruhrkohle AG, and Siemens.

— Apart from government sources, funds are provided by the Dresdner Bank, Deutsche Bank, Allianz-Versicherungs (Insurance Alliance) and Allgemeine Verwaltungs Gesellschaft für Industriebeteiligungen (General Administrative Society for Industrial Affairs).

— Nuclear reactors are manufactured by Brown, Boveri and Cie (BBC) and Kraftwerk Union (KWU), owned by the Interatom (Internationale Atomreaktorban). Nukem and RBG (Reaktor Brennelemente) are producing the reactor fuel from imported uranium and the enrichment process is carried out by the Essen-based energy concerns Steag and Uranit. (See the chart in photographic section).

— The global reach of the West German nuclear industry is demonstrated by agreements on nuclear co-operation with Third World countries such as Brazil (18 November 1975), India (19 May 1972) and Iran (4 July 1976).[71]

— The DM17 billion spent by the government on nuclear research has yielded considerable profits. The results of the research are being converted into projects which produce jobs and both domestic and international sales. The question of the countries involved does not seem to have concerned the Ministers in Bonn, who do not allow politics to interfere with business. The President of the Bundesverband der Deutschen Industrie (BDI, the

Federal Association of German Industries), Hans-Gunther Sohl, goes even further: 'Every effort to improve trade should be welcomed in the interest of politics. Misunderstandings, though, are to be expected'.[72]

'The German government,' wrote *Der Spiegel* on 15 March 1976, 'is thankful for every big order which industry receives from abroad at a time of world-wide economic recession, and shrinks from controlling the freedom of trade and industry. What applies to the arms trade — export restrictions controlled by the Cabinet — has no validity in atomic trade. Here, private industry is uncontrolled, and even the *Bundestag* is by-passed.' *Der Spiegel* went on to describe how the Bonn government operates. 'The business people prepare the deal inconspicuously in advance, then, as in the case of Brazil, present to the government virtually ready-to-sign contracts; Bonn, in the interests of employment and a booming economy, then has to agree. "They put pressure on us, and we like to be pressed" ', said Herr Dieter Vogel, adviser to former Economics Minister Hans Friedrichs. As a safety valve the Bonn Government deliberately avoids Cabinet decisions on controversial trade matters and appears to leave business to industry and a couple of State Secretaries in the Ministries. If a deal goes wrong, an excuse is at hand: the Cabinet did not know anything about it. (An example of this can be seen in the case involving Dr Eppler, on pp. 69 ff.).

An illuminating example of how far the industry is prepared to go in order to deceive the government and the German public is the case of the intended delivery of compressors made by the German company GHH (Gutehoffnungshütte, the parent company of MAN) for the South African enrichment plant at Pelindaba.

At a meeting held on 24 November 1975, at the Federal Ministry of Economic Affairs, Professor Fiedler of Steag argued that the compressors in question were normal compressors and should therefore be exempt from the export regulations whereby special permission must be sought from the government for the export of nuclear equipment. During the discussion, the issue of the export of the jet-nozzle enrichment system came up. The debate developed into a sharp exchange between Professor Fiedler of Steag and Herr Heil of the Federal Ministry for Research and Technology. Herr Heil bluntly stated that he would use all means at his disposal to see to it that the entire jet-nozzle system would be placed under strict export regulations. The following day Professor Fiedler reported on the meeting to Herr Wenzel, Head of the Nuclear Installation Department of Steag, who decided that he had to take action on the matter personally.

In a confidential inter-office memo of 25 November 1975, a copy of which is reproduced in the appendix (Document 24), Wenzel recommended to the Director of Steag, Dr Völcker, that the inventor of the jet-nozzle system, Professor Becker, should be informed as quickly as possible and that it should be seen to that a 'credible description of the unimportant military significance of the jet-nozzle system is produced without delay.' As Head of the Nuclear Installation Department of Steag Wenzel must have known that jet-nozzle enrichment of uranium is, like any other enrichment process, a route to nuclear weapons capability.

The export of the compressors *was* stopped — but only for a time. The company found a different way round ministerial opposition. According to *Der Spiegel:* 'The order was switched to the coal–processing company Matla. For compressors delivered to Matla no export licence is required. The Bonn export controllers were satisfied and the South African Atomic Energy Board got what it wanted.[73]

Wenzel's idea was repeated some eighteen months later by the *Auswärtiges Amt* (Foreign Office), which produced its own 'credible description' of the jet-nozzle system. In a Memorandum of February 1977 issued to deny charges of nuclear and military collaboration with South Africa, it justified itself as follows:

The rise of West Germany as a nuclear power

'The German separation nozzle process for uranium enrichment is not suitable for the manufacture of material usable for weapons. Hence it is not covered by secrecy regulations. It does not involve any risk of proliferation.'[74]

But this is not how the enrichment process is viewed by the nuclear scientists at the Stockholm International Peace Research Institute: 'Enrichment facilities give the physical capability for a country to produce uranium-based nuclear fission weapons and so far are the preferred way of producing the nuclear weapons,' wrote Dr Peter Boskma in 'Nuclear Proliferation Problems', published in 1973. He then described the jet-nozzle process, at that time still used in pilot plants only, as one of the 'second generation enrichment methods.'[75]

The jet-nozzle system was also sold openly to Brazil as part of the biggest nuclear deal in history: the DM2 billion agreement signed on 27 June 1975. The West German sale to Brazil and the French sale of a reprocessing plant to Pakistan in 1976 were described in the 1977 SIPRI Yearbook as representing '. . .the first exports to non-nuclear weapons countries of plants capable of producing material in a form suitable for direct use in atomic bombs.'[76] So much for the 'unsuitability of the jet-nozzle system for the manufacture of material usable for weapons', as claimed by the West German Foreign Office.

Perhaps the officials of the Foreign Office really were unaware of the facts, and a Minister can be misinformed. *Der Spiegel* has an interesting story on the subject: 'Soon after his entry into office in May 1974, Research Minister Matthöfer stopped state subsidies worth DM50 million to the Nuclear Research Institute at Karlsruhe for the exceptionally expensive German jet-nozzle enrichment process. Fifteen months later Matthöfer discovered that the funds had been granted on the instructions of his State Secretary Hans-Hilger Haunschild.'[77] It is not without interest that the same Herr Haunschild had been one of the chief contacts between the

South African Atomic Energy Board and the West German nuclear industry.

The aggressive West German sales tactics for nuclear technology have support from Chancellor Schmidt himself. An example of his attitude towards the selling of nuclear hardware was his interview on the American television programme 'Agronsky at Large', shown on 29 April 1977 — only a week before the Summit of seven Western powers in London where nuclear proliferation was expected to be one of the toughest items on the agenda.

Chancellor Schmidt was reminded of President Carter's concern over nuclear proliferation and '. . .his concern most importantly about Germany selling to Brazil that group of nuclear processing plants which have the capacity to produce plutonium — plutonium being the material with which nuclear bombs are made.'[78] Asked if it was still his position that West Germany would stand by its treaty with Brazil, Schmidt replied: 'Of course. And we are acting on the basis of the Non-Proliferation Treaty, which we have subscribed to and ratified at the end of the 60s, early 70s, as many other countries did.' When asked if West Germany, or he as its Chancellor, would be willing to postpone or cancel the Brazilian transaction, Schmidt said: 'The answer is no.'

Agronsky then said: 'Now, you've certainly made very clear your position on the Brazil deal. Germany is in the process of a number of other negotiations for exporting nuclear reactors with Portugal, with Spain, Saudi Arabia. I understand that you're even in negotiation with Iran in competition with an American firm which is trying to sell, I believe, eight or nine nuclear reactors to Iran. Is that true?'

To this, Schmidt said, 'Well, roughly speaking, it's true to say that we're exporting nuclear reactors to any country which wants to buy one, or even more than one. In Germany, it's a future industry and it, at present, does employ about 100,000 people, skilled labour. We don't have the same chances for instance in the field of civil aviation where the world is dominated by American corporations, nor do we have the same chances in the field of computers where, more or less, the world market is in the hands of American corporations.'[79]

In terms of nuclear capability, West Germany had already acquired the status of a nuclear Great Power. With 4,855MWe (megawatts of electricity) produced in 1976, it ranked fourth in the world after the US, Japan and France; and the projected capacity of 35,116MWe in 1984 will move it to second place after the United States. West Germany has also broken into commercial uranium enrichment, as the dominant partner in the joint venture with Holland and Britain to establish the enrichment plant at Almelo, in Holland.[80]

In 1971, a West German plant for reprocessing low-enriched uranium dioxide went into operation at Karlsruhe. Established by leading members of the West German 'nuclear family' — Farbenfabriken Bayer, Farbenwerke Hoechst, Gelsenberg and Nukem — it has been operated by Kewa (Kernbrennstoff-Wiederaufbereitungsgesellschaft GmbH) and has a projected capacity of 1,400 tons of light water reactor-type oxide fuel in 1987. For comparison, the projected capacity for the same year of the British plant is 1,000 tons and of the French 800 tons.

The West German drive to secure its own reprocessing facilities was motivated by the anticipated shortage of enriched uranium during the coming decade and by the resolve to terminate German dependence on the American reprocessing facilities. The suspension of the enriched uranium deliveries by the Carter administration in 1977, with Canada following suit, added urgency to the Kewa project, despite the fact that the separating and recycling of plutonium as a reactor fuel has not yet been shown to be economically justified.[81]

Prodded by a West German nuclear industry lobby, supported by about seventy German professors insisting that without atomic energy the West German economy would collapse, the West German Government has continued to push ahead with its domestic and foreign nucelar policy apparently with little concern about the possible consequences.[82]

3
Nuclear Conspiracy

General Rall takes a trip

'Pilots in all armies of the world are fine people, extravagant animals; they enjoy a special status not only because their toys and training cost a lot of money but also because of the romanticism, occasional bravado and the undeniable spirit of "sportsmanship" of most of their doings.'[1]

Heinrich Böll

On 28 September 1943 *Der Adler*, the official magazine of Hitler's *Luftwaffe*, had on its cover a photograph of Captain Rall who, with Lieutenant Hrabak, was awaiting the return of his comrades-in-arms from a mission to the German-Soviet front. Captain Rall was then Commander of an air squadron and was only the 34th soldier of the German armed forces to be personally decorated by Hitler with the highest military order of the Third *Reich*: the Knight's Cross. Captain Rall had earned it by shooting down 200 enemy aircraft. Thirty years later Captain Günther Rall had become a three-star General, Chief of the West German Air Force and German military representative to the Military Council of NATO in Brussels.

Other *Luftwaffe* officers also had outstanding careers. One of them, Walter Scheel, became President of the Federal

51

Republic of Germany. Less lucky was the most decorated hero of the Second World War: Colonel Hans-Ulrich Rudel. He had carried out 2,530 missions and held a record for destroying 519 Soviet tanks. Rudel switched from the controls of his Stuka-bomber to the pen, and continued to fight the *'Barbaren mit dem Pferdemagen'* ('barbarians with horses' stomachs') — as he used to describe the Russians — in a political campaign for the neo-nazi *Sozialiste Reichspartei* (SPP), its successor, the *Deutsche Reichspartei* (DRP), and finally the National Democratic Party (NDP) led by Adolf von Thadden.

In 1951 Rudel wrote a pamphlet entitled 'Dolchstoss oder Legende?' ('A stab in the back or a legend?'). In it Rudel said of Hitler: 'Have not past and present events clearly proved that of all statesmen of our time the German state leader Adolf Hitler alone had correctly assessed the world situation and knew how to handle it?' [2] Rudel's evocations of the past became an embarrassment to many of his former colleagues who continued their military profession in the *Bundeswehr*.

In October 1976 Rudel addressed a veterans' meeting of officers of the Stuka-bomber squadron 'Immelmann', the most-feared *Luftwaffe* unit of the Second World War, at Baumgarten near Freiburg. Two of his former colleagues were fired from their jobs because they attended the ceremony and defended Rudel's right to take part in it as well. They were the Commander of the Air Force, General Walter Krupinski, who had flown many missions with Rudel (and had 197 enemy planes to his credit) and General Karl Heinz Franke, Commander of the 'Nike' and 'Hawk' rocket units.[3] The spokesman of the Defence Ministry, Arwin Halle, explained that '. . .while Minister Leber has nothing against veteran meetings, men like Rudel, who will never learn, and never improve, have no place in the *Bundeswehr*.'

General Günther Rall lost his post through another World War II Stuka pilot — Kurt Dahlmann, who had settled in Windhoek as the editor of the *Allgemeine Zeitung*. It was Dahlmann who was used by Ambassador Donald Bell Sole to provide a cover for General Rall's secret visit to South Africa. Sole was carefully cultivating his top military contacts in West Germany. When, for example, South African Airways inaugurated jumbo-jet flights from Frankfurt to Johannesburg in 1972 Sole arranged a free ride for General Grüner, Commander of the Federal Border Guards. Defence Minister Georg Leber and his wife, Erna, were also friends of the Soles.

From 1972 Ambassador Sole tried to arrange for Rall to visit South Africa. He knew Rall and his wife, Hertha, well; on 14 June 1972 he confided to L.B. Gerber, Director-General of the South Africa Foundation, that Rall '. . .has always been interested in South and South West Africa'. Sole approached General Jim Verster, Chief of the South African Air Force, with a request that the Defence Ministry should bear the costs of the Rall trip. However, the Ministry appears to have been rather hesitant about arranging a private visit for so senior an officer of the NATO army. No doubt the consequences of a possible leak were being considered, hence the excuse about the 'economy drive' when it declined to accept Sole's proposition. The officials of the Defence Ministry were apparently unaware of Ambassador Sole's ingenuity. Indeed, when he later presented them with his plan for covering up the visit they spared no expense in entertaining Rall during his South African tour.

The trip was planned originally in 1973 and details of the programme were discussed between Sole and Rall at the Hotel Königshof in Bonn on 30 May 1973. The occasion was a reception to celebrate South Africa's National Day. However, six days later Rall wrote to Sole that he had to postpone the visit due to his preoccupation with weapons planning and decisions which had to be made on the changes in the defence structure of the Federal Republic.[4]

Rall apologized for the 'unforeseeable events' which had prevented him from making the journey, assured Sole that he had obtained his Minister's approval and gave him a firm promise to visit South Africa in the autumn of 1974.

In his dispatch to General Laubscheer of the Defence Ministry in Pretoria, Sole commented approvingly on Minister Leber's attitude: 'If he (Rall) had another Minister, who was not as well disposed to us as Georg Leber, he might not have got permission to visit South Africa.'

Sole made sure of it himself. At the *Luftwaffe* Ball in February 1974 Sole was seen talking to Leber. Whether the two men were discussing Rall's trip cannot of course be established, but a few weeks later, on 4 April 1974, Sole wrote to State Secretary van Dalsen in Pretoria's Foreign Ministry: 'I was able to arrange for the issue of a visiting permit on an informal basis by Defence Minister Georg Leber, with whom I likewise discussed the matter.'

When General de Villiers succeeded General Laubscheer in handling the foreign visitors he too was assured by Ambassador Sole in a dispatch of 6 August 1974 that Rall was travelling '. . .in his private capacity, as guest of Kurt Dahlmann. The German Defence Minister Herr Leber knows that the visit involves more than this.'

A secret telegram from Sole to the South African Embassy in London, dispatched on 13 August 1974, tells it all:

> 'I have arranged for Lieutenant-General Günther Rall and Mrs Rall to visit South Africa as the guests of our Department of Defence, but ostensibly as the guests of Mr Kurt Dahlmann, editor of the Windhoek Allgemeine Zeitung. General Rall was until 31 March of this year Head of the Luftwaffe and is now German Military Representative to the Military Council of NATO. No publicity whatever is being given to this visit and for security reasons General and Mrs Rall will travel as Mr and Mrs Ball.
>
> Details of their flight arrangements are as follows:—

SN 609	05 October	dep Brussels	16.05
		arr London	17.05
SA 233	05 October	dep London	18.15
	06 October	arr Johannesburg	10.35

> You will note that there is only a limited transfer period

available between the arrival of the flight from Brussels and the departure of the flight to Johannesburg, but South African Airways Frankfurt has been assured from London that the time is sufficient.

I should be grateful if you would arrange nevertheless with South African Airways that General Rall is received suitably on arrival on the Sabena flight and transferred immediately to the South African Airways flight. It would perhaps also be an appropriate gesture if your Air Attaché (in civvies) could also be on hand to welcome General and Mrs Rall.

With reference to the invitation to General Rall ostensibly coming from Mr Kurt Dahlmann, I should perhaps explain that Mr Dahlmann and General Rall were comrades in arms in the same squadron during the war.'

On their arrival at Jan Smuts airport General Rall and his wife were greeted by General J.P. Verster, Chief of the Air Force, and his wife. From then on Rall had a chauffeur-driven limousine at his disposal and was flown around by South African Air Force planes. An army helicopter took him to pay a visit to Kurt Dahlmann with whom the Ralls spent three days in a game reserve. For the rest of the three-week-long visit Rall had business to attend to. On his itinerary was an inspection of the Devon Air Force unit, Waterkloof military aerodrome, the Atlas aircraft factory, the Transhoek and Stellenbosch military academies and the Navy Head-quarters at Simonstown. For 'security reasons' two versions of the programme were prepared. The visit to Pelindaba and a dinner with the West German Ambassador, Dr Sträthling, were marked on the confidential copies only.

On Thursday 17 October a special programme was arranged for Hertha Rall while her husband had the following agenda:

9 a.m. Concluding discussions with the Chiefs of Defence Staff about the strategic importance of the Republic of South Africa for the West.

11 a.m. Visit of the Atomic Energy Board at Pelindaba.

11.30   Discussion with the President of the Atomic
a.m.   Energy Board,Dr A. J. A. Roux followed by lunch
           at 12.30.

After the lunch Rall was shown the nuclear establishment at Pelindaba.

After his return General Rall sent (on 2 November 1974) a hand-written letter of thanks to Ambassador Donald Sole for what he described as a trip 'rich in experience'. However, by that time there was already a leak. Reporter von Jochen Raffelberg of the *Frankfurter Rundschau* attributed it to Elisabeth Sole, who allegedly said in the International Women's Club in Bonn that Rall's South African journey was made at the invitation and expense of the South African government. This had reached the ears of the Inspector-General of the German Armed Forces, Admiral Zimmermann, who talked with Rall four days after his return on 29 October at the meeting of the Commanders in 'Damp 2000'.

Admiral Zimmermann had good reason to be worried – it was he who had signed the authorisation for Rall to go. But the public was alerted only after the news about Rall's trip to South Africa and his visit to the uranium enrichment facilities at Pelindaba was broken by a Reuters news despatch on 26 September 1975. It quoted the African National Congress of South Africa as its source. Questioned by the press, Minister of Defence Georg Leber denied the whole thing. He maintained that he had no knowledge of Rall's South African 'holiday'. Rall supported him by repeating emphatically to editor Werner Heine of *Der Stern*, 'The Minister knows nothing at all', and when confronted with his own letter to Sole he stuck to his version: *'Es war eine Urlaubsreise. Absolut.'* ('It was a holiday trip. Definitely.') 'How then, Herr General, can you explain your mention of having secured the Minister's consent?' asked Heine. 'That is incorrect. I cannot remember that,' answered Rall.[5]

The assertion that it was a private visit was also defended by the Defence Ministry: one of their officials, Colonel Kommer, told the *Frankfurter Rundschau*[6] that Rall had made a private visit at the invitation of a friend, for the purpose of a holiday. No official duties were connected with the visit, and there were no grounds for speculation.

However, when the 'private visit' cover was blown by *Der Stern* and *Der Spiegel*, Minister Leber began to sweat. Still insisting that he had not known anything about Rall's journey until 26 September 1975, Leber appeared on television on 1 October 1975 with a solution: he announced that he was acceding to General Rall's wish to be retired. Speaking about Rall as '... the highly decorated officer, honoured soldier, courageous officer', he lowered his voice to express his regret that Rall had concealed from him 'the special circumstances' of his visit. However, despite his admission that '... the interest of our country could have been jeopardized by the nature of such a visit', he said that it was a case '... which does not call for disciplinary measures.' That was the end of it.

One year later he was asked about the Rall affair by *Der Spiegel*:[7]

> *Spiegel:* And the trip to racist South Africa by the former Inspector of the *Luftwaffe*, Günther Rall?
> *Leber:* The circumstances of this trip became known to me only long afterwards.
> *Spiegel:* Rall allowed the *apartheid* regime to pay all his expenses.
> *Leber:* Herr Rall drew the only possible conclusion and asked me to retire him. It had nothing to do with politics.

If Defence Minister Georg Leber was telling the truth, that he had never consented to Rall's trip to South Africa, then Ambassador Sole would have been deceiving Pretoria when he wrote that he had secured Leber's permission. If Sole did this on the basis of Rall's assurances that Leber *had* agreed to the trip, then it would have been General Rall who was lying. Georg Leber's case is rather thin, considering that Admiral Zimmermann talked to Rall soon after his return from South Africa in 1974, and that the German Ambassador in Pretoria wined and dined Rall at his residence on 13 October 1974. It would be very surprising if neither of these two gentlemen had said anything about it to their superiors.

The most alarming aspect of the Rall affair is the fact that it hardly stirred the *Bundestag*, where nobody seriously challenged the glaring contradictions in the various statements by spokesmen of the Defence Ministry and by Georg Leber himself. One could sense a deep relief when Rall's retirement was announced and the unpleasant incident was closed.

Rall was not the only leading West German military officer to visit South Africa's secret nuclear installations. Apart from General Grüner, whose visit to South Africa was arranged by Sole, documents in our possession show that other German Generals made similar trips: the Inspector-General of the *Bundeswehr*, de Maizière, in 1972 and again in 1974 after his retirement; General Graf Kielmannsegg in 1971 and 1975; and General Schnez. None of these has denied visiting South Africa, nor has it been explained what they were doing during their visits to Pelindaba.

The Karlsruhe-Pelindaba Connection

'The technocrats, the sort of people you find not only among the engineers and businessmen but even among the teachers and lawyers, constitute one of the greatest dangers to mankind. This because of their narrow and one-sided "specialization". They are totally absorbed in their own jobs into which they put all their efforts and zeal in order to be successful. They do not care about the consequences of their work, nor whether it may harm society.'
Albert Speer, Hitler's Minister for Armaments[8]

In December 1968 South Africa's Governor on the Board of the International Atomic Energy Agency in Vienna, Donald Sole, was back in Pretoria to prepare himself for the most important assignment of his career. He was appointed South

Africa's Ambassador in Bonn, one of his special tasks being to secure for the South African Atomic Energy Board the technology necessary for the enrichment of uranium – the most secret part of the nuclear energy process. Donald Sole had excellent qualifications for the job. He had been involved in nuclear diplomacy since 1954 when South Africa was invited, together with Belgium, Australia and Portugal, to join the United States, Britain, France and Canada in the negotiations which subsequently led to the establishment of the International Atomic Energy Agency (IAEA) in 1957.[9]

Since 1957 the South African Mission to the IAEA in Vienna has served as an external wing of the South African Atomic Energy Board, through which scientific information is channelled and the exchange of nuclear experts arranged. Sole had filled the post since 1959 as a Member of the Board of Governors of the IAEA. However, since the events at Sharpeville, followed by South Africa's departure from the British Commonwealth in 1961, things became more difficult. The African diplomatic offensive against South Africa at the United Nations gathered new momentum in 1963 when the Organisation of African Unity, established in the same year, made the elimination of *apartheid* one of its principal aims and began to press the UN to follow suit. This had an impact on all the UN agencies, including the IAEA which has a special link with the Security Council.[10] South Africa gradually found itself isolated in international affairs, and the benefits derived from participation in the IAEA for international co-operation in the field of atomic energy were jeopardized. After the Security Council passed its first resolution on the embargo of arms supplies to South Africa, nuclear co-operation with South Africa could have become embarrassing even for such countries as the United States and Britain, which had originally ushered South Africa into the nuclear world.

South Africa's refusal to accede to the Treaty on the Non-Proliferation of Nuclear Weapons, signed by the United States, the Soviet Union and Great Britain on 1 July 1968,[11] did not help matters. Article II of the Treaty imposes *inter alia* an obligation on the part of non-nuclear-weapon states '. . . not to manufacture or otherwise acquire nuclear weapons or other nuclear explosive devices; and not to seek or receive any assistance in the manufacture of nuclear weapons or other nuclear explosive devices.' South Africa was not prepared to promise anything of the sort. The Treaty also binds its signatories '. . . not to provide a source of special fissionable material[12] or equipment or material especially designed or prepared for the processing, use or production of special fissionable material, to any non-nuclear-weapon state for peaceful purposes, unless the source of special fissionable material shall be subject to the safeguards under the Treaty.'

In 1968 a three-man committee headed by Dr H. J. van Eck, Chairman of the Industrial Development Corporation, (who was also a delegate to the First International Atomic Energy Conference in 1955 mentioned above) was instructed by the Atomic Energy Board to explore the possibilities of establishing a uranium enrichment plant.[13] This required uranium enrichment technology, very substantial industrial capacity, and the money to finance the project. In order to proceed with its ambition, South Africa had to find a partner who would overlook its refusal to sign the Non-Proliferation Treaty (NPT) and provide the technology, the industrial base, and if possible also the finance. The Federal Republic of Germany was willing on all three counts, and Donald Bell Sole was the man who arranged it. As *Der Spiegel* put it: '. . . since 1969 the South African Ambassador in Bonn has made his Embassy the most effective centre for military and nuclear interests his country has. And West German authorities have helped willingly.'[14]

Sole approached his new job with such enthusiasm that his predecessor, Ambassador J. K. Uys, had to calm him down. On 5 December 1968 he received a draft of the speech Sole proposed to deliver on the presentation of his credentials. In this draft Sole expressed the hope that as South Africa's Governor of the IAEA he would be able to give special attention to co-operation between South Africa and the Federal Republic of Germany in the field of atomic energy. Ambassador Uys gave the following advice to Sole in a personal letter dated 19 December 1968:

'Regarding the speech, I should be grateful if you would permit me to comment on the inclusion in the speech of a reference to nuclear energy and the production of uranium – *vide* the penultimate paragraph. As you know, the East Germans have for many years accused the Federal Republic and South Africa of close co-operation in this particular field and of secretly producing atomic weapons. I fear that the reference to nuclear energy – even though you specifically mention the peaceful uses of such energy – and South Africa as a major uranium producer, and the fact that you specifically express the hope, as South Africa's Governor on the IAEA, to be able to give special attention to this aspect of the relations between us, could be seized upon by our enemies as further proof of the collaboration of which we have been accused for so long. This we should avoid. Moreover, from the German side it may prove difficult to prepare a proper reply in this connexion for inclusion in the Federal President's answer at the presentation of credentials ceremony, especially as both your speech and the President's reply will be published in the official bulletin which enjoys wide circulation. I feel that the less said in public at this stage about this aspect of our relations with the Federal Republic, the more success we shall be able to achieve behind the scenes. It is therefore strongly recommended for your consideration that the particular paragraph in the speech be omitted.

I should be grateful to receive your views on the above comments, before handing a copy of the speech and translation to Dr von Rhamm in Protocol.'(Document 2).

Ambassador Sole presented his credentials on 12 February 1969. Eight months later he arranged the visit of the first important visitor from the South African Atomic Energy Board, Dr W. L. Grant, then its Director General.[15] Grant was known for his excellence in nuclear engineering (Gold Medal of the Institute of Mechanical Engineering in 1958) and for his hobby of target shooting. Accompanied by Sole, Grant visited the *Gesellschaft für Kernforschung* (Society for Nuclear Research) at Karlsruhe on 10 and 11 November. The visit marked the beginning of discussions aimed at obtaining the nuclear enrichment process for South Africa. Two months later Ambassador Sole pursued these discussions with Professor Leussing, then the Federal Minister of Education and Science, by inquiring into the possibility of South Africa's participation in Urenco/Centec, the joint enrichment project of West Germany, the Netherlands and Britain. With the expensive jet-nozzle system still at laboratory stage, the Almelo plant was to use the gas centrifuge technology then believed to be the most economically viable system of uranium enrichment.

This is how Sole described his conversation with Minister Leussing in a secret letter of 4 February 1970 addressed to the Secretary for Foreign Affairs in Pretoria:

> 'I referred to the interest shown in participation in this tripartite project by the Belgians and the Italians, also to the fact that Mr Wedgwood Benn had made it clear in his statement that collaboration with other countries interested could include States outside of Europe as well as in that continent. South Africa, as a major producer of uranium, was naturally interested in the success of such a project, not only from the point of view of providing an additional market for uranium producers, but also because it was the natural trend in any important uranium producing country to improve and expand its own technology – where possible also in collaboration with other countries. I outlined to the Minister our interest in promoting sales of uranium to the Federal Republic of Germany, referred to our traditional sales relationship in this field with the United Kingdom, sketched the thinking now going on in South Africa about our own future atomic power programme and expressed to him the personal view that there would be considerable interest in South Africa in exploring the possibilities of future collaboration with the Federal Republic and its partners in the

production of enriched uranium by centrifuge techniques. I enquired how the Minister himself would view the possibility of South African participation, emphasising that I was well aware that much more development work was necessary among the three countries concerned before the prospect of participation by an extra-European state would be ripe for consideration.'

According to Sole, Leussing was 'cautious in his reply', and he summed up his answer as follows:

'He said that at the present time the Federal Republic and the Netherlands were discussing the whole project in Euratom. The Euratom hurdle had first to be cleared and that would take some time. There was still a lot of work to be done on the development side and he stressed that as far as enrichment was concerned West Germany was keeping all her options open. Work was proceeding for example also on the "nozzle" process. But he agreed that participation in the tripartite project might well be extended to a country or countries outside of Europe.'

Although Sole promised Pretoria that he would come back to the issue a month later, nothing more emerged. South Africa was, no doubt, an attractive partner because of its uranium resources; however, the political implications of participation in the Almelo project, notably its refusal to sign the Non-Proliferation Treaty, outweighed the economic considerations.

South Africa did not press the issue for reasons of its own. From the South African point of view participation in Almelo did not offer any guarantee of the transfer of enrichment technology for uncontrolled use at home. This would have been an open violation of the Euratom and IAEA safeguards to which the project was subject. In the end the West German jet-nozzle system turned out to be the best option for the following reasons:

1. South Africa with its cheap electric power was in a position to test the jet-nozzle enrichment technology on

63

a scale that the Karlsruhe Research Centre could hardly afford. This in turn would enable the Germans to perfect the method and perhaps make it more viable commercially.

2. By providing South Africa with enrichment technology the Germans could reasonably expect to be rewarded by a steady supply of enriched uranium, which would reduce their dependence on the United States and the Soviet Union.

3. West Germany, which was one of South Africa's best trading partners and politically one of its most understanding allies, could be relied upon to keep the co-operation secret in order to avoid accusations that South Africa was acquiring the atomic bomb.

4. By providing South Africa with the 'know-how', West Germany would ensure that contracts for supply of equipment for the enrichment plant would go to German companies.

These and other aspects were, no doubt, discussed by the German visitors to Pelindaba and South African visitors to Karlsruhe and Essen. However, there were problems: in the early 1970s West Germany was still waiting to be admitted to the United Nations, and since 21 August 1969 it had had a Social Democratic Government whose Chancellor, Willy Brandt, was rather concerned about West Germany's standing in world affairs.

On the basis of the secret documents in our possession, it can be seen that co-operation between West Germany and South Africa in the field of uranium enrichment proceeded broadly as follows:

On 18 April 1972 the Secretary of State at the Ministry of Education and Science, Dr H. Haunschild, who was also Chairman of the Society for Nuclear Research at Karlsruhe, led a delegation to Pelindaba for talks with the South African Atomic Energy Board. It should be recalled that 90 per cent of the shares of the Society for Nuclear Research, which is a parent body of the Institute for Nuclear Processing Techniques where the jet-nozzle system for uranium enrichment was developed, are owned by the Federal Government. In 1972 both research and technology still fell under the competence of the Federal Ministry of Education and Science. In 1973 a separate Ministry for Research and Technology was established with Professor Horst Ehmke as its first Minister. Herr Haunschild was accompanied by Dr Dietman Frenzel of the Technological Research and Development Department of his Ministry, Dr Martin Nettesheim of the Department for International Co-operation, and Dr Rainer Gerold.

South African Ambassador Donald Sole, who had called several times on Minister Haunschild in connection with preparations for the visit, conveyed his impressions to Pretoria in a letter of 24 February 1972 in which he wrote: 'It is quite clear from my discussions with Dr Haunschild that there is a great deal of official interest in this visit. It is not simply a courtesy gesture in response to an invitation from Dr Roux, hence the inclusion of other experts in the party, although the invitation was issued originally only to Dr Haunschild and his wife.' (See Document 11, below.)

The details of Dr Haunschild's talks with Dr Roux at Pelindaba are not known, but that they concerned the 'exploratory phase' of collaboration in uranium enrichment between Karlsruhe and Pelindaba, which South Africa was anxious to keep secret, is clear from a confidential letter from Haunschild to Roux on 12 July 1972. Haunschild referred to the conversation he had with Dr Roux at Skukuza in South Africa, in the course of which Dr Roux asked Herr Haunschild to enquire into how, in German law, it could be ensured that '... the employees of an industrial company involved in the exploration stage of the enrichment process would keep the information a secret even prior to the decision on mutual co-operation.' Haunschild promised to look into the matter and, after explaining to Dr Roux the relevant provisions of German law, wrote: 'In our discussion we agreed that to conclude an agreement between our two countries on keeping our dealings secret is at present not advisable.' However, he pointed out '... the German private companies are of course free to decide themselves which information about their foreign transactions they wish to treat as confidential.' (See Document 14, below.)

Steag and Ucor make a deal

Two months after a visit to Pelindaba on 11-13 January 1973 by Dr K. H. Bund, Chairman of the Executive Committee of Ruhkohle AG and Chairman of the Board of Steag (which had acquired the exclusive patent rights to the uranium enrichment jet-nozzle system), Dr A. J. A. Roux arrived in West Germany for further discussions. The importance of his visit can be deduced from his itinerary, which shows that Roux met the three key people from Steag concerned with the jet-nozzle system, namely Dr K. H. Bund, Dr H. Völker, Director of the Department of Nuclear Energy of Steag, and Herr Hugo Geppert, who was in charge of the Steag-Ucor project for co-operation in the field of uranium enrichment. All of them also attended social events in Dr Roux's honour. Secretary of State H. Haunschild and Dr Nettensheim, who handled West German/South African relations at his Ministry, came to the first dinner for Dr Roux.[16] On 15 March, after a lunch at the restaurant *Die Mühle* in the Sheraton Hotel in Munich, the afternoon was devoted to discussion about co-operation.

On 1 May 1973 Dr H. Völker and Herr H. Geppert of Steag flew to Pelindaba for further talks with Dr Roux. Six weeks after their return, on 13 June 1973, Völker sent Roux the following telex:

'Dear Dr Roux,
With reference to Article 14 of the Memorandum of Understanding between Ucor/AEB and Steag we have

the pleasure to inform you as follows:

1. Steag Board of Management has agreed to the Memorandum.
2. State Secretary Haunschild was informed by Dr Bund about the Memorandum and has agreed to proceed as planned.
3. GFK (the Society for Nuclear Research at Karlsruhe) has given approval to the Memorandum in principle. GFK has recommended an equivalent wording with respect to points 2a and 2b in the final agreement.
4. With respect to the political situation we refer to the recent letter exchange between Prime Minister Vorster and Chancellor Brandt.
5. We kindly ask you to proceed in this matter as agreed during our meeting.'

In the evening of the following day (14 June 1973) Dr Völker received Ambassador Sole, who handed him four sealed envelopes which had been sent by the Uranium Enrichment Corporation of South Africa in the diplomatic bag. They were addressed to Herr Hugo Geppert, and Ambassador Sole was instructed by the Secretary for Foreign Affairs to deliver them personally. When Dr Völcker gave Sole a copy of his telex to Dr Roux (quoted above), Sole was rather upset by the fact that a communication of this kind was transmitted on an open line.[17] The telex revealed that there existed a 'Memorandum of Understanding' between Steag and the South African Uranium Enrichment Corporation (Ucor) to which both Secretary of State H. Haunschild and the GfK (Society for Nuclear Research of Karlsruhe) had 'agreed'; and also that there had been an exchange of letters between Prime Minister John Vorster and Chancellor Willy Brandt in connection with nuclear co-operation.[18]

'We operate on the basis that both telex and telephone communication to South Africa is periodically monitored,' Sole pointed out to Dr Völcker. 'We had warned other German firms engaged in undertakings of a confidential nature to exercise care in the use of open telex and telephone lines. These firms had been told that confidential communications to, for example, the Armaments Board could, if they wished, be sent through the South African Embassy utilizing its cipher facilities for this purpose.'

Ambassador Sole then suggested to Dr Völcker that Steag too would be welcome to use this secret communications link with the South African Atomic Energy Board.

The agreement on co-operation between Steag and Ucor, under which Steag was to arrange the legal transfer of rights to the jet-nozzle system, was signed at Pelindaba on 15 August 1973 by Dr Bund and Dr Völcker on behalf of Steag and by Dr Roux for Ucor. The text of the agreement is not known, but the fact that it contained Steag's approval for sublicensing its rights to the jet-nozzle system to Ucor is revealed in a letter written by Dr H. Völcker and H. Geppert of Steag to Dr A. Roux of Ucor on 2 October 1973. The letter refers to Article 3 of the Agreement, according to which Steag was obliged to seek the German Government's permission to conclude the deal:

'Dear Dr Roux,
Referring to Article 3 of our agreement we have requested formal approval for sublicensing our rights according to our contract with Gesellschaft für Kern-forschung Karlsruhe (GfK). GfK has agreed to our request in principle but needs approval of Staatssekretär Haunschild as chairman of GfK supervisory board.

We have unofficially been informed that on the request of Mr Haunschild Staatssekretäre of the Ministry of Economy, the Ministry of Foreign Affairs and of Chancellor Brandt's office met on 27th of September to discuss this matter. They have given a positive reaction to the GfK position. However at the request of the Ministry of Foreign Affairs a legal investigation to find out whether the Aussenwirtschaftsgesetz[19] is applicable in this case is still necessary. The expert of the Ministry of Economics, who has meanwhile contacted us, has already unofficially confirmed that this law is not applicable and that he cannot see a reason to withhold Mr Haunschild's approval.

We have once more underlined the urgency of this matter and expect the final decision within the next few days. We are extremely sorry for this delay and shall keep you informed about any further steps.

Yours sincerely,
Steag Ltd.
Aktiengesellschaft
H. Völcker   H. Geppert

STEAG GOES TO THE CABINET: 'GROSS DECEPTION'

'I never knew of the existence of the letter from Steag to Dr A. Roux of 2 October 1973', Dr Erhard Eppler, former Minister for Development Aid and member of the German Cabinet (from 1968 to 1974), told the London *Sunday Times* correspondent Anthony Terry in October 1977,

> '... but it makes perfect sense in the light of the sequence of events. Obviously the officials concerned had to go to Haunschild and present him with a proposal for Haunschild to decide. Haunschild did not want to decide as the whole matter was a bit too hot for him to handle. So he presented it immediately to the Cabinet as a joint proposal from the three Federal Ministries concerned. I know that the presentation was rushed onto the Cabinet agenda so suddenly that my personal aide for Cabinet affairs in my Ministry did not even have time to make me any advance notes on the subject as is normally the case. It meant I had to read the documents through myself on the spot and it was then that the idea occurred to me that there was something fishy about the whole project.'[20]

The proposal, as put forward on 17 October 1973, was for a joint study for a nuclear enrichment plant by Steag and Ucor. 'There was nothing in it about actually building the plant but as far as I was concerned the whole operation only made sense if the plant was built,' observed Dr Eppler. He described the Cabinet session as follows:

> 'The Steag proposal put before the Cabinet contained a sentence which said that the South Africans regarded it as an "essential pre-condition" for the start of the comparative study that Steag should grant an option for a sublicence for the manufacture of the jet-nozzle system. Though the whole operation was supposed to be "the study" this little sentence was slipped in and made it clear what it was all about.'

Dr Eppler opposed the proposal at the Cabinet meeting, and argued that '. . . to link the Federal Republic of Germany with South Africa in the nuclear field would effectually destroy West German relations with other African countries.' He was supported by Werner Maihofer, then the Minister without Portfolio, while the present Chancellor Helmut Schmidt, then the Minister of Finance, the present Foreign Minister Hans-Dietrich Genscher (then Minister of Interior) and Ministers Hans-Jocken Vogel (Housing), Hans Friederichs (Economy), Josef Ertl (Agriculture and Forestry), Walter Arendt (Employment and Social Welfare), Katharina Focke (Health, Youth and Family), Egon Francke (Domestic Affairs), Egon Bahr (Relations with the GDR) and Georg Leber (Defence) were in favour of the Steag proposal. Chancellor Willy Brandt, according to Dr Eppler, 'appeared to oppose the idea' and suggested that they should postpone the matter until the next Cabinet meeting. At this meeting, on 24 October 1973, Dr Ehmke, the Research Minister, announced that Steag had asked for the proposal to be withdrawn. 'I naturally assumed from this,' said Dr Eppler, 'that the matter was dead and buried, as in the whole six years of my Parliamentary and Cabinet experience I have known that when a proposal was put forward to the Cabinet and then withdrawn, it was because it was dead and buried.'

This was also the official version of events put out by the Federal Ministry of Foreign Affairs, which assured *Sunday Times* correspondent Anthony Terry, who investigated the matter on 6 October 1977, that 'there was no further co-operation between Steag and Ucor'. This claim is contradicted by the documents and also specifically denied by Dr Helmut Völcker of Steag. Dr Völcker also revealed that the agreement about the co-operation and feasibility study was signed in August 1973, but, as he put it, '. . . the work only got under way in April 1974'. This appears to fit the chronology: the public press release of 9 April 1974 about the Steag-Ucor co-operation was issued after a secret visit of Steag principals to Pelindaba described below. Dr Völcker did not explain how Steag got around the Cabinet nor the fact that when Steag, in October 1974, asked the Cabinet for the approval of its agreement with Ucor, *the agreement was already signed.*

Dr Eppler did not know how this could have happened but he made the following guess:

'It was not until much later, in early summer 1975, when I had long since dropped out of the Government, that I learned by pure chance that the whole affair had not ended at all but had been kept going despite the absence of Cabinet sanction. It had evidently continued on the lower level of Secretaries of State, with the two State Secretaries Dr Hans-Hilger Haunschild in the Ministry for Scientific Research and Dr Detlev Rohwedder in the Economics Ministry as the driving force behind it.

I naturally do not know, and it is none of my business, how far the Government Ministers who brought the proposal up at the Cabinet meeting gave their State Secretaries a free hand afterwards. But normally in a Ministry the State Secretary does not do anything without the green light from the Minister himself. I do not know whether in this case the Secretaries of State plotted to get the matter through themselves claiming that it did not need to be submitted to the Cabinet or whether they got a private nudge from their ministers to go ahead with it below Cabinet level.

But in the Cabinet there was no decision in favour and we were left in the belief that the matter was on the shelf for good.'

He then continued:

71

'When the Cabinet failed to take action the State Secretaries, including Rohwedder, obviously then got together to sidestep the Cabinet and decided to put the whole operation through themselves with Haunschild playing the leading role and, I fear, someone from the Foreign Ministry going along with it too.

What I fail to understand, because it is contrary to the official procedure of the West German Government, is how something that reached Cabinet level and was left undecided can then be continued at State Secretary level. This was nothing more or less than a gross deception of the Cabinet. If I had been West German Chancellor and had noticed this was going on I would have sacked both of them immediately. Otherwise the Government makes itself ridiculous if it discusses something which is then withdrawn and the State Secretaries then go on and do what they like – it's an impossible procedure.

It is clear now that the bureaucrats in the Economics and Research Ministries were in favour of the whole project anyway and they merely wanted Cabinet's blessing for it. I can say this now quite openly because I believe this scandal should be brought into the open and both of them should be for the high jump.'[21]

The opposition of Dr Eppler to nuclear co-operation with South Africa, as well as against West German participation in the Rössing uranium-mining project in Namibia (he stopped Government credit guarantees of the West German investments in Rössing) has made him many enemies. Shortly after the showdown in the Cabinet over the Steag proposal to pass the uranium enrichment technology on to South Africa, the right-wing press unleashed a campaign against Eppler which persisted for six months until he resigned. 'I certainly did notice that the real campaign started against me immediately after this affair – in the winter of 1973 and beginning of 1974 – in the Springer newspaper *Bildzeitung* and elsewhere... This campaign took me by surprise', Dr Eppler recalled, 'it lasted about six months and they used all sorts of methods. I do not want to say that the two things are directly connected but it cannot be excluded.'

SECRET DEALS GET UNDER WAY

The first move in covering up the transfer of the 'licence' for the uranium enrichment technology from Karlsruhe to Pelindaba, was to replace the words 'sub-licensing' by 'comparative study' in the official pronouncement on the relationship between Steag and Ucor. However, in their secret communications the words 'agreement on co-operation' were still used. What the two companies were now saying in public was that they were interested only in making a comparison between the South African enrichment process and the West German jet-nozzle technology, in order to find out which system was more feasible technically and more viable financially.

In the absence of the official Cabinet support which is an essential prerequisite for a Government credit guarantee, funds also had to be found to finance the Steag-Ucor deal. Dr N. Diederichs, the South African Minister of Finance, took care of this problem. On 31 October 1973 he arrived in Frankfurt for talks with the West German bankers involved in the financing of the West German nuclear industry. His two-week programme included meetings with a number of leading West German financial heavy-weights. On the afternoon of 30 October he met Dr Paul Lichtenberg, Chairman of the Board of the Commerzbank. That evening he met Jürgen Ponto, Chairman of the Dresdner Bank and Helmut Hönsgen, a Board member; the Vice-Chairman of Metallgesellschaft AG, Casimir Prinz Wittgenstein; and Paul Ungerer, Chairman of the Degussa Board. On 8 November there were discussions with Dr Wilfred Guth of the Deutsche Bank Board and that evening with Dr Karl Klasen, President of Deutsche Bundesbank, the National Bank of the FRG. On the following morning, 9 November, discussions were held with the Honorary Chairman of the Board of Deutsche Bank, Herman Joseph Abs.

The new phase of co-operation between Steag and Ucor was finalized in March 1973. In a secret coded cable of 20

February 1974 Ambassador Sole, using military communication channels, notified Pretoria of the impending arrival of Professor E. Becker accompanied by Dr Völcker and Dr Schulte of Steag. The text of the message was as follows:

OUTGOING SECRET TELEGRAM
To: CHIEF OF THE ARMY PRETORIA
From: THE SOUTH AFRICAN EMBASSY COLOGNE
Date: 20 February 1974

Please convey the following message to the Secretary of Foreign Affairs as our normal channels have broken down. Begins.

Secret
Nr. 38.
'Professor Becker of Karlsruhe (the uranium enrichment specialist) accompanied by Dr Völcker and Dr Schulte of Steag departing Frankfurt by SA 231 on Friday, March 1, arriving Johannesburg Saturday at 1335 hours. This visit was arranged by Dr Roux on his last visit to Germany. Please, ensure appropriate entry facilities and endeavour to avoid publicity re Professor Becker.'

OFFICIAL ATTEMPTS TO EXPLAIN THE DEALS: A MASS OF CONTRADICTIONS

The South African claim that their uranium enrichment process is of their own invention caused considerable embarrassment to the Germans, who suddenly found themselves accused of passing nuclear secrets to a non-signatory of the Non-Proliferation Treaty, and one suspected of wanting to use the acquired uranium enrichment technology for military purposes. Naturally, the Germans tried to cover it up.

When *Sunday Times* correspondent Anthony Terry asked Professor Becker about his visits to Pelindaba, the latter was careful in his answers. It became clear that Becker was not

aware that the secret coded cable about his visit to South
Africa in March 1974 had been made public, in a front page
story by Andrew Wilson in the London *Observer* of 14 August
1977; the German press had maintained a notable silence
about the whole affair. Professor Becker still maintains that
he himself never saw what the South Africans were really
doing.

> 'As scientists and technicians developing a process we
> would naturally like to discover that something some-
> body else is doing is the same as ours or at any rate has
> similarities in some respects. But you no doubt know how
> politically explosive this subject is at the present time,
> when for political reasons it is stated that two things are
> quite different. This is why I cannot make any statement
> on the matter as I genuinely do not know how the South
> African process works. The best way to discover the
> similarities between the two processes is in the documents
> of the Nuclear Energy conference in Paris in April 1975.
> Dr Roux from South Africa disclosed to the conference
> then what he felt he wanted to make known about his
> process and a public discussion followed in which I also
> joined. As a result we arrived at the following conclusion
> about the similarities between the two processes: namely
> that both are gas dynamic processes using a light
> additional gas together with UF6 (uranium hexafluoride).
> But there are also differences. At least the South Africans
> claim that there are considerable differences. That
> means that there is a difference in the way the separation
> is established when using the UF6 together with a light
> additional gas. The South Africans have described the
> process as a centrifuge with a stationary wall.
> This means that they have something like a tube and
> they introduce a mixture of UF6 and an additional gas in
> such a way that a vortex is formed and there is separation
> between the isotopes through the movement of the gas.
> That is the similarity. In our process we also expand UF6
> in a mixture of light additional gas and we have curved
> flow passages. I cannot answer the direct question of how
> close the two processes are as we simply do not know how
> the South African one actually works. There are certainly
> marked differences between them but we do not know

where. We can only draw conclusions and make assumptions. The South African development has taken place under conditions of total secrecy while ours is entirely open and public. This means that South Africans know what we are doing but we do not know anything about what they are up to apart from the little they were prepared to disclose at the 1975 conference in which they dealt solely with points of similarity but not dissimilarity.'

Dr Völcker, of Steag, used stronger words when trying to explain the curious similarity between the West German and the South African uranium enrichment systems: 'Steag had no secrets to hide about its own enrichment process. The blackest Zulu kaffir can find out all what he wants to about it.'

'Were they [Ucor] further than Steag?' Terry asked Völcker. 'No comment,' was the reply.

It was shortly after Professor Becker's visit to Pelindaba that Steag announced (on 9 April 1974) the existence of a '. . . joint comparative economic feasibility study between our Karlsruhe jet-nozzle uranium enrichment process and the secret South African process.' It is not clear how the Germans could participate in a comparative economic study of a system which is secret and, according to Becker, 'inaccessible'. None of the Steag representatives has explained anything beyond saying 'no comment'.

This is how Professor Becker described his visit to Pelindaba in 1974, which he maintained was the only one he made:

'They never let me inside. Reports which have appeared that I worked there or that we had South Africans here for training in this field are wrong. Naturally, the Karlsruhe Research Center has had visiting South Africans. But they worked on quite different things and not on the jet-nozzle process. In Pelindaba I only saw what has already appeared on photographs of the installation, some of them showing it from inside, but I was never allowed in myself. I did meet Dr Roux and Dr Grant whom I knew anyway, but they never let me in.'

76

Dr Völcker supported Professor Becker's explanation of the 'tourist visit' of Pelindaba by saying: 'When we got there they already had a lot. They could have achieved a final result entirely on their own. They were an efficient team.' But this claim contrasts oddly with Becker's protestations of total ignorance about South Africa's achievements.

Völcker also contradicted the Foreign Office's repeated claim that all co-operation between Steag and Ucor had ceased in 1973. He told Anthony Terry on 7 October 1977:

> 'One of the many lies that have been circulated is that we ended our collaboration in 1973. And I cannot do anything to help someone who persists in clinging to this nonsense. When we agreed to carry out the feasibility study in 1973 it was agreed that it would be concluded as soon as possible but would end finally at the latest by March 31, 1976. We would have had to be bananas to have ended the study in 1973 and then to issue a press statement in 1974 saying we were still working on it.'

The President of the South African Atomic Energy Board, Dr A. Roux, reacted irritably to the suggestion made by Professor Becker to the correspondent of the *Washington Post* on 16 February 1977 that as Dr Roux and other South African scientists had had free access to the results of his research, it was possible that they had adopted them. Roux retorted: 'Becker may be a friend of mine but already before he has talked so much nonsense that this is just another proof. All we have attained, we attained by ourselves.'[22] The official explanations, it seems, have not yet been as well co-ordinated as the original deals.

German-South African co-operation on the testing and improvement of Professor Becker's system was conducted by Steag and Ucor, and it is curious that Steag, which claimed that the technology of the South African enrichment process was not the same as the German jet-nozzle system, did not choose to publish the results of the study comparing the two systems which was the ostensible basis for the whole deal.

Obviously the German system was considerably improved at Pelindaba. It also seems that the uranium enrichment technology sold to Brazil in 1975 as a part of the DM2 billion deal is based on the jet-nozzle system improved through testing in South Africa.

After South Africa announced that it was ready to embark on a large-scale commercial enrichment programme, the question of further co-operation between Steag and Ucor was no longer confined to the 'comparative study' but involved investment by Steag in the Ucor project. This opened the door to West German suppliers of technical equipment. While Steag's technological assistance was still indispensable, as far as capital was concerned, there turned out to be another party very keen on co-operation with South Africa: Iran. On 13 October 1975 the *International Herald Tribune* reported that South Africa had negotiated an agreement with Iran according to which Iran would invest in the South African enrichment project in return for the enriched uranium needed for the Shah's own ambitious nuclear programme.

CO-OPERATION CONTINUES ...

New ideas on West German-South African nuclear co-operation were discussed at Pelindaba in April 1975 between the Secretary of State of the Federal Ministry of Economics, C. D. Rohwedder, the Vice-President of the South African Atomic Energy Board, Dr J. W. L. de Villiers, and the Deputy General Manager of Ucor, Dr R. S. Loubser. Rohwedder, who spent ten days in South Africa, visited Pelindaba on 22 April and ended his visit with talks with the Minister of Mines, Dr P. G. S. Koornhof, in Pretoria. After his return home, Herr Rohwedder wrote to de Villiers and Loubser expressing his enchantment at the beauties of the country and thanking them for an 'interesting afternoon at Pelindaba.' He continued:

'It would be nice if we could come to a long-term and mutually fruitful co-operation in your field or in the general field of peaceful uses of nuclear energy. I have already started to put together suggestions and the contents of our talks and brought them to the attention of those who are interested in a closer exchange and who would be the right type of partners.'[23]

Herr Rohwedder's view was fully shared by the economics expert of the Free Democratic Party group in Parliament, Count Lambsdorff, who visited South Africa in February 1975 at the expense of the South African government. On 24 April 1975, during the debate on energy in the *Bundestag*, he recommended to Parliament that West Germany should participate in the uranium enrichment project in South Africa.[24]

Perhaps the best insight into the 'understanding' between Sole and Rohwedder is provided by Ambassador Sole's own report on the conversation with Rohwedder after the latter's return from South Africa. Sole summed it up in his confidential letter (of 2 May 1975) to Brand Fourie at the Department of Foreign Affairs:

CONFIDENTIAL   2 May 1975

Dear Brand,

Visit of Secretary of State Rohwedder

Since my telegram no. K23 (copy attached) was despatched I have had an opportunity for a discussion with Rohwedder in some depth, lasting about an hour and a half, and what he told me more than confirmed my telegraphic report of his impressions based on my earlier conversation with him and with one of his advisers. I give below a summary of the cardinal points which emerged.

Rohwedder described his trip to South Africa as the most instructive overseas visit he had ever undertaken. He had learned a tremendous amount, not only about South Africa and its economic and political potential but perhaps even more of the nature of its people. Apart from the traditional hospitality for which he said we were

79

deservedly famous, he was struck by the warm and genuine friendship for West Germany; it was friendship such as he had encountered in hardly any other country of the world. He considered himself privileged too to have had interviews with so many Ministers, including even a brief call on the Prime Minister, but he had especially appreciated his talk with Dr Diederichs where he had felt that he as a junior economist of some standing had been able to share the thoughts and ideas of a wise leader and elder statesman who ranked amongst the world's foremost practical economists.

On the political side he referred to the insight which especially you, but also Dr Koornhof had given him into our relations with other African countries. This had been a complete eye-opener and he was most grateful for the openness with which the information had been imparted and the confidence which had been reposed in his discretion. In private talks with his colleagues and with German Ministers he would be able to make most valuable use of the impressions he had gained with respect to this particular aspect of South African foreign policy.

I had not known beforehand that his itinerary would include a visit to Silvermine. This visit, and the talks which he had on that occasion, have convinced him of the necessity for the German Government to take a fresh look at its policies with respect to the supply of equipment to South Africa which can be regarded as important for the defence of Western Europe. He has returned with a completely new concept of what South Africa means, strategy wise, with respect to oil supplies to Western Europe from the Persian Gulf. He mentioned in this context that Chancellor Schmidt had recently instructed that a fresh study be undertaken of how best the Federal Republic can be assured of receiving oil supplies from the Persian Gulf. His impressions, arising out of his visit to Silvermine, would enable him to make a valuable contribution to that study and he had already arranged for a detailed discussion with his opposite number in the German Ministry of Defence. He had been surprised to learn how little contact there was, even informally, between South African defence experts and the top officers of the NATO command in Brussels. And he wondered whether something could be done about it, although this was not his field. On the question of the supply of certain types of defence equipment, however, he would take an early opportunity of initiating a re-appraisement of what was politically feasible. Obviously he could not make any promises. He also emphasized that it is German policy to base its defence equipment deliveries on its commitments to its NATO allies but he said for example that with respect to such items as fast patrol boats (schnellbooten in German), which were delivered to countries such as Peru and Venezuela as well as to NATO countries, it might be worthwhile studying whether a case could not be made out for delivery of this kind of item to South Africa if this kind of item could be identified as a contribution to Western defence. Of course, there would be the United Nations embargo to be contended with. Of course, West Germany could not act as independently of United Nations injunctions as France could do. But the matter should be looked into. He repeated that he could promise nothing. But I think myself that if he can get the ear of Chancellor Schmidt with respect to this particular problem, progress may be made. There will be objections from the Foreign Ministry. Of that I have no doubt. But he will have Defence Minister Leber on his side. Perhaps I should add too that this is a development with *long* term rather than short term prospects which will be important if it is Schmidt who is returned to power as Chancellor in next year's elections and the CDU/CSU opposition fails to make the grade.

We had a long discussion about co-operation in the field of energy resources. There are one or two points which are worth recording in this context. He said that although he had nothing against STEAG it was important that we should press STEAG to agree to widen the area of co-operation on the German side by bringing into a new consortium other companies with a major interest in this field. He specifically mentioned Kraftwerk Union and suggested that we also secure the collaboration of Prof Mandel of RWE (the Professor is of course an old friend of ours). I said that Dr Koornhof and I had already broached this matter in our discussions with Dr Bund during our working breakfast with Ruhrkohle in Essen

81

last month. I had received the impression that Ruhrkohle was not opposed. He said that this might be so but we *must* put on pressure from our side.

We talked about the possibility of trilateral collaboration with respect to the long term nuclear energy requirements of Iran. I told him of recent discussions which I had with Mr Beitz of KRUPP in this regard. He referred to the close liaison already established between the Iranians and the Americans and between the Iranians and the French but pointed out that it was only the Germans who so far have got off the ground in the field of providing Iran with reactors. In the case of the French and the Americans it was still a lot of talk plus of course letters of intent.

With respect to German/South African co-operation in the uranium enrichment field he was most grateful for the insight provided by his visit to Pelindaba and by his discussions with the top people of the Atomic Energy Board during the Kruger Park weekend as well as with Dr Koornhof. But he stressed that there would be tricky political problems to be overcome inside his Government before a firm final agreement would be reached. He mentioned his intention to talk to Haunschild in this respect. Although he did not say so explicitly, this may be one of the reasons why he advocated a wider base than just STEAG alone as the German side of the tie-up with South Africa.

Fundamentally, he said, there were two broad options open to the Federal Republic in its relations with South Africa.

The first option was for Germany to align itself with popular ideologies (which also had overwhelming support within the two governing parties) and to curtail German contacts with South Africa to what was regarded as absolutely essential or vital for Germany's continuing economic stability and prosperity; for the rest the Federal Republic could play along with the Third World and with the United Nations majority.

The second option was to recognise that South Africa is a country with tremendous problems, a country which is extremely well disposed towards Germany, a country with which Germany *in her own interests* and in the interest of the Western world can develop increasingly close

82

economic ties. From this recognition would flow a further recognition that South Africa politically is moving in a direction which can command understanding and sympathy, and also respect. Furthermore, the Federal Republic should do what it can, through its own international contacts, to assist South Africa by promoting a better appreciation of what South Africa is trying to achieve, as also of what South Africa can contribute in various fields towards the promotion of stability and security, not only in Africa but in Europe as well.

He said that I would have no doubts as to where he stood as between these two options. But I should not underestimate the strength of opinion within the two parties making up the Government which would prefer to see the Federal Republic adopt the first option.

Rohwedder is of course an SPD man. He occupies a post of Secretary of State which is a civil service post but this does not make him any less SPD. It is therefore the more encouraging that he has returned from South Africa with the impressions summarized above.

May I repeat to you and to others concerned my warmest thanks for what you did to ensure that his visit produced such positive impressions.

Yours sincerely,

DBS/svdb

Mr Brand Fourie,
Department of Foreign Affairs,
CAPE TOWN.

... BUT STEAG BACKS OUT

However, at the beginning of 1976 the negotiations between Steag and Ucor about Steag's participation in the commercial uranium enrichment plant in South Africa ran into difficulties.

On 31 March 1976 co-operation between the two companies was officially terminated. In an internal confidential note addressed to Dr Bund, Dr Völcker summed up the main points of disagreement over Steag's participation in the Ucor project as follows:

1. Ucor had refused to commit itself to guaranteeing that it would supply Steag with uranium from South African sources, so that Steag could meet its obligation towards the project.
2. No interest would be paid on the capital invested by Steag if Steag failed to provide its obligatory quota of uranium.
3. Steag was to be financially responsible for any failures in supplies of equipment caused by the political situation.
4. Secretary of State H. H. Haunschild had made it clear to Steag that it could not expect any government guarantees for its investment.

The document clearly shows that the reasons for termination had nothing to do with the key to the whole commercial enrichment project – the jet-nozzle enrichment process – since the technology for that had been handed over long before 1976. Steag declined to put money into Ucor's project simply because the conditions were unacceptable. It would have been interesting to know from where Steag was supposed to get the extra uranium for the South African enrichment plant, which South Africa alone could not supply.

In March 1976 Steag withdrew completely from financial participation in the South African enrichment plant. The order for compressors for the enrichment plant placed by South Africa with the Maschinenfabrik Augsburg-Nürnberg (MAN) was official cancelled.[25] On 29 May 1976 Kraftwerk Union of Karlsruhe lost to the French Framatome consortium the contract to deliver two nuclear reactors to South Africa.[26] However, the most crucial element, the technology of the enrichment, had already been 'licensed' to South Africa by the Federal Republic of Germany.

West Germany and South Africa – a special relationship

In the 1950s West Germany's first Chancellor, Konrad Adenauer, won the struggle for the Federal Republic to have equal status with its Western allies. During the 1970s, while the economies of France, Britain and other countries have had severe problems, West Germany has hardly faltered in its climb up from the bottom of the ladder which, in the words of Chapter XVII of the UN Charter, was reserved for 'the enemy of the signatories of the Charter.' It is now second only to the United States in economic power in the Western World. Its high rate of growth, combined with low inflation, a strong balance of payments and full employment have remained consistent since the post-war recovery.

West Germany is now an important banker for the Western world in general and Western Europe in particular. In mid-1976 its foreign assets stood at £89,487 m., as against liabilities of £61,794 m.

West Germany's African policy has had two faces. One is the posture of supporting the political and economic aims of African countries, including the condemnation of *apartheid*, and providing development finance. When talking about West German-African relations, Bonn politicians like to dwell at length on the West Germany aid programme and the claim that Bonn has actually exceeded the UN target for industrial nations to allocate one per cent of their GNP annually for development finance. West German politicians say that their country's contribution is now 1.18 per cent. This assertion is true only if one accepts, as the Germans claim, that private investment and commercial loans by private banks at high interest rates (exceeding 10 per cent) count as a real contribution to 'development'.

Real development aid provided by the Government – technical assistance, low-interest loans and grants – amounts to only 0.3 per cent of West Germany's GNP, a rather poor

85

record in comparison with that of many other European countries.

The other face of Bonn's African policy is manifested in its collaboration with the Government of South Africa. For several years now African States have expressed their concern at the growing economic and political ties between West Germany and South Africa, exemplified by the steadily increasing volume of trade, financial investment, nuclear co-operation and supply of arms to Pretoria: West Germany and South Africa clearly have a 'special relationship'.

The true size of the West German economic involvement was recently revealed by the controversy arising from a public pronouncement by Egon Bahr, the Manager of the Social Democratic Party (SPD). On 5 July 1977 he said that West German investments in South Africa had fallen to practically nothing.[27] He was soon corrected by a press release from the South African Embassy in Bonn issued the following week, stating that the opposite was true. According to the South Africans, West German private investment totalled DM576.2 m. at the end of 1976, a rise of DM 38 m. over 1975. About 6,000 West German firms had direct or indirect contacts with South Africa and in the industrial and trading sectors more than 300 German firms had subsidiaries in South Africa. The South African Embassy estimated total German financial involvement in South Africa to be in the region of DM12 billion, including indirect investment through subsidiaries and export credits. 'All major German concerns in the electro-technical, chemical, automobile and steel industries are intending to go ahead with their investments in South Africa,' said the statement.

The most startling response to Egon Bahr's claim was a revelation made by the Information Service on Southern Africa (ISSA) in Bonn, which is linked with anti-*apartheid* groups throughout the Federal Republic of Germany. Quoting secret documents from the Government-controlled insurance company Hermes Kredit-Versicherungs AG in Hamburg, Peter Ripken of ISSA stated that during 1976 there had been a dramatic increase in Government credit guarantees for exports to South Africa. Over the previous 18 months, up to mid-May 1977, the credits totalled DM 2,775.13 m., almost four times the credit accorded to South Africa in 1975.[28] As a result South Africa received about 70 per cent of the total German government export credits and guarantees earmarked for Western industrialised countries in 1976. Among the deals supported by the guarantees is DM515m.-worth of steam-generating equipment supplied by Deutsche Babcock for the Sasol II project, manufacturing oil from coal: 14 million tons of coal are to be reduced to 2.7 million tons of oil annually in order to reduce South Africa's dependence on oil imports. Peter Ripken's report, which the Federal Government reluctantly acknowledged to be correct, caused serious embarrassment and led to an official investigation. It transpired that the purpose of the guarantees is to safeguard German firms which do business with companies in 'politically insecure' parts of the world. Documents relating to such guarantees are classified as secret, and are not normally referred to Ministers unless there are doubts about the credit-worthiness of the importing country; the credit-worthiness of South Africa was never in doubt, hence the decision about the credits was left to the civil servants. Neither Foreign Minister Genscher nor Economics Minister Friederichs was told about the massive South African commitments by Hermes, the official export credit agency of the West German Government.

African criticism of West German ties with South Africa, often expressed in OAU resolutions and the African media as well as in private conversations between West German and African representatives, had very little impact on public opinion in West Germany itself and even less on the Bonn Government. The Tanzanian *Daily News* described the West German Government as being '... probably the most in-sensitive to progressive international protest'.[29]

There are various reasons for this. Most Germans are relatively ignorant about Africa. For the German marketing personnel representing the South African travel agency

Satour, *apartheid* is a 'tourist attraction' and until recently the catalogue of the Tourope-Scharnow travel agency appealed to travel-happy Germans to visit South Africa and be 'thrilled by its brutal contradictions'. The ignorance of the German public about Africa has enabled business interests to seek their share of the South African market, in which cheap black labour has provided ideal conditions for the investment of German capital.

Good relations with the Nationalist Government which date from the Nazi era, as well as a sizeable German community there, are of considerable importance. Business interests have also been given active aid and advice by successive Bonn governments, which have been more concerned with sustaining Germany's 'economic miracle' than with waging a battle for human rights in foreign places. At present the ruling coalition of Social Democrats (SDP) led by Chancellor Helmut Schmidt, and Liberals led by Foreign Minister Hans-Dietrich Genscher, feel the same way as the conservative Christian Democrat opposition: that the best foundations for healthy international relations are free enterprise, increased trade and investment.

The other side of the argument has been put by Dr Uwe Holtz, one of the SPD members of the *Bundestag*: 'Our present policy towards South Africa is short-sighted. We have more to gain if we press for majority rule in South Africa and have good relations in the future than if we do business with *apartheid* at the expense of the credibility of our African policy, with the risk of losing African markets altogether. After all, South Africa ranks only 17th on the West German list of trade partners. The volume of trade with South Africa represents only 1.2 per cent of Germany's total imports and 1.5 of its exports'. Unfortunately, such arguments are being drowned out by the well-organised lobby clamouring for ever-increasing commitments to the South African economy.

South Africa's image in West Germany is probably more favourable than in any other Western country. The South African lobby in the *Bundestag* and in the West German business community is perhaps unparalleled in Western Europe. It has in its ranks a number of eminent West German politicians such as Franz Josef Strauss, the leader of the Bavarian Christian Social Party (CSU), who visited South Africa in 1966, 1971, 1973 and 1975. During his first visit to Cape Town on 2 May 1966 he explained his understanding of *apartheid* as follows: 'The policy of *apartheid* stems from a positive, religious awareness of responsibility for the development of the non-white population. It is therefore wrong to speak about an oppression of the non-whites by a super-race.'[30] The *South African Digest* quoted Herr Kai-Uwe von Hassel, Vice-President of the *Bundestag*, as saying after a two-week visit to South Africa that '... in no part of Africa were blacks better cared for than in South Africa'.[31]

Among other prominent political visitors to South Africa during the last six years are former Minister of Defence and Foreign Affairs Dr G. Schröder (in 1971 and again in 1972), former Minister of Finance Alex Möller (in 1973), and Dr Gerhard Stoltenberg, Prime Minister of Schleswig-Holstein and former Minister of Science (in 1973 and again in 1974). CDU *Bundestag* members Horst Schröder, Peter Willi Sick and Ursula Besser attended the ceremony for Transkei 'independence', despite a protest from their own party executive. Hans Klein, another member of the *Bundestag* (CSU), led a parliamentary delegation of CDU/CSU members (Dr Herbert Hupka, Peter Petersen, Gerhard Kunz, Prof. Dr Hans Hugo Klein, Dr Hans Stercken and Dr Fritz Voss) who visited South Africa in 1977, including the Transkei and Namibia – which is still referred to in the West German media as *Sudwestafrika* (South West Africa). When they returned they told *Deutschland-Magazin* that 'South Africa must not fall'[32]: the Transkei was a step in the right direction, and they had been impressed by Vorster's 'reforms' in racial policy, such as the opening of beaches, parks, museums, libraries and art galleries to blacks.

THE OFFICIAL COVER-UP

The Federal Government in West Germany has put severe pressure on domestic critics of its foreign policy. One example of this is the silencing of the German Africa Society when it started to issue embarrassing information and comment on official policy towards Africa.

On 16 October 1974, following reports of West German supplies of arms to South Africa and military support for the former Portuguese regime's colonial wars, the *Times of Zambia* issued an editorial containing one of the sharpest critiques of the Bonn Government ever published by an African paper. It read as follows:

> 'If the document published by the German Africa Society and reported in this paper yesterday is correct — and we have no reason to doubt its accuracy — then the truth is laid bare that West Germans will not hesitate to plunge their hands into the blood of black people if by so doing they will earn a few miserable coins for Germany.
>
> Of all the dastardly deeds committed by selfish money hunters, the deeds of West Germany in Southern Africa against African freedom just revealed surpass all.
>
> It is clear that, but for the exposed military involvement of West Germany in support of fascism, Caetano and his lot should have been flushed out long before.
>
> Any person, any country that supports *apartheid* we condemn. For any country that arms racism in southern Africa against freedom we can find no word of condemnation strong enough to express our feeling.
>
> Over the last ten years the indigenous people of southern Africa have been fighting a hot war of liberation. During this period the German Africa Society has revealed that West Germany has placed against African freedom 400,000 guns and over 200 military planes.
>
> These German weapons have been used against freedom fighters in Mozambique, Guinea-Bissau and Angola and are still maiming black people in the unliberated areas of southern Africa.
>
> We refuse to agree that the West German government

or for that matter any government worthy of the name, is powerless over commodity manufacture and trade by its nationals. The excuse is capitalist soft talk which amounts to nothing.

Black people are not that stupid to know that, if a factory in West Germany embarked on producing poison gas to kill Germans, the West German government will not hesitate to take swift action to stop it and bring the morons responsible to book.

As far as Africa is concerned, the leaders of West Germany have to take the responsibility for German poison and German planes that have [sic] and may still poison crops in southern Africa in aid of the futile attempt of racism to halt the freedom of the African people.

Western Germany has to accept the condemnation of Africa for the training its nationals are giving to racist minorities in biological warfare and other military resistance against the black majorities.

Africa is aware that Germans were responsible for ushering a Hitler into the world, also for Nazism, for the First and Second World Wars. Africa does not want a Germany-inspired African genocide.

West Germany has shown some fine gestures to Africa but unless German military involvement against Africans ceases, Africa will understand this first action as merely a cover for devilish work in the area.'

A few hours later the text of the editorial, cabled verbatim by the West German Embassy in Lusaka to the Foreign Office in Bonn, was brought to the Secretary of State, Hans Jürgen Wischnewski, one of the principal Africa policy-makers of the Social Democratic Party (SPD) and its highest official in the Foreign Ministry, which was then headed by the Liberal Party (FPD) Minister Walter Scheel. Nicknamed by *Der Spiegel* as 'a Dr Bloodless dealing in software' and described as a man who can talk like a gun — 'fast, slippery and round as a bullet'[33] — he enjoys the reputation of being able to make friends and influence people. Unfortunately, his charm had no effect on the source of the report about West Germany's collaboration with South Africa, the German Africa Society and its

Secretary-General, Dieter Habicht-Benthin, one of the most perceptive critics of West Germany's Africa policies.[34]

Wischnewski rose to his position as an 'authority on the Third World' through the office of trade union secretary at IG Metall, through chairmanship of JUSO (the Young Socialists) and through membership of the SPD and *Bundestag* Committees on Foreign Relations. He earned a reputation as a radical in African affairs by making friends in the Algerian FLN, whose struggle he supported. That was some time ago, however, and since then he had become a part of the 'establishment', furthering his claim to expertise on Africa by jetting around the world at the rate of 160,000 km a year. Less well known than his trips, which he says he makes to meet interesting people, is a fact revealed by the German weekly *Der Stern* of 28 November 1974: in a story about General Gehlen it named State Secretary Hans Jürgen Wischnewski as a close collaborator of the German Secret Service (*Bundesnach-richtendienst*). According to *Der Stern*, whenever he travels abroad Wischnewski is first briefed by the Secret Service on the country he is to visit and the people he is to meet there. On his return he reports back to the headquarters of the Secret Service at Pullacher.[35]

In 1970 a group of young rebels in the German Africa Society ousted the conservative Board of the Society and transformed it from a docile appendage of the Foreign Ministry into a forum for criticism of Bonn's Africa policy. The 'Zambian affair' was not the first of the problems that Wischnewski attributed to dissidents in his own country. Wischnewski summoned Habicht-Benthin and bluntly told him that this would have to be the last. 'Rather than have a German Africa Society which causes us difficulties, it is better to have none at all.'[36] A cable was dispatched to the head of the Federal Republic's UN Mission in New York instructing him to deny the arms supplies story.

On 28 October 1975 Herr Rudinger von Wechmar, West Germany's Ambassador to the United Nations, dutifully inserted the following statement into his contribution to the debate on *apartheid*: 'The Federal Republic of Germany adheres strictly to the resolutions of the Security Council and does not supply any arms or other military equipment to South Africa. Anyone maintaining the contrary view does not speak the truth either unintentionally or even deliberately.' 'The question really is,' remarked Zambian Foreign Minister Vernon Mwaanga after hearing von Wechmar's speech, 'was *he* really speaking the truth?'[37]

The following day, on 29 October 1974, Wischnewski and his colleague at the Foreign Ministry, Karl Mörsch, summoned the Board of the German Africa Society to tell it that the Society had lost the Ministry's confidence and its financial support and was given five months to dissolve itself. When Board members objected to the fact that such a move would have a negative impact on Germany's image in Africa, which the Society had tried to improve by identifying with progressive forces in Africa and by advocating Africa's cause in the Federal Republic, Mörsch replied, 'No African country can afford to get angry at the Foreign Ministry of the Federal Republic of Germany.'[38] The government subsidy was terminated.

Today, nothing has really changed. Herr Wischnewski has moved up to become a high official in the office of Chancellor Schmidt, and since masterminding the capture at Mogadishu Airport of the terrorists who had hijacked a Lufthansa plane from Mallorca, he has become virtually irremovable.

The Foreign Ministry continues to follow Mörsch's line based on the assumption that African goodwill can be bought. The critics, represented largely by the German-Arab-Africa Bureau (the successor to the German Africa Society), the German Anti-Apartheid Movement, ISSA and other organisations, remain a constant thorn in their side.

WEST GERMANY AT THE 1977 LAGOS CONFERENCE ON APARTHEID

It was expected that West Germany might use the

opportunity offered by the World Conference on Action Against *Apartheid*, held from 23-26 August 1977 in Lagos, to give a detailed refutation of the evidence of its nuclear and military co-operation published by the African National Congress of South Africa (ANC). This evidence was contained in a pamphlet entitled 'The conspiracy to arm apartheid continues', distributed at the Lagos Conference. The leader of the West German delegation to the conference, Dr Klaus von Dohnanyi, Minister of State in the Foreign Office, mounted the rostrum only to repeat what had always been his government's reaction to any such criticism: 'To defeat *apartheid* by peaceful means, the Federal Republic has observed along the lines of its general policy of extreme limited arms sale to other countries, absolute rigid arms sale embargo against the Republic of South Africa for decades [sic]. And contrary to unfounded, uninformed and unfortunately sometimes malicious allegations, there is *no* co-operation between the Federal Republic of Germany and South Africa in the military or nuclear field.' He then continued: 'We are grateful to anybody who seriously points out to us where our strict anti-*apartheid* policies are not adhered to.'

DR ENGELTER VISITS WEST GERMANY

Dr Adolf G. Engelter, Senior Chief Research Officer of the South African National Research Institute for Mathematical Sciences, is stationed at the Simonstown Naval Base. At one time he was involved in *Tyrant*, the code name for a minefield belt secretly laid by the South Africans along parts of their coast. Dr Engelter's particular interests are electronic instruments used for monitoring the signatures of ships of the South African Navy, together with the electronics of modern mines such as sensors, decision-making circuitry and power sources. On 28 August 1969 the Department of Military Intelligence gave him a top-secret security clearance.

Considering the sensitivity of his job, and the fact that he often travels overseas on behalf of the South African Navy, this is not unusual. However, Dr Engelter is a German national, born on 18 August 1927 in Marburg, now in West Germany. On a number of occasions he has been permitted to visit West German scientific establishments engaged in classified military research similar to that carried out at the Simonstown Navy base.

In order to visit these West German institutions he needs, and apparently gets, security clearance from West German Military Intelligence. The fact that Dr Engelter is employed by the armed forces of a foreign power (with which any suggestion of military co-operation has always been publicly and vigorously denied) does not seem to worry the West German military authorities. Official denials are made on the assumption that any co-operation would remain secret, protected by the 'scientific' cover provided for visiting researchers by the Scientific Counsellor of the South African Embassy. The true nature of the visits is conveyed by secret correspondence sent through the diplomatic bag.

On 5 November 1969 the Secretary-Treasurer of the South African Council for Scientific and Industrial Research (CSIR) — which controls, among other organisations, the National Institute for Defence Research (NIDR), the National Mechanical Engineering Institute (NMERI), and the National Physical Research Laboratory (NPRL), all of them conducting research of military significance – wrote a confidential letter to Dr F. le R. Malherbe, then the Scientific Counsellor of the South African Embassy in Cologne, which included the following: 'Dr A. G. Engelter, a Senior Chief Research Officer of the National Research Institute for Mathematical Sciences will shortly be visiting various establishments in Germany and Italy on behalf of the South African Navy.' Enclosed with the letter was the itinerary and classified explanatory appendix, security certificate and a copy of the security clearance from the Directorate of Counter Intelligence.

The purpose of Dr Engelter's trip was described as follows:

'Instrumentation is required for monitoring the signatures of South African Navy ships, namely pressure, magnetic and low and audio frequency acoustic signals. This would include ultrasonic aspects for auxiliary tasks, for example the position determination of South African Navy ships while their signatures are being measured during checking the effectivity of their degaussing.' His task was then specified in an appendix of 14 items which show that France and Italy were offering mines to the South African Navy.[39]

Dr Malherbe passed the letter to Brigadier General Hamman, who scribbled at the bottom of the letter a note dated 10.11.1969 which read: 'The necessary application is being submitted to the Bundeswehr for the clearance of Dr E. for his visit to Germany.' The first thing which had to be done was to fill in a form called '*Anmeldung von Ausländischen Besuchern*' ('notification of foreign visitors'). The South African Embassy keeps a special file called 'Instructions and guidelines for compliance by members of the South African Defence Forces (SADF) visiting Germany to attend courses or on other duty visits.' The instruction concerning such visits reads as follows: 'Write a letter in triplicate (in German) to the *Bundesministerium der Verteidigung* (Federal Ministry of Defence) to reach them at least eight weeks prior to the commencement of the visit.' The Embassy keeps a stock of blank forms for these applications, addressed to the Minister of Defence S 11 8 (22) (the Department of Military Intelligence), 53 Bonn-Duisdorf. Dr Malherbe duly filled in this form and the Military Attaché countersigned it. Permission was requested for a visit to the Vacuumschmelze plant at Hanau on 22 December 1969, for Dr Engelter to discuss with Dr Assmus of Vacuumschmelze the requirements for instruments used in measuring weak magnetic signals. The other visit requiring security clearance was to the Hewlett-Packard computer system installed at Stuttgart airport, which Dr Engelter proposed to inspect on Christmas Eve in the company of Herr Dieter Egerman and Herr Fred Schröder of the Hewlett-Packard Company.

Item G of the form, on the grade of security clearance accorded to Dr Engelter, was filled in with the words 'top secret'. Attached was the security clearance from South African Military Intelligence. The German authorities cleared Dr Engelter and the visit went ahead as planned, and he also took three weeks' holiday in his home town of Marburg, where he called on Professor F. H. Müller of the Institute for Polymerics. On 22 December he visited the Institute for Oceanography at Kiel to inspect the oceanographic equipment and discuss projects related to his own work at Simonstown.

Dr Engelter's 'scientific visit' to Germany was completed on 25 December 1969 and he went back to Simonstown. But he returned later on behalf of the South African Atomic Energy Board. A number of Dr Engelter's colleagues have been through the same procedure and were also cleared by the West German military authorities to visit classified institutions and companies involved in weapons production and other military work.[40]

One of them, Dr L.L. van Zyl, Chief Research Officer at the National Institute for Defence Research, also visited the British Royal Radar Establishment at Malvern and the Defence Operational Analysis Establishment (DOAE) at West Byfleet, Surrey in 1974. News of his visit caused a major political row in the ruling Labour Party.[41] It has been established that on 11 October 1974 Dr van Zyl visited Eltro Engineering in Heidelberg, whose production includes laser range-finders, homing devices and military electronic equipment. The firm is partly American-owned and there is extensive co-operation with French companies.[42]

The visit of two more South African scientists should perhaps be mentioned. Their names are G. Lampen and V.C. Wikner. They are both attached to the South African National Institute for Defence Research as specialists in missile guidance systems. From 15 July to 13 August 1974 they attended a course at the Siemens factory at München-gladbach to learn how to operate a ciné theodolite, an instrument used for measuring angles and horizon lines when aiming at a target, which Siemens was to install at the test range at St Lucia in South Africa.

The St Lucia missile base and test range is sited on the Natal coast, 50 km north of Richards Bay and 200 km north of Durban. Close to the Indian Ocean and Lake St Lucia, the range is South Africa's main rocket-testing site as well as one of its most strategically located missile bases. Its northern boundary lies only 50 km from the Mozambique border and less than 200 km from its capital city of Maputo. The base is fully operational and manned by a special unit of the South African Army, while the test range is used by various weapons manufacturers, both South African and foreign, and by the South African Navy and Air Force. Guided missiles have been tested there since December 1968.

As in the case of the other 'scientists', the visit of Lampen and Wikner was handled by the scientific counsellor of the South African Embassy, who at that time was Dr W.T. de Kock, at the request of the Director of the National Institute for Defence Research.[43]

There have been other instances of West German/South African military co-operation, for example concerning the Lindau neutron monitor supplied to a South African installation at Tsumeb. It was to be transported free of charge by the Gesellschaft für Kernenergienutzung in Schiffbau and Schiffahrt (Society for the Employment of Nuclear Energy in Ship-Building and Ship-Propulsion), which operates the West German experimental nuclear freighter, the *Otto Hahn*.[44] The story is revealed in a telex sent from Bad Godesberg by Professor P.H. Stoker, a member of the West German delegation to the 14th International Cosmic Ray Conference held in Munich in August 1975. On 15 August 1975 Professor Stoker sent a telex to the Council for Scientific and Industrial Reseach (CSIR) in Pretoria. The telex reads as follows:

'Bonn 15/08/75   tlx948
CSIR Pretoria
   Professor Stoker requests to inform Dr Hewitt of the
South African CSIR in Pretoria that the Gesellschaft für
Kernenergienutzung in Schiffbau und Schiffahrt mbH
who will transport the Lindau neutron monitor by the
Otto Hahn free of charge from Rotterdam to a harbour
in South Africa for Tsumeb, was advised not to unload in
Walvis Bay because of possible political demonstrations
against any foreign vessel entering the harbour. Conse-
quently they decided to unload in Durban, which
harbour the Otto Hahn has to visit on this trip.
   The monitor has to be dismantled and packed virtually
immediately for delivering in Rotterdam in time for
transportation to South Africa in September. We have to
meet the costs of unloading the monitor at harbour and
of the transportation to Tsumeb. The monitor comprises
29 tons of lead, 3.5 tons of paraffin wax in boxes, 18 large
neutron counter tubes and electronic recording equipment. Please inform whether the additional costs of
transportation from Durban to Tsumeb are acceptable.
P.H. Stoker
c/o SASLO[45] Bad Godesberg

The South Africans were happy to pay the extra transport
costs, although they did not like the open telex communication
about it.
   Since 1975 West German firms have been involved in the
production of what SIPRI calls 'the main battle tank for South
Africa'. The tank's hull is supplied by A. Jung, a company
based at Betzdorf; the transmission was developed at the
university laboratory of Bochum, and other components by
Thyssen-Rheinstahl.
   Aviation Week reported the delivery of the German BO-105
helicopter to South Africa.[46] Helicopters are a crucial weapon
in anti-guerrilla operations, and although the Bonn Government claims that '. . .there exists no military version of the
BO-105,' *Aviation Week* reported that these helicopters can be
armed with both conventional bombs and air-to-air missiles.
   The delivery of Daimler-Benz Unimog military vehicles to

South Africa is not being denied by Government officials; they argue that it is only '. . . a cross-country vehicle constructed for civilian purposes'. The West German magazine *Wehrtechnik* ('Defence Technology'), which is financed by the Federal Ministry of Defence, takes a somewhat different view. It has commented: 'Ever since the Portuguese colonial wars in Angola, Mozambique and Guinea, the military version of the Unimog is considered to be the best all-round military transport in Africa.'[47]

THE GENERALS ARE COMING TOO

Military co-operation between West Germany and South Africa, as represented by the visits of Dr Adolf Engelter and other scientists, is covered by a Cultural Agreement signed in 1962. Article 1 reads as follows: 'The contracting parties shall strive to facilitate, (a) the interchange of university staff, lecturers, teachers, research workers, students, journalists and other approved persons, (b) co-operation between scientific and cultural institutions and societies of the two countries.'

The closeness of the relationship between the *Bundeswehr* and the South African Defence Forces is shown in a letter in which it appears that Brigadier General D.J. Hamman, Military Attaché of the South African Embassy, and Admiral of the Fleet Rolf Steinhaus in the Federal Ministry of Defence, had discussed the '. . .expansion of NATO in the South Atlantic' on the telephone (30 November 1972). They were talking about the delicate subject of South Africa and the NATO defence system. The letter by Admiral Steinhaus reveals that NATO planners were far from agreed on the issue of South African involvement in NATO, and several controversial recommendations had been submitted on which discussion was continuing. However, Steinhaus wanted to keep General Hamman in the picture and immediately after putting down the phone he sent him '*mit freundlichen*

Grüssen' ('with warm regards') the draft of the final formulation by the Federal Ministry of Defence to be submitted to the NATO Military Council in Brussels.

What Admiral Steinhaus and his colleagues were recommending to NATO is not known and can only be guessed. However, one can assume that by November 1974 NATO had to acknowledge the loss of Angola, and this in turn encouraged South African ambitions to fill the defence vacuum created by the departure of Portugal from Africa. It would appear to be most irregular for an Admiral of a NATO country to provide a General of a foreign power with documents destined for NATO meetings.[48]

When the new building of the South African Embassy at Auf der Hostert in Bonn was inaugurated in September 1975, a senior official of the Federal Minister of Defence, Helmut Fingerhut, and Brigadier-General Günther Schneider attended the ceremony. This was not surprising. Sixteen months earlier on 30 May 1974, Colonel Paul F. Strauss of the Protocol Department of the Federal Ministry of Defence (formerly Brigade Commander of the 12th Brigade at Amberg) had written a letter to the Military Attaché of the South African Embassy, General P. E.K. Bosman, telling him that a visit by a South African military delegation of computer experts to look at the traffic computer control systems of the *Luftwaffe* would be welcomed. The South Africans wanted to see a Full-Duplex installation, a device consisting of two traffic control computers: when one breaks down the other automatically takes over. 'As far as we know,' wrote Herr Strauss to General Bosman, 'a Voll-Duplex-Betrieb is used only by civilian air-safety installations.' He told them that '. . .what could be shown to your delegation is a multiple computer device (*Mehrrechnerbetrieb*) where 2 computers are fed by the same data system and where the linked decision-making instruments can be operated by either of the two computers.'[49]

The string of South African military visitors to the Federal

Republic of Germany has included General R.C. Hiemstra, when he was Chief of the South African Air Force, General J.P. Verster, Lt. General H.P. Laubscheer, Brigadier A.F. Erasmus, Major-General H. de V. du Toit of South African Military Intelligence, and last but not least General H. van den Bergh of BOSS, the Bureau of State Security.

What should also be added is the visit of a team led by Commander H.J. Jooste of the South African Navy which, accompanied by members of the South African Armaments Board, visited West Germany for two weeks in May 1974. The purpose of the visit was to undergo instruction in the operation of project *Advokaat*.

Advokaat is probably the biggest single example of West German military hardware delivered to the Republic of South Africa. It has been justified by the Bonn Government as a civilian installation. According to an official statement by the Ministry of Foreign Affairs of January 1977, *Advokaat* is a system that helps to maintain the safety of shipping around the Cape, which has increased to about 25,000 vessels a year. The West German government claims that it can no more be blamed for the military uses to which the system can be put than for selling automobiles which, as the memorandum puts it, may be used for military purposes. *Advokaat* is a military radar surveillance installation built at Silvermine, near Cape Town, at a cost of £16.2 million; it is described by the South Africans themselves as the 'new headquarters of the Navy'. Its range stretches far across the South Atlantic and Indian Oceans and it monitors all maritime traffic in the area. The installation is staffed by South African Navy personnel headed by Commander Jacobus Brink, who personally took over the project from the West German companies Siemens and AEG-Telefunken.

The level of overtly military co-operation which we have outlined above, routinely covered up by blanket denials from Government officials, reflects the combination of mercenary and military ambitions which lies behind West Germany's collaboration with South Africa in the nuclear field. For the German companies it is indeed good business; but the collusion involved in passing secret NATO documents and codes to the South Africans clearly goes beyond the purely commercial relationship which the West Germans so vociferously claim to be the basis for their deals.

PART TWO SOUTH AFRICA

Introduction

When the South African Prime Minister announced, in 1970, a new process for enriching uranium as '... an event of unequalled importance for South Africa,' he was ostensibly referring to a process that might increase South Africa's export earnings from uranium and contribute to a very modest nuclear programme, not expected to comprise more than one power station, to come on stream in the 1980s. In fact, the application of the nuclear programme is far more important than these relatively modest goals would suggest. Seven years after the announcement came the warning from Soviet satellites, confirmed by intelligence sources in France and the United States, that South Africa had a nuclear-weapons testing site.

To construct a nuclear weapon South Africa would need either a certain quantity of highly-enriched uranium, for construction of a Hiroshima-type atomic bomb, or a small amount of this material plus a larger amount of plutonium (obtained by reprocessing spent fuel from a reactor) to produce the more sophisticated and deadly hydrogen bomb. For both types of device, sophisticated technology is required not only for the construction of the explosive, but also for the detonator that will allow the operator control of the explosion.

The technology itself, however, according to a variety of

studies, is increasingly widely available, and both atomic and hydrogen bombs have been designed by nuclear physics students at undergraduate level. South Africa certainly has a higher level of expertise than this, its own and that of foreign nationals working in South Africa. What this means is not so much that South Africa will have no technical problems, but that they will not be a real obstacle to production of nuclear weapons.

As more and more of the technology is published, international debates about the danger of proliferation have focused increasingly on the imposition of safeguards on nuclear materials and equipment. Ideally, all reactors and associated nuclear plants such as enrichment and reprocessing plants would be subject to trilateral safeguards agreements between the supplier, the purchaser and the International Atomic Energy Agency (IAEA), allowing for regular inspections by the IAEA to ensure that they are not used for military purposes; in addition, all nuclear materials from natural uranium to enriched uranium and plutonium would also be safeguarded, thus providing a full guarantee against diversion of the equipment and materials for production of atomic weapons.

The real threat to the international safeguards system lies in the availability of uranium from a supplier more interested in politically advantageous deals with other countries than in the ideal of non-proliferation, and in the existence of nuclear installations which are not subject to inspection under a safeguards agreement. South Africa is a threat to non-proliferation on both counts. Firstly, it is a major uranium producer, with a strong interest in dealing with countries such as Israel, Iran, Brazil, France and others which place high priority on obtaining an 'independent' nuclear capability. Secondly, it has the world's only unsafeguarded enrichment plant outside the existing nuclear powers. Not only is the plant itself unsafeguarded, thanks to the understanding between the South Africans and their West German suppliers, but there can be no question of safeguards on the raw material fed into it for enrichment since the South Africans produce their own uranium and can now carry out every stage in its processing, from extraction out of the ore to final enrichment.

In Part II we therefore examine South Africa's nuclear establishment, from its origins as a supplier of uranium for the nuclear weapons programme of the United States and Britain immediately after the Second World War — a connection which has helped the South Africans to obtain technical assistance and know-how for the South Africans' own nuclear programme — to the development of the 'unique' enrichment process which is in fact derived from West German technology. Finally we look at indications that South Africa is basing its strategy for survival on the development of its own nuclear weapons. We conclude that these weapons, relying as they do on delivery systems supplied by France, West Germany and other Western countries, are in fact 'aimed' not only at the heart of Africa and the 'communist menace' there, but also at the West itself, in a calculated manoeuvre to force it into guaranteeing the future of white minority rule in South Africa.

As a means to evaluate the South Africans' claim that their nuclear programme is for commercial purposes only, we examine in Chapter 5 the economics of the programme, and argue that the enormous investment of capital in commercial personnel cannot be justified in commercial terms. The facts and the arguments can be summed up as follows:

(a) *Koeberg power station*
— To provide an electricity supply flexible enough to meet demand at an economic rate, nuclear power is not justified. The station at Koeberg will probably require a very expensive pump storage system, making it economically even less viable.
— The financing of Koeberg will greatly exacerbate the problem of rising electricity prices, eroding the advantage to South African industry of extremely low power costs.
— The contract for Koeberg was granted to the French consortium, almost certainly the least competitive bidder.

Intense political pressure was applied to the electricity authority, ESCOM, in a deal with strong military overtones.

— There are no guarantees that Koeberg will be built on schedule — and delays will mean massive cost overruns. The French consortium is inexperienced and cannot offer the guarantee that ESCOM considered of critical importance — i.e. that the construction consortium should have an identical reactor underway elsewhere to provide a model for Koeberg.

— To sum up: efficiency, quality control and value for money were *not* the criteria applied to Koeberg. The French consortium won the contract, as we shall show, for political and military reasons.

(b) *Valindaba enrichment plant*

— There has been intense controversy in South Africa about the proposed enrichment plant on a commercial scale. The cost is likely to be staggering, at a time when South Africa is having great difficulty financing other major projects. External observers consider it impossible for South Africa to compete in price terms on the world enrichment market.

— If the pilot plant were for purely commercial purposes, South Africa would have allowed the application of safeguards and international inspection: the suspicion attached to the existence of an *un*safeguarded plant would be too damaging.

(c) *General*

— If the nuclear programme were for purely commercial purposes, South Africa would not be taking an increasingly hard line against signing the Nuclear Non-Proliferation Treaty, which stand antagonises the United States and other important trading partners.

— If the work at Pelindaba were subject to economic criteria, it would not mechanise menial jobs and employ whites at much higher pay to do the work normally done by blacks in South Africa.

— The secrecy legislation covering all aspects of the nuclear establishment goes far beyond that of other countries with commercial nuclear programmes.

4
South Africa's uranium

The origins of South Africa's nuclear capability are to be found at the very beginning of the atomic age. In their race to stockpile nuclear weapons aimed at their wartime allies, America and Britain needed South Africa, which played a key role in the provision of the raw material – uranium. The post-war Smuts government gave this top national priority, and when the right-wing National Party claimed victory in 1948 they too saw the embryonic uranium industry as of crucial importance to the future of the new, *apartheid* South Africa.

We therefore examine the building up of South Africa's uranium mining and processing capacity, backed by the capital and technological expertise of the Anglo-American Combined Development Agency. Uranium and gold – the basis of the white South African economy – are shown to be intimately linked; moreover, the gold-mining and other companies are expanding their exploration and marketing of uranium. South Africa is shown, on the evidence of secret

108

Australian documents, to be central to the whole international uranium cartel which has come under attack in the United States for price-rigging on a massive scale.[1] We also take a detailed look at the importance of the Rössing uranium mine and the place of Namibian uranium in South Africa's ambitions.

Finally, we examine the South Africans' uranium-marketing strategy, the obsessive secrecy which surrounds it, and the hints that South Africa cannot always be relied upon to apply 'peaceful uses' safeguards to its uranium exports.

South Africa's uranium: protégé of the nuclear arms race

The existence of uranium in Witwatersrand gold-bearing ores has been known since the 1920s, but its significance was not appreciated until the end of the Second World War, when the United States and Britain, through the Combined Development Agency (CDA), launched a world-wide search for the raw materials with which to manufacture atomic bombs. In 1945 the two governments approached South Africa about the possibility of extracting uranium from gold ores, and at the same time two of their top uranium specialists, Dr G. W. Bain and Dr C. F. Davidson, visited South Africa.

The South African Government, with Field Marshal Smuts as Prime Minister, took the keenest interest in the uranium question and Smuts informed the President of the Transvaal Chamber of Mines that this was a matter of vital importance to South Africa.

Regular exchanges of visits were established in the early years, setting up close working relations between the CDA experts and the South Africans – a pattern of co-operation which has been of continuing importance to South Africa's

109

own nuclear ambitions. Even after the National Party takeover in South Africa in 1948, there was undiminished enthusiasm on both sides of the Atlantic in favour of making South Africa a leading uranium producer.[2]

The South African Atomic Energy Act, passed in 1948 to replace certain War Measures relating to uranium, placed the sole right to prospect, mine, produce and dispose of uranium and thorium in South Africa, as well as the right to concentrate, refine or process them or to produce atomic energy, in the hands of the South African Government. This legislation was later extended to cover the occupied territory of Namibia, formerly South West Africa, as well. The Government's monopoly powers over this industry are exercised through the Atomic Energy Board (AEB).[3]

The industry's rate of production, after massive inputs of capital and technology from the United States and Britain through the CDA, rose from zero in 1950 to over 6,400 tons of uranium oxide in 1959, carrying an annual export value of over R100 million.[4]

When the original 1950 agreement proved insufficient and the CDA showed great urgency in stepping up uranium output, the AEB agreed to their request that many more gold mines be brought into the uranium industry; at the same time, a higher price per pound was negotiated for all South African uranium producers.[5]

Between 1951 and 1957 R140 million was invested in the infrastructure in special plants – including a new power station and sulphuric acid plants, as well as the calcining plants for extracting the uranium oxide from the ore.[6] The whole of this investment originated in the United States and Britain. About 2,000 white employees were trained in various specialised fields (some of them very useful for the subsequent nuclear programme) and a large building scheme was inaugurated for them. The acid and pyrite plants also had to be staffed and operated, and African Explosives and Chemical Industries (a subsidiary of Britain's ICI and the Anglo-American Corporation of South Africa) made a valuable contribution to this.[7]

Orders for the materials and products involved provided a major stimulus to South Africa's engineering, steel, rubber, chemical and other industries, and were doubly important because, according to Mr R. B. Hagart, a key participant, they represented '. . . new money injected into the South African economy; and by far the greater part will have been spent in South Africa. The benefit to this country is permanent.' Uranium oxide exports for the atomic weapons stockpiles of America and Britain formed '. . . an extremely important source of foreign exchange to South Africa, particularly valuable because a major portion is paid for in dollars.'[8]

From the start therefore, uranium mining has been important to post-war development in South Africa, contributing to the economic boom, increasing prosperity for the white minority and generating inflows of foreign capital to strengthen the *apartheid* structure.[9] As the definitive paper on the subject describes it, the South African mining companies were:

> '. . . amazed also at the expeditious way in which delivery of material to the project was made, due largely to the extremely high priority given by all of the three Governments directly concerned with the South African production of uranium. . . It can safely be said that no major industry in the history of South Africa has been developed as rapidly as the uranium industry.'[10]

URANIUM AND GOLD

Uranium mining has been of particular benefit to the gold mines, which are still crucial to the economy and the regulation of black labour. Exports of gold, although less important than in the past, are still the major factor in allowing South Africa to spend more on imports than the value of their other exports. The fluctuating world price of gold has a considerable impact on the state of the South African economy as a whole.[11]

111

Until the opening of the Rössing mine and the extraction of uranium from the residues of copper-mining at Palabora (a mine owned largely by Rio Tinto-Zinc of Britain and the Newmont Mining Corporation of America), uranium has been exclusively a by-product of gold-mining in South Africa. Uranium oxide is extracted from treated ore at a very low concentration, and a sophisticated purification process, the Purlex process, makes a final product of high quality. It is able to do this at at relatively low cost, since the basic costs of mining the ore are carried by the gold-mining operation.

Uranium production is about one-fifteenth the value of gold produced. The dependence of the level of uranium production on gold is evident from the fact that a sharp rise in the world price of gold in 1974, while it allowed South Africa to stockpile uranium instead of having to sell it, also had a detrimental effect on uranium output. This was because Government policy, including the use of appropriate subsidies and taxation, requires mines to exploit the more marginal ores while profits are high, leaving the richer ones (which also contain more uranium) for a time when margins will be leaner; this stretches out the life of the mines, many of which are on the verge of closing down.[12] In fact the close relationship between gold and uranium is beginning to go into reverse: some of the gold-mines are kept viable because of the extra revenue provided by uranium oxide production. At least one worked-out gold mine, with substantial reserves of uranium still underground, is being kept open in the hope that it may become commercially exploitable.[13]

On the whole, South Africa's uranium deposits are of such low concentration that, until Rössing, it was not economically feasible to mine them in the absence of a gold-mining operation. Towards the end of 1976, mining executives were looking hopefully toward uranium–mining as a possible solution to the squeeze between rapidly rising costs and a lower-than-hoped-for market price for gold. Dennis Etheredge, Chairman and Managing Director of Anglo-American's gold and uranium division, announced: 'Uranium will no longer be a by-product of the gold mines. It will be a co-product.'[14] The *Rand Daily Mail*, which has very close ties with Anglo and the other mining companies, commented that 'almost in fairy godmother fashion, uranium has come to the rescue of many gold mines at a time when the gold price has slumped.'[15]

THE MINING COMPANIES AND THE FUTURE OF URANIUM

The gold-mining companies have from the start been extremely important partners for the Government, not only in developing uranium mining and extraction but also in giving a major boost to the whole nuclear industry. The companies decided at an early stage to start uranium production after the war with the least possible delay. The project was dealt with at the highest level with the most experienced people in the industry taking personal responsibility.[16] The Nuclear Fuels Corporation (NUFCOR), the uranium producers' consortium, was set up in 1967 to act as a 'mine-to-enrichment' service. It collects ammonium diuranate slurry from the mines and, by the Purlex process, dries and calcines it to produce concentrates containing about 92% uranium oxide. NUFCOR has spent considerable sums on research to improve the purity of the uranium oxide which would enable South Africa to produce its own uranium hexafluoride – the final process before enrichment. By 1969 it was already producing uranium tetrafluoride, an intermediate step towards 'hex', and is now bringing its own 'hex' facilities into operation.[17] NUFCOR therefore has been important in extraction and in 'hex' conversion; it is also a substantial financial contributor to research by the Atomic Energy Board.[18] In the early days of uranium production in South Africa leading personalities in the mining industry — including C. S. 'Stowe' McLean, former President of the Chamber of Mines (the gold-mining body), and the late R. B. Hagart – urged the Government to plan for a full-scale South African nuclear programme. The advice was heeded and the result was the construction of the SAFARI-1 reactor, and all the development leading from there. In the first ten years of the AEB's research programme the uranium-mining industry contributed R8 million; and it has remained a substantial contributor.[19]

The research into techniques for extracting uranium from the low-grade South African ores continues, based on the initial achievements of what Dr A. J. A. Roux, President of the AEB, has called the 'highly successful international programme' which first set up the industry after the war.[20] In 1968-69, as announced by Dr Roux, the AEB's extraction metallurgy division, working in collaboration with the National Institute for Metallurgy, developed a new process – Purlex – for the extraction of uranium from heavy metal concentrates. Investment and research from abroad continued to be crucial, with Bechtel-W.K.E. of West Germany assisting in the design of the full-scale plant.[21]

Assistance has also been provided by Britain in the development of uranium hexafluoride production. This is the important final stage of processing before enrichment. The United Kingdom Atomic Energy Authority has had a lucrative agreement with NUFCOR, which gave it first refusal on the conversion of any NUFCOR ore into hexafluoride, or 'hex'. In 1970 the British Government had to decide whether to allow South Africa access to the technology of hex conversion.[22] Although the decision was never announced, Dr Roux stated in October 1970 that research on the hex process had been carried out by South Africans in collaboration with 'overseas interests', and that South Africa was now in a position to build its own hex plant. It seems probable that the 'overseas interests' were in fact British, and that the technology for South Africa to build its own hex plant was handed over by the UKAEA. A pilot hex plant was commissioned in South Africa in June 1975, and production on an industrial scale began in early 1976.[23]

South Africa's uranium mining industry is at present at a cross-roads. Demand for uranium from the US and Europe for military purposes declined in the 1960s, and the long period of stagnant prices during that decade and the early 1970s discouraged most of the mines from producing uranium. From a peak, when twenty-nine mines were producing uranium at the height of the military contracts, the number dwindled to eight in 1975. Now, with 'new vistas opening' as a result of rapid price rises, as Jean-Pierre Hugo of the AEB puts it, the country is reassessing its uranium assets. Further advances depend greatly on a new extraction process to replace Purlex: this is the CCIX (counter-current ion-exchange) process, under development since 1970 as a way of processing dilute slurries and successfully tested in 1974.[24] It is apparently this process which is being used at Palabora and at Rössing.[25] It is also a key factor in the expansion of the uranium-mining industry on a new basis, now no longer in conjunction with gold-mining. Dr Roux told visiting American investors at the end of 1976 that uranium oxide production in South Africa and Namibia combined would reach 10,850 tonnes a year in 1978 – of which the Rössing mine would supply half – and that this could rise to 16,270 tonnes in 1985.[26]

Dr Roux has said that prospecting for new uranium deposits has been intensified, involving a R3-million aerial radiometric survey and the mapping and sampling of target areas by joint teams of the US Geological Survey and the AEB. An automated activation analysis facility at Pelindaba has been set up to handle thousands of samples a month.[27] Two large American corporations, Union Carbide and Utah Mining, have already undertaken intensive prospecting expeditions, while two others – Newmont Mining Corporation and US Steel – are also seeking areas to begin prospecting. It appears likely that geological surveys by the US ERTS (Earth Resources and Technology Satellite) prompted the interest of the American corporations in the deposits involved.[28] Of the South African companies, Rand Mines, Afrikander Lease, General Mining, Gold Fields, Anglo-American and Johannesburg Consolidated Investment Co. (JCI) are also in the

115

picture; so is Compagnie Française des Pétroles and Aquitaine, both of France.

Apart from the Karroo Desert, north of Cape Town, the search has been concentrated in Namibia, particularly in the vicinity of Rössing, which seems the most likely economic prospect[29] (see the map on p.118). Rio Tinto-Zinc is reported to have staked claims all over Namibia, in addition to its existing operations at Rössing.[30] It is probably no accident that in 1973 South Africa substantially improved the conditions of mining- and prospecting- grants applying to overseas companies, and this provided a major incentive for them to operate in the occupied territory.[31] If substantial new uranium mines are started up in Namibia with foreign backing, it would increase South Africa's intransigence in the face of repeated rulings by the International Court of Justice and the United Nations that it is obliged to withdraw its occupation of the territory.

Meanwhile, existing plant capacity at the gold-mines is being brought back into production, improved and enlarged.[32] Anglo-American has announced plans to build a big uranium complex in the Orange Free State gold-fields area.[33] Ambitious plans are also being formulated to extract uranium from the 'slimes dams', the mounds of mine-tailings which have accumulated all over the gold-mining area since the late 1890s; this could boost South African uranium reserves overnight by 10-20%. The problem is not so much one of chemistry as of mechanical handling of the vast heaps of finely-ground tailings, which are dispersed over a very wide area.[34]

Namibia: the Rössing connection

The most significant development of uranium production in recent years has taken place not within South Africa itself but in Namibia, the vast, mineral-rich territory controlled by Pretoria since 1919. Despite the mounting pressures from the international community for Namibia's decolonisation in line with resolutions of the United Nations, and the finding by the International Court of Justice at the Hague in its advisory opinion of June 1971 that South Africa's presence is illegal and contrary to international law, South Africa refuses to abandon its occupation of the territory.

Apart from strategic and political considerations, a major factor in South Africa's determination to maintain its occupation of Namibia is its desire to retain access to Namibia's extensive resources of minerals and other primary commodities. This economic dimension has been perhaps the most important motivation behind South Africa's attempts since 1975 to devise an 'internal solution' for Namibia which would exclude from power the South West African People's Organisation, recognised by the United Nations and the Organisation of African Unity as the sole authentic representatives of the people of Namibia.[35]

Even before uranium production began at Rössing at the end of 1976, Namibia had already become the fourth-largest African mineral producer – after South Africa, Zaire and Zambia – due to the development of its deposits of alluvial gemstones (the world's richest source of diamonds) and a variety of base minerals including copper, lead, lithium, silver, tin, vanadium and zinc. Production provided an estimated mining turnover of £250 million a year, almost all of which was exported along with most of the profit. The importance of mining activity can be gauged by a comparison with the gross domestic product (GDP), which excludes imports and exports; in 1976, total GDP was estimated at £250-300 million.

The extensive low-grade uranium deposits at Rössing, a desolate area in the Namib Desert some 65 kms north-east of the coastal town of Swakopmund, are generally reckoned as amongst the largest anywhere in the world. There exist no official figures for the size of the reserves, but they have been estimated by most outside sources as 100,000 metric tons of assured reserves exploitable at less than $15/lb. This is

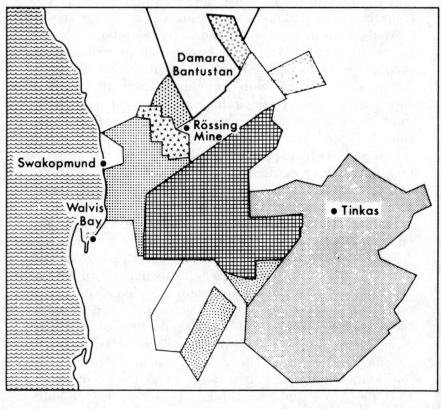

PROSPECTING CONCESSIONS FOR URANIUM IN NAMIBIA

Damara Bantustan

Rössing Mine

Swakopmund

Walvis Bay

Tinkas

General Mining

Aquitaine

Union Corporation

Anglo-American

Falcon-bridge

RTZ

Jhb Consolidated Investments

Gold Fields of South Africa

Westwind Ventures (Union Corporation)

equivalent to about half of South Africa's reasonably assured reserves of just over 200,000 tons. Apart from increasing the deposits under South African control by around 50%, the Rössing deposit is of great value to them because it can be mined alone, rather than as a by-product of gold which ceases when gold mines are closed down. If it were an independent producer, Namibia would have the potential of becoming the third largest non-communist uranium producer after the United States and Canada; however the immediate political and strategic significance of Rössing has been to provide South Africa with uranium on a scale surpassing any of its actual or planned domestic production.

From the outset the Rössing project has involved close collaboration between the South African Government, the British-based Rio Tinto Zinc Corporation Ltd (RTZ), and a number of other transnational and South African mining corporations.

The project is the largest single mining venture ever carried out in Namibia, and RTZ's single most important new mine in the 1970s. The total costs of development involved a pre-production investment of £100-£120 million, most of it raised in the form of loan-finance outside the RTZ group, the remainder through equity-participation by other companies. Partners in Rössing Uranium Ltd, the company established in 1969 to operate the mine, include: the Industrial Development Corporation of South Africa Ltd (IDC); the South African mining company, General Mining/Federale Mynbou, a uranium producer in its own right and a subsidiary of the powerful Federale Volksbeleggings Beperk; Rio Algom, RTZ's Canadian associate, also a uranium producer; and the French-owned Total Compagnie Minière et Nucléaire, part of the Government-owned Compagnie Française des Pétroles.

The IDC, a wholly Government-owned parastatal which is involved in a number of other mining operations in Namibia, is reported to have provided R60 million in loan-finance for the metallurgical and treatment plant at Rössing.[36] The availability of this substantial finance from the public sector

119

indicates the priority given to the mine in South Africa. The remaining outside financing has probably come from loans to RTZ and the IDC arranged by the American and European banks participating in the Euro-American Banking Corporation, EABC, headed by the Deutsche Bank which has provided finance for a number of state-supported capital investments in South Africa. In 1974, RTZ obtained a German-secured, variable-interest loan of DM 59 million (£10.4 million) for the period 1979/85. The RTZ Annual Report for 1974 also revealed that construction and pre-production work at Rössing was by then accounting for much of the group's current expenditure. This had increased by £3.7 million, most of it due to Rössing, where work was completed on basic facilities at the mine in that year.[37]

As the result of a partnership and sales agreement concluded in 1973 between RTZ and Total, the French company received a 10% equity stake in Rössing in exchange for a long-term contract involving the purchase of a substantial quantity of uranium concentrates during the 1980s.[38] The deal presumably provided a pre-payment element, although this was not disclosed. In other RTZ projects, advance customer payment has been a financing method widely used. In 1975 Total's interest in Rössing was taken over by Minatome SA, a newly-established energy consortium jointly owned by Total and Péchiney-Ugine-Kuhlmann, a giant chemical and metallurgical concern involved in all aspects of the uranium cycle.[39]

Rio Algom took up a 10% holding in Rössing during 1974 through the exercise of an equity option; according to the company's 1976 report this was valued at 8.1 million Canadian dollars. Rössing is not a publicly-quoted company, therefore details of its structure are not generally available, but its status appears to be that of a subsidiary of RTZ's South African holding company, Rio Tinto Management Services South Africa (Pty) Ltd, based in Johannesburg. Most of the senior management at Rössing are also directors of Rio Tinto South Africa. In 1975 it was reported that Mr John Berning, General Manager of Rössing, had been appointed Deputy

Managing Director there and would also remain a Director of Rio Tinto South Africa.[40]

But while RTZ seems to have controlled the majority of Rössing's equity from the outset, subsequent arrangements provide for major South African influence via the IDC and General Mining. This is in fact greater than their actual equity holding would indicate.

In 1974, Rössing's equity structure was as follows:

RTZ	48.5%
Rio Algom	10.0%
Total	10.0%
IDC	13.2%
General Mining	6.8%
Other interests	11.5%

As of 1976 the beneficial interest of RTZ in Rössing was given in that year's annual report as 45.5%, a decline of 3%. By normal commercial criteria this would still be sufficient to ensure overall control by RTZ at Rössing. But the company's accounts also revealed the significant feature that the RTZ group's holding carried only 25.8% of the total voting rights.[41] There were apparently two categories of shares, 'A' and 'B', with disproportionate voting power; RTZ holds 53.0% of the 'B' class shares, but the accounts made no mention of any holding of 'A' class shares. Assuming that a majority of the 'A' class shares are held by the IDC, this provides for effective control of basic Rössing policy decisions by the South African Government. At the 1977 annual general meeting of RTZ in London, the chairman, Sir Mark Turner, in response to persistent questioning on the issue, denied that there was any veto exercised on the Rössing board by the IDC or South Africa. But he admitted that RTZ would have to take account of the views of its South African partners in coming to decisions. He also stated that the Uranium Enrichment Act of 1974 allowed the South African Government to obtain any amount of uranium for any use it required.[42]

RTZ has consistently denied any intention to supply

uranium to South Africa from the Rössing mine.In 1976 Mr Alexander Macmillan, Deputy Chairman of Rio Tinto South Africa and with overall responsibility at Rössing, told the London *Financial Times* that South Africa would receive no uranium produced at Rössing until at least the 1980s, since every pound of the output had already been sold abroad. In 1977 this was reiterated by a spokesman at RTZ's London headquarters who said: 'It would not be our intention to supply South Africa with uranium. There is no contract whatsoever between Rössing Uranium and the South African Government.'[43]

The existence of legislation covering all aspects of the uranium industry has enabled Pretoria to shroud developments at Rössing in secrecy. Almost no official information has been provided as to the ore-body and other technical features of the mine, and commercial details have been restricted to those essential to enable the recruitment of expert mining personnel. Section 36 of the South African Atomic Energy Act of 1967 applies the Act's provisions to the 'Territory of South West Africa, including the Eastern Caprivi Zipfel (Strip).' It defines any substance containing uranium in excess of 0.006% as being a prescribed material, the sole right to search, prospect, mine and process any such material being vested in the state. The Act also provides the South African State President with powers to prohibit the export of restricted material. This clearly provides South Africa with the ability to direct operations at Rössing in the interests of South Africa's own uranium and nuclear requirements. No formal contract between RTZ and South Africa would therefore be required for South Africa to obtain any uranium it might require. The Uranium Enrichment Act of 1974 increases the powers of the Uranium Enrichment Corporation, providing for Government intervention in the production and marketing of 'source material' and 'special nuclear material', as well as the financing of its production. South Africa has been careful to make this Act fully applicable to Namibia too.[44]

The degree of secrecy surrounding Rössing is considerably more stringent than it is for Palabora or the gold mines producing uranium as a by-product. The 1967 Act provides for the imposition of heavy fines or prison sentences for the unauthorised disclosure of information, and no journalists have been allowed into Rössing. Security has been so tight that even the highest South African officials in Namibia, including the Pretoria-appointed Administrator, require written permission from the Atomic Energy Board to visit the installations. This possibly reflects the illegal character of the operations at Rössing in terms of international law as well as a strategic importance, apart from the fact that it is located in an occupied territory.

The refusal to provide information about the project extends equally to the companies involved; the 1967 Act has been invoked frequently by RTZ, for example as the main reason for its own refusal to supply more information. This attitude also derives from attempts to protect companies investing in Rössing from the political implications of an exposure of their interests. Mainly for this reason governments known to have approved delivery contracts from Rössing have remained extremely tight-lipped on the subject of details in their possession about developments at the mine. The French Government has refused to provide any details about Minatome's contract with Rössing; this circumspection has been even more marked in the case of Britain, where the Cabinet approved a major deal in 1968 between the UK Atomic Energy Authority and RTZ for the supply of 7,500 tons of uranium between 1977-82 from Rössing (see below, pp.142 ff.). The existence of these forward contracts was essential to RTZ to ensure the commercial viability of the project and enable RTZ to raise the necessary capital. For these reasons the British deal has come under intense criticism, both domestically and at the United Nations. But the Foreign Office has continued to defend the contract, and on its return to office in October 1974 the Labour Government refused to honour its pledge to cancel it.

Japan is the third country known to have arranged contracts with Rössing. Namibia and South Africa are providing 43% of all contracted supplies between 1975 and 1985, a total amount of 83,800 tons, of which 8,200 short tons are for delivery from Rössing to nine electric power companies in Japan. Following criticism of the deal in the Japanese Diet and at the United Nations, government spokesmen acknowledged the existence of the contract, and pledged themselves to deal with it in accordance with UN resolutions on Namibia. The deal has now been suspended. In addition Japan has two further contracts with RTZ companies, including Rio Tinto South Africa, which manages RTZ's copper/uranium mine at Palabora, and Rio Algom of Canada.[45]

Although there has been no mandatory ruling on Rössing or any other specific foreign investment in Namibia, the clear tenor of the series of resolutions adopted by the UN Security Council has been to emphasise the invalidity and detrimental effect of foreign investment in South-African-occupied Namibia. In 1966 the UN General Assembly unilaterally revoked South Africa's old League of Nations Mandate over the territory, following Pretoria's strenuous opposition to the idea of placing Namibia under the UN's Trusteeship system as a first stage in the decolonisation process. The revoking of the Mandate was subsequently endorsed by the Security Council; in 1970 it formally requested all member states of the UN to ensure that their nationals, both individual and corporate, ceased all dealings with respect to commercial enterprises located in Namibia, under rights and concessions purportedly granted by South Africa.

In a resolution adopted in October 1971, the Security Council warned that all 'franchises, rights, titles or contracts relating to Namibia granted to individuals or companies by South Africa after the adoption of Resolution 2145 (of 1966) are not subject to protection or espousal by their states against claims of a future lawful government of Namibia.' However, South Africa's major trading partners, with one significant

124

exception, have sought to maintain a distinction between
continuing private commercial interests by their nationals in
Namibia, and the issue of the legal and diplomatic non-
recognition of South Africa's administration of Namibia. The
one exception has been the United States, which in 1970
announced in the UN General Assembly that it would hence-
forth 'officially discourage' investment by American corporations and individuals in Namibia, would not supply credit
guarantees, and would withold protection of investments
made since 1966 against claims made by a future lawful
government of Namibia. A decision to invest would be left to
the company concerned, but the existence of this policy seems
to have been instrumental in deterring any direct US involvement in the Rössing project, despite other existing interests in
the Tsumeb base-metal mine and elsewhere.

The illegal exploitation of Namibia's natural resources is
also subject of a special Decree for the Protection of Namibian
Natural Resources, issued by the UN Council for Namibia in
1974. This declares all mining, processing and export of
Namibian resources to be illegal, and subject to seizure by the
United Nations on behalf of the people of Namibia. SWAPO's
attitude to foreign investment has also become tougher over
the years. In a statement issued in October 1976 by the
organisation's Publicity and Information Secretary, Peter
Katjavivi, SWAPO warned: 'It is our view that foreign
investment in Namibia is one of the major factors contributing
to the continuing presence of South Africa's illegal occupying
forces... Foreign companies are taking advantage of the
immediate political situation in Namibia and it is necessary to
emphasise that all mining titles and temporary prospecting
rights granted after 1966 are illegal, and constitute moreover,
a criminal exploitation of irreplaceable natural resources,
which rightfully belong to the people of Namibia.' The
statement called on investors to take a 'long-term view' of
their prospects, and recognise that their activities were
jeopardising any future role in Namibia after its
independence.[46]

RTZ held secret discussions with SWAPO in 1975, at which the gap between the two became clear. RTZ apparently sought a private understanding guaranteeing its future in an independent Namibia, with subsequent discussions on taxes, royalties and wages; SWAPO wanted immediate discussions of these issues, and acknowledgement of the unlawful status of the Rössing project whose construction postdates the revocation of the South African Mandate in 1966. The implication of SWAPO's subsequent public comments has been that to attempt negotiations the day after independence would be too late, and companies such as RTZ would have to take the consequences.[47]

RÖSSING'S DEVELOPMENT

Extensive radioactive mineralisation was discovered in the central Namib Desert area before the First World War, during the period of German colonial rule. There are reports that radioactive properties were first suspected in the area because of the sterility of African women after exposure to the atmosphere: they apparently visited the area especially for the purpose of becoming sterile.

In 1928 pieces of davidite were found in the vicinity by Peter Louw, an army captain turned prospector. No attempt was made to mine the ore, either then or for many years subsequently, because of the low concentration of uranium, the absence of an economic mining method, and the low world price for uranium. But by 1960, systematic mapping and aerial photography carried out by the Geological Survey Division of the South African Department of Mines had established the potential of the ore-body, its extensive nature and its suitability for open-cast mining. The actual decision to initiate mining probably followed publication in 1964 of the report of the official *Commission of Enquiry into South West African Affairs*, which had spent two years surveying Namibia's economic potential for the South African Government. It recommended the rapid development of the territory's mineral sources, in order to provide revenue for infrastructure projects such as the Cunene hydro-electric scheme then on the drawing-board. Existing mines were reported likely to become exhausted in the 1980s.

The Rössing deposits are regarded as geologically unique, since they are the first porphyry uranium finds in Alaskite and granite intrusions. Similar deposits have subsequently been located elsewhere in Namibia, as well as areas of southern Brazil and Uruguay. The problem of developing a suitable mining process for the vast amounts of rock and waste ores in which the uranium is located was solved by RTZ (developing their own extraction process at Palabora), who decided to initiate a full-scale prospecting programme with a view to the immediate commercial development of the deposit.

In July 1966, mining rights were obtained from G. P. Louw (Pty) Ltd, and a programme of exploration, including metric surveys, geological mapping, percussion and diamond drilling, mineralogical and metallurgical investigations, was carried out prior to bulk-sampling and pilot-plant test-work to establish the economic viability of the project.[48] The rapid rate at which the mine was subsequently commissioned reflected the crucial role the project had come to occupy in RTZ's bid to become a leading world uranium producer and so lessen its heavy dependence on earnings from copper. Through Rio Algom RTZ already controls one-third of the Canadian reserves, and a major stake in Australia, but the potential for expansion in these two countries is limited.

Underground bulk-sampling was completed by the end of 1971, and preliminary test-work on the material obtained by April 1972. In that year the management contract for the design, engineering, procurement and construction work of the mine was awarded to a joint venture comprising Power Gas Ltd of London, part of the Davy-Ashmore Engineering group, and the Western Knapp Engineering Division of Arthur G. McKee, a US company. Basic engineering took place in San Francisco, with general design and construction handled by Power Gas Ltd.[49] Another US company, Interspace Inc., was awarded a contract in 1973 by the South African Department of Water Affairs for the manufacture and supply of pipes for the water needs of Rössing and Swakopmund. Construction of the mine's basic facilities, road and rail connections to Walvis Bay (Namibia's main harbour and only deepwater port), and the building of a special all-gravel airstrip at Arandis near Rössing, had all been completed by the end of 1974. Work on the main processing plant and the commissioning of mining equipment began in 1975, with the first ore deliveries initially scheduled for mid-1976.

Shortly before work on the main plan began, RTZ decided to double the mine's rated output from 2,500 to 5,000 tons of U_3O_8 a year, which meant an increase in the ore and waste handled from 60,000 to 120,000 tons a day. This decision could well have been related to the increase in the world market price of uranium following the 1973 energy crisis, in which RTZ itself played a leading role. However, the increase resulted in a number of unexpected difficulties at the mine which seriously delayed production targets. In its 1975 Annual Report, RTZ forecast that full-scale mining would be started in the second half of 1976, with a rapid build-up to the rated output of 120,000 tons. In October, rumours of trouble at Rössing which had spread to London and affected RTZ's share price on the Stock Exchange forced the company into one of its rare public disclosures on Rössing. This revealed that '. . . problems mainly due to mechanical and design weaknesses' had been encountered in some sections of the plant; although the problems were said to have been identified, certain modifications had to be made. This meant a delay of up to 18 months before full production could be reached. The implications were amplified by the mining correspondent of the London *Times*, who reported the view in mining circles that Rössing might run at only 50% capacity until early 1978. The process of separating uranium from the host-rock had apparently been plagued by a number of faults, possibly resulting from the crash programme to double capacity after work on the project had already started.[50] The extent of the problem was more fully revealed at RTZ's 1977 Annual General Meeting. In his statement to shareholders, Sir Mark Turner said that it had been the 'most disappointing aspect' of the group's 1976 operations. Design weaknesses and the abrasive nature of the ore, '. . . surprisingly not apparent at pilot plant stage', had meant heavy maintenance costs and frequent component replacement.[51] During 1976 only 771 short tons of U_3O_8 was produced, well below the initial production target and insufficient to enable RTZ to fulfill its delivery contracts with major customers – the company revealed that it was now seeking to renegotiate these for later delivery.

The technical problems also had the effect of increasing the capital costs of the project and forcing a revision of RTZ's optimistic forecast of its growth prospects, based as they were on the expectation of an early return on investment at Rössing. Under South African tax legislation, extremely generous provisions allow for up to 100% of capital expenditure to be set off against taxation, while the promise of a high level of pre-interest profit to set against development costs suggested that Rössing was expected eventually to become RTZ's largest profit-earner. But at the 1977 Annual Meeting Sir Mark Turner said that RTZ had been obliged to inject a further £20 million into the project to cover running costs incurred in the period up to the end of June 1977, and shareholders were told that further funds would be required, involving discussions with other partners in Rössing. In an almost classic piece of corporate understatement, the RTZ chairman emphasised that the profit flow would remain unsatisfactory, and unlikely to contribute much to the balance sheet before 1979. But this is likely to prove dangerous for RTZ if in the meantime the political situation changes radically in Namibia through the installation of a SWAPO-orientated, majority-rule government. A stockbroker's survey concluded that '. . . the main risk for Rössing is the political one.' It also accurately predicted that 'local opposition' would intensify once the mine was in production.[52] RTZ has itself admitted that labour conditions at Rössing were not good, while the mine has acquired a bad reputation among the African work-force for health- and working-conditions.

The importance of Rössing to South Africa as a future source of uranium has been obscured by the refusal on the part of Rössing Uranium to disclose any precise figures. But *Nuclear Active*, the official publication of the South African Atomic Energy Board, has reported that, on the basis of new ore discoveries including ones at Rössing, South Africa's reasonably assured reserves exploitable at under $15 per ton, a standard international measure, were revised upwards from 205,000 tons to 300,000 tons in December 1972.[53] An IAEA report of 1974 suggested that porphyry uranium deposits such as those at Rössing could become the chief source of the world's uranium. This followed reports of exchange visits between Brazilian and South African geologists in which the very similar stratigraphy of Namibia and Brazil was studied.[54]

The most specific figure for the grade of ore at Rössing was provided in 1970 by the US Department of the Interior, which estimated the average grade as 0.03% of U_3O_8 or 0.8 lbs per ton of ore. More recent estimates have suggested an average ore grade of 0.045%, capable of being upgraded to 0.07% by selective mining techniques. The total size of the ore-body indicated is enormous, and on the basis of these grade estimates it is placed at around 250 million tons. Some sources regard even this as too conservative a figure, and interpret the geology of the deposit as indicating an ore-body in excess of 500 million tons.[55]

Further uncertainties about the real potential of Rössing were caused by the discovery that RTZ was planning underground mining in addition to open-pit mining. Projections have been made by some brokers of a production increase to between 8,000-10,000 tons of U_3O_8 a year through the introduction of underground mining and the treatment of lower-grade ore found further beneath the surface. Until 1976 there was no public reference by RTZ to any plans to introduce underground mining; the first hint was provided in an advertisement for mining personnel in the London *Mining Journal:* the job was to involve responsibility for the planning of a 'fully mechanised underground operation'.[56] Subsequent advertisements referred to the use of 'underground trackless methods' at Rössing. RTZ's *1976 Annual Report* merely stated that ' . . . during the year, development work commenced on the underground mine which complements the open pit.' In the following year's report, development was said to be 'proceeding satisfactorily' with underground operations planned to start in the second half of 1977.[57] One effect could be to increase the expected life-span of the Rössing mine from 25 years to 80 years.

Mining analysts found themselves puzzled by the timing of RTZ's construction of underground mining capacity at Rössing since underground mining costs at least twice as much as open-pit mining. Usually a shaft is sunk after several years of open-pit mining, once all the financing loans have been paid off and the pit has become so deep that it is cheaper to mine underground. RTZ's explanation that it was simply the best way of getting the ore out was not thought convincing. One report found there to be two possible explanations – either the underground ore-body was surprisingly rich or RTZ was going for pockets of rich ore immediately, bearing in mind the political uncertainties surrounding Rössing.[58]

The extent of Rössing's contribution to South Africa's overall uranium production is clearly considerable, and official projections of increased output in this period up to the mid-1980s are widespread. A 15-year forecast of production in 'Southern Africa' (an official euphemism in the context for South Africa and Namibia) was presented to the 1976 international symposium of the Uranium Institute in London by R. E. Worrall, the influential uranium advisor to the South African Chamber of Mines. He forecast an increase in production from 2,800 metric tons in 1975 to approximately 10,000 metric tons in 1978, due 'in large measure' to production from Rössing. Output would subsequently attain 12,000 tons in 1980 and 15,000 tons in 1985. At the end of 1976 these figures were revised upwards by Dr A. J. A. Roux, president of the Atomic Energy Board, to 10,850 metric tons in 1978 and 16,270 tons in 1985.[59]

Despite RTZ's denials that any of the Rössing production is to be used by South Africa, the prospects of a huge new source of uranium which it opens up have probably been a decisive factor in the initial planning of South Africa's enrichment capability. The projected expansion of uranium output inside South Africa (excluding Namibia) would be insufficient to supply the feed necessary for the full-scale enrichment plant planned for the 1980s. The throughput required for a production of between 4,000-6,000 tons of enriched uranium has been estimated by one source as at least 20,000 tons of uranium oxide a year.[60] Only Rössing (or other uranium deposits yet to be developed commercially in Namibia) has the potential to provide uranium on this scale. If the Namibian deposits are not available, the Republic would have to import the raw material from overseas.

A NAMIBIAN URANIUM RUSH?

By the mid-1970s the development of Rössing, together with a high uranium price on the world market, had stimulated a widespread prospecting boom in Namibia of gold-rush proportions. The search for uranium was centred on prospects in the Namib Desert inland from Swakopmund and Walvis Bay, with large concessions staked out by many of the biggest South African and multinational mining firms. Little information on the scope of this prospecting activity has been revealed by Pretoria or the companies concerned, but in 1977 the Johannesburg *Financial Mail* published a map of the concessions, which cover thousands of square kilometres, showing that the main thrust of exploration activity is south-east of Rössing.[61] (See map on p.118).

Companies involved in prospecting include South Africa's General Mining, Gold Fields of South Africa, the Anglo-American Corporation and others, as well as the French companies Minatome and Aquitaine, Canada's Falconbridge Nickel Mines, and Union Carbide of the United States.

Secrecy over the prospecting activity has been assisted by a new technique of exploration known as the Track Etch process, a patented system owned by a subsidiary of General Electric of the United States, which greatly reduces the time and expense involved in uranium exploration. It can be used in ways which protect proprietary rights, so that only the customer is aware of the results.[62] There is no information available as to the financing of these developments, and many potential investors are taking pains to keep their identity secret. However, it is known that representatives of banking and other financial institutions have been paying visits to Windhoek, while sources in certain OPEC countries are reported to have shown interest.[63]

Apart from the political uncertainties, a major question-mark over the development of the Namibian deposits is the long delay in commissioning the Cunene hydro-electric scheme on the Angola/Namibia border. This was started in 1969 as a joint Portuguese/South Africa venture involving a series of dams and pumping stations inside Angola and a hydro-power station at Ruacana, just inside the Namibian border. The project was financed mainly by South Africa at a total cost of around R230 million, and was due to be completed in mid-1977 with an initial power output of 240MW. Namibia is without any indigenous fuel sources of its own, and power has been supplied by thermal stations using coal imported at considerable expense from the Republic; so the Cunene scheme formed part of a comprehensive programme to meet the anticipated power needs of Rössing and other new mines. By 1976 ESCOM, the State electricity programme, had completed a power grid linking all the major mining centres with a direct transmission line to

133

Ruacana.

Protection of the Cunene installations was South Africa's pretext for its intervention in the Angolan civil war of 1975/76. Although the MPLA Government agreed not to interfere with work at sites inside Angola, in March 1977 it ordered all civil engineering personnel to leave the Calueque dam, which was only partly completed owing to continuing border tensions with South Africa. Luanda alleged that the South Africans were continuing to provide assistance to Unita guerrilla forces in southern Angola. The Calueque dam was to regulate the flow of water to the turbines at Ruacana, and unless it is complete the power station will be able to generate only 70% of its installed capacity. At the start of 1977 South Africa was forced to provide an additional R20 million for the provision of power from alternative sources, mainly through the expansion of existing thermal capacity. This enabled Rössing to begin production; however, only the Cunene project is likely to be able to provide the additional power needed for further mining operations.[64]

Secret West German involvement in Rössing

In Part I we described the deliberate misleading of the West German Cabinet by government officials and commercial interests allied with the South Africans over Rössing. The Federal Government's close involvement in the Rössing project derives from its financing of the prospecting activities of Urangesellschaft. This was part of Bonn's programme of providing subsidies to German concerns engaged in uranium exploration activity around the world, in order to meet the anticipated needs of the West German nuclear energy programme in the 1980s. A major cover-up operation was launched by Bonn and Pretoria over the degree of West German involvement at Rössing. Among the highly confidential

134

documents from the South African Embassy are a number relating specifically to Rössing, made available to us by the SWAPO office in London.

A major reason for the anxious cover-up was an earlier political row over Cabora Bassa. Opposition to the official financing of West German consortia involved in the construction of the Cabora Bassa dam in Mozambique had forced Bonn to announce a halt to further involvement in the project in 1971. African opposition had been particularly intense, and was conveyed personally to Chancellor Brandt by Zambia's President Kaunda during a State visit to Germany in October 1970. In the case of Rössing the diplomatic position was even more tricky, because of West Germany's application for full membership of the United Nations in 1972. A further worry was that a diplomatic row over Rössing would prematurely blow the lid off the whole process of collaboration between the two countries in the nuclear field.

The documents reveal that hidden government subsidies for Urangesellschaft's activities in Namibia, in the form of extra subsidies on operations in Niger, may have been crucial in enabling Urangesellschaft to continue its involvement in Rössing. The Federal Science Ministry announced in 1972 that it was withdrawing further financial support for Urangesellschaft's prospecting activities in Namibia, although it admitted that the Government contribution had accounted for 75% of the total DM8 million prospecting costs incurred by Urangesellschaft until then. The official announcement gave as one reason for the decision the availability of alternative supplies from Australia and Canada, reducing the need for any uranium from Rössing; however, a spokesman admitted that '. . . in these things there is always a political context.'[65] Informed sources in West Germany confirmed that the Cabinet had decided to withdraw its backing, largely due to the embarrassment caused by its earlier commitment to Cabora Bassa, as well as the fear that continued support could jeopardise West Germany's application for membership of the United Nations, where the involvement of Urangesellschaft had been publicly mentioned in the context of criticism of the Rössing project in general.

Urangesellschaft was ostensibly replaced by General Mining, but in fact the company has retained a financial interest in Rössing, together with an option on some of its uranium output. Its share of the equity is between 10% and 15% apparently reduced from 25% following the allocation of 10% to Total in 1973. An interview with Urangesellschaft revealed that the company was hoping to become involved in actual production at Rössing.[66]

The major advocate for a withdrawal from Namibia was Dr Erhard Eppler, then Minister for Economic Cooperation, whom we cite in Part I (see pp.69 ff.). In 1971 he stated that the Government should not provide a financial guarantee for Urangesellschaft's involvement in Rössing, but it was not clear if this was any more than a personal opinion on the political risks. A letter dated March 1971 from another Government source was less positive, merely stating that since no guarantee for the mining of the deposit had so far been requested by Urangesellschaft, the issue of Federal Government backing had not arisen. The letter made it clear that at that time Urangesellschaft was still involved with RTZ in investigating one of the '. . . largest deposits in the world'.[67]

During a debate in the UN General Assembly on measures to end South African rule in Namibia, the West German spokesman claimed that incentives granted to encourage investment in developing countries were not given to concerns investing in Namibia, while contracts promoted by the Government Development Agency all excluded Namibia. In reply to a questionnaire sent out by the UN Council for Namibia in 1974 to all UN member states, the Federal Government said that it in no way encouraged economic activities in Namibia. But in an answer to a question in the Federal parliament in the same year, a government spokesman stated that the Federal Government did not feel itself bound by the UN Decree of 27 September 1974 on the Natural Resources of Namibia, which banned export of Namibian minerals without UN approval.[68] He indicated that West Germany would be purchasing between 500 and 750 metric tons of U_3O_8 from Urangesellschaft's share of the production quota. This coincided with the uranium enrichment agreement concluded by Steag (which has a one-third stake in the consortium controlling Urangesellschaft) and strongly suggests that German participation in the Rössing project was an integral part of the whole nuclear co-operation agreement.

The documents relating to Rössing consist of a series of cypher telegrams and secret reports, mainly between Ambassador Sole and the Department of Foreign Affairs in Pretoria, between 1970 and 1974. When viewed chronologically their significance is self-evident.

1/ *Outgoing Pro Cypher Telegram, 28/10/1970 from SA Embassy, Cologne, to Secretary for Foreign Affairs, Pretoria*

'Your 105. I have already reported on this trend brought to your notice by German Ambassador. See my 8/19 of 9th October entitled "Is there a threat of a second Cabora Bassa" also my 9/2/8/28/1/1 of 20th October entitled "German participation in the Rössing Project" and my report on Kaunda's visit 8/5/1 of 20th October. It is for this reason I expressed reservations about Rössing in my 115. As regards the Urangesellschaft it is necessary for me to talk to one of their *top* men and this has now been arranged for November 6th. I must defer my final recommendation until I have spoken to Urangesellschaft and at this stage would merely remark that the German Government was persuaded to stand firm on Cabora Bassa and that it is important that it should stand equally firm on Rössing, although it may be more difficult to do so. . . It is an open question whether publicity at this stage would be wise or unwise and it is here that I consider that we should be guided by Urangesellschaft.'

2/ *Letter to Andries Mare, Department of Foreign Affairs, 6/12/1971 re 'Uranium Deposits' from D. B. Sole*

'It is not clear from your 137/23(5) of 2 December 1971

whether you are interested in the reference to South West Africa or whether you are interested in the overall programme of the Federal Republic.

As far as South West Africa is concerned I am sure that Mr Fourie [Secretary for Foreign Affairs] is aware of the developments with respect to Urangesellschaft's participation in Rössing, although this is being kept very hush hush. As to other German interests in South West Africa I am aware of other negotiations proceeding in respect of the possible exploitation of an area not far from Rössing but here the interested South West African partners are still awaiting clearance from Pretoria.

As regards the policy of the German government with respect to Uranium prospecting, as in the case of Japan, the government is concerned to ensure the continued availability of supplies to meet the expected requirements of the German nuclear power programme in the 1980's. Where this is feasible the German government encourages German companies such as Urangesellschaft or Uranerzbergbau in Bonn to enter into agreements for prospecting of Uranium ore...

As regards Dr Klaus von Dohnanyi's reference to South (West) Africa in his reply to a parliamentary question, I am a little surprised that this reference slipped through because since the earlier propaganda campaign against German participation in Rössing, the practice has been to keep this kind of activity in wraps.'

3/ *Letter to D. B. Sole from the Secretary for Foreign Affairs, 11/1/1972*

'I passed your letter to Hein van Niekerk who also asked me to thank you for the information, particularly as regards the German Government's policy in respect of Uranium prospecting. What interested me in the first place was that Dr Klaus von Dohnanyi had mentioned South West Africa by name, in the Uranium prospecting context. I do not think I need further information at this stage.'

4/ *Confidential letter from D. B. Sole to Secretary for Foreign Affairs, 28/3/1972, re 'Sales of SA Uranium: Rössing'*

138

'More publicity has appeared on the question of the withdrawal of German Government sponsorship of uranium prospection [sic] in South West Africa; see for example the attached article from "German International" the English language monthly published in Bonn.

According to Secretary of State Haunschild of the Ministry of Science which normally contributes 50% of preliminary prospection surveys, the issue with respect to Rössing has been somewhat eased for Urangesellschaft by the fact that in the case of a Central African State, which he did not identify but which was obviously Niger, the Ministry will contribute 75%. Herr Haunschild's version was that Urangesellschaft would be in receipt of a subsidy in respect of all its approved prospecting activities overseas and the Ministry would not inquire too closely into how precisely Urangesellschaft allocated its money. The inference was that Urangesellschaft could use some of the money for Rössing.

However, when I spoke to Dr von Kienlin of Urangesellschaft in Frankfurt last week, he denied that the extra payments to the Central African State were being received and said that this was no more than a hope and a possibility, on which he did not place much reliance. My impression from my talk with Dr Kienlin was nevertheless that the financial problem was not the principal cause of concern on the part of Urangesellschaft. The real issue was the present uncertainty with respect to supplies from South West Africa and their availability over a long-term period in the future. Neither he nor his company doubted our ability to deliver – the doubts lay with the consumer. Since the uranium would be required for use ultimately in nuclear power stations operated by public utilities or under the supervision of Länder Governments, Urangesellschaft had to take into account that in the minds of many responsible persons concerned in this particular sector, there was a fear that delivery of the uranium would be banned or become impossible because of resolutions adopted by the United Nations and unwillingness on the part of the German Government to act counter to those resolutions. These fears are being played upon by the radical anti-South African elements in this country who have a spokesman in the Cabinet on this question in the person of Herr Eppler. This was a problem about which very little could be done at the moment (although he did not say so he was obviously inferring that a change of Government would lead to an easing of the situation).

The question of Rössing will certainly come up in one form or another in the course of Secretary of State Haunschild's visit and I should be grateful to be informed of any discussion which may take place on this topic.'

5/ *Immediate Outgoing Cypher Telegram, 14/4/1972, from South African Embassy Cologne to Secretary for Foreign Affairs*

'I cannot understand why the report which I sent to you should have been sent back to me. Surely it should have been sent to Dr Roux. Since Dr Haunschild arrives in South Africa on Sunday suggest you give this matter your immediate attention.'

6/ *'Secret' Letter 24/7/1973 from D. B. Sole to Secretary for Foreign Affairs*

'As far as negotiations on enrichment are concerned I am reasonably in the picture, but should be glad to know what the present picture is with respect to the German interest in Rössing. I was asked for example about a report date-lined Windhoek appearing in the *Handelsblatt* of 11 July according to which France has now taken over the share originally envisaged for Germany in the development of the uranium deposit at Rössing.

What I would like to know please is whether the German interests concerned have completely surrendered the option which they had with respect to Rössing.'

7/ *'Personal and Secret' Letter 27/8/1973 from Brand Fourie, Department of Foreign Affairs to D. B. Sole, Re: German Interest in SA Uranium Developments*

'In response your minute 9/2/30, 9/2/8/28/1 of 24 July 1973, the Atomic Energy Board has reminded us that matters such as uranium sales and the allocation of equity in the share capital of Rössing, although done with the approval of the Minister and/or the Board, are the sole functions and responsibility of that company in the normal course of business practice, and information pertaining thereto must be treated as commercial secrets unless the buyer wishes to make a public or press announcement at the appropriate time, as in the case of Total, to which you refer. For these reasons it is *not possible for the Board to divulge secret information to outside persons who are not directly concerned − even if they were to make enquiries through one of our Ambassadors.*

However for your personal information I may mention that Total, having concluded an agreement with Rössing for a substantial supply of uranium U_3O_8, has acquired an option to take up 10% of the equity of Rössing. This deal has nothing to do with Urangesellschaft and "the German interests concerned" have therefore not surrendered the option which they have had with respect to Rössing to the French company.

I may also mention that Urangesellschaft has in fact renegotiated a somewhat smaller quantity of U_3O_8 than originally planned,* and as a result the percentage equity to which they are entitled has been reduced.'

'*The Total contract is far in excess of this reduction – it is even larger than the quantity originally contracted for by Urangesellschaft.'

8/ *Personal Telex from Sole to Roux, Atomic Energy Board, Pretoria, 22/3/1974*

'Dr Hampel of Urangesellschaft Frankfurt has applied to me for assistance in connection with facilities for representatives of Urangesellschaft to visit Roessing [sic] next week. Persons concerned are − Dr Albrecht von Kienlin, Managing Director, Dr Karlfreig Knolle, prokurist (commercial), Dr Hans-Michael Jonndorf, prokurist (legal) – they fly to South Africa tomorrow March 23 for three days discussion with NUFCOR in Johannesburg commencing Monday but have been asked to visit Roessing before returning to Frankfurt SA 250 on Friday 29 from Windhoek. I understand that normally six weeks advance notification has to be given to the AEB in order to obtain permission to visit Roessing but in the special circumstances applicable in this case I hope that it will be possible for you to authorise the visit.'

The visit was obviously important to the establishment of secret participation by Urangesellschaft and the Federal Government in Rössing.

The British Government dragged in

West Germany is not the only country where the Cabinet has been tricked into a secret involvement in Rössing. In 1968 the UK Atomic Energy Authority (UKAEA), a government agency then under the jurisdiction of the Ministry of Technology, was authorized by the Cabinet (under a Labour Government) to conclude a contract with Rio Algom, RTZ's Canadian subsidiary, for supplies of uranium from the mid-1970s to the early 80s. The Cabinet decision relating to this deal specifically stated, in the light of information from the UKAEA that the contract might have to be switched to a RTZ subsidiary in South Africa, that approval of the deal with Rio Algom was conditional on the supplies being from Canada; if there were any question of purchasing uranium oxide from South Africa the Cabinet should be informed immediately, and no switch should be made without authorization by the Cabinet. The deal with Rio Algom was described at the time as probably the most significant new uranium deal since the initial military contracts.[69]

In January 1970 it was learned in the Foreign and Commonwealth Office, following an apparently routine, low-level notification from the Ministry of Technology, that two major contracts were about to be signed between the UKAEA and Riofinex, a South African subsidiary of RTZ, for the purchase of uranium oxide from the deposits to be exploited at Rössing, Namibia. The exchanges which followed made it obvious that Ministry of Technology officials had authorized the UKAEA to switch its purchasing contract not only to South Africa – in defiance of a specific Cabinet decision – but, to make matters worse, the occupied territory of Namibia, a possibility which had never even been suggested to the Cabinet. The Ministry of Technology officials maintained that the distinction between South Africa and 'South West Africa' was insignificant, and that they had no obligation to refer the matter back to the Cabinet because of the switch. It seems possible that even the Minister of Technology, Mr Anthony Wedgwood Benn, had not at that point been informed of the issue. The Foreign and Commonwealth Office therefore decided that, although it was probably too late to stop the deal, the Cabinet should be informed because of the defiance of its earlier decision.

A Cabinet Committee discussed the contract two months after the discovery of the unauthorized switch. The Government departments represented included the Foreign and Commonwealth Office (FCO), Ministry of Technology, Ministry of Defence, Board of Trade, Treasury, Ministry of Overseas Development, and the Attorney General's Office. The issue was presented to the Cabinet as a *fait accompli*, with the contract already signed. In fact, this was misleading since the first contract, for 6,000 tons, was claimed by RTZ to be inoperable unless a second contract, for an additional 1,500 tons, was also signed; this was necessary to ensure a sufficient scale of production to make the operation profitable and therefore to raise the necessary finance. This second contract was signed only after the Cabinet, on the assumption that there was only the first contract at issue, had reluctantly decided that there was nothing they could do about it, and that it should therefore go ahead.

A memorandum on the original contract by the Attorney General pointed out that there was a *force majeure* clause in the contract, and that any policy decision by the Government to stop the deal would amount to *force majeure*, allowing the deal to be terminated without penalty. This was hardly discussed, however, in the light of two further arguments put forward,

143

without any evidence, by the Ministries of Technology and Defence. The first was that there were no readily available alternatives to purchasing from Rössing. This was patently untrue, since in 1970 there was a glut of uranium oxide on the world market, and producers everywhere were either closing mines or stockpiling uranium which could not be sold. The falsity of this information was made obvious when, at about the same time as the contract with Rössing was signed, the Rio Algom mines at Elliott Lake, Canada – from which the UKAEA purchase was originally to have been made – were closed down owing to the lack of customers.

It seems, therefore, that the second argument was in reality the valid one: that the Canadian Government, having been approached by the French Government for the sale of Canadian uranium to France for its nuclear *force de frappe*, had been able to refuse the deal only by imposing a policy of applying 'peaceful safeguards' clauses to all export deals for Canadian uranium. This would involve international inspection of the uses to which the material was being put at all stages, to ensure that it was not used for the manufacture of nuclear weapons. Although it was not specifically stated, the clear implication was that the UKAEA purchase in question was specifically intended for the British atomic weapons programme. The fact that uranium oxide for commercial purposes, contrary to the claim made to the Cabinet, was widely available at relatively cheap prices from many sources, including the original one in Canada, supports this conclusion.

Given this combination of arguments and misleading information, the Cabinet members felt it impossible to terminate the deal with Rössing. However, considerable dissatisfaction was expressed at the failure of government officials to inform the Cabinet before the crucial decisions had been made.

Fully appreciating the sensitivity of this major new commitment to South Africa's occupation of Namibia, instruction was sent out that no word of the deal should be allowed to leak out to the British press and public before the General Election, in June 1970. It was only after the defeat of the Labour Party in that election that news of the Rössing deal emerged in the press: a total of 7,500 tons to be delivered between 1976 and 1982, at a cost of about £40m, which has subsequently increased to about £120m. The disclosures prompted a considerable debate within the Labour Party, then in opposition. It was pointed out, for example, that despite the abundance of cheap uranium on the market the UKAEA contract had never been put out to tender to any company except RTZ subsidiaries.[70] RTZ, apart from being a major British multinational with powerful influence in London, had an important political asset in Sir Val Duncan, its Managing Director, who among other things was entrusted with a significant report on the British Diplomatic Service, published, by coincidence, at approximately the same time that the Rössing deal was being debated; its major recommendation was that the British foreign service should be drastically overhauled so as to concentrate representation in areas relevant to British multinational business interests. As a result of the controversy, Mr Wedgwood Benn took public responsibility, defended the deal with Namibia, and finally wrote in a letter to *The Guardian* in September 1973 that he had in fact made a mistake:

> 'We have already decided to terminate the AEA-RTZ contract. That particular case, in which neither the AEA nor RTZ were altogether candid with the last Labour Government, points to the need for even greater vigilance than has been shown in the past. As the Minister responsible at the time, I certainly learned that lesson.'[71]

The decision referred to was that of a Labour Party policy study group, subsequently endorsed by the National Executive and by the Party Conference of 1973. This reads, *inter alia*: 'Labour will terminate the atomic-energy contract with Rio Tinto Zinc for uranium in Namibia.'

Under the Conservative Government of 1970-74, not

145

surprisingly, there was no question of revoking the contract with Rössing Uranium. The Foreign Secretary, Sir Alec Douglas-Home, had himself visited Namibia in 1968, and commented then that South Africa was '. . . the natural administrator of South West Africa. . . It is difficult to see how the situation could be otherwise.'[72]

A move relating to the UKAEA at that time may have been connected with the sensitive Rössing deal. Under the United Kingdom Atomic Energy (Weapons Group) Act of 1973, the Weapons Group of the AEA was transferred to the jurisdiction of the Ministry of Defence as of 1 April 1973.[73] As described by the Secretary of State for Defence, Lord Carrington, in the House of Lords debate, the intention was to transfer the activities of the AEA's Weapons Group, including the Atomic Weapons Research Establishment at Aldermaston and its outstations at Foulness and small ones elsewhere, from the AEA to the Ministry of Defence:

> 'The main task of the Group, some $\frac{1}{5}$ of its total effort, is research, development and production work on explosive nuclear devices for use in nuclear weapons . . .'[74]

This followed another Act which split off the AEA's radio-chemicals and nuclear fuels business into new companies.[75] The contracts with Rössing are therefore probably transferred from the AEA to the new governmental corporation for nuclear fuel. There was also a rumour in London at the time of the 1973 Act that the incorporation of the Weapons Group into the Ministry of Defence was to protect it from legal action or too many questions about its reliance on Namibia as a source of uranium.

The question arises, with a Labour Government in power authorized to cancel the Rössing deal, whether it will in fact do so. Mr Wedgwood Benn (now Tony Benn) has not honoured his 1973 pledge to cancel the contract. A statement of policy on Namibia in terms of a recent 'policy review' on Southern Africa slightly restates British policy on Namibia, admitting that South Africa's occupation of Namibia is illegal.[76] However, for the first time it gives it *de facto* authority. This is a retreat from the policy stated in 1966 by then British representative at the United Nations, Lord Caradon, that 'We recognize that, by her actions, South Africa has forfeited the right to administer the Mandate.'[77] The British retreat from observing international law, as established by the International Court of Justice, on Namibia is closely related to the Rössing deal.

An interesting commentary on the deal in the London *Financial Times*, citing the arguments of senior British Government officials, confirms the importance of Rössing for the British Government. The commentator, David Fishlock, writes that the full dimensions of the dilemma posed by its deal with South Africa over Rössing are rarely admitted in public, but that the Government '. . . would dearly love to associate itself with the growing international pressures building up against South Africa. Yet it almost certainly cannot, *because of a major contract* signed about six years ago which is not merely with South Africa but, to compound the embarrassment, concerns the exploitation of resources in the hotly disputed territory of Namibia. . .'[78] (Authors' emphasis.)

South Africa's uranium marketing strategy

Most of the Western uranium suppliers, led by the United States, are imposing highly restrictive conditions on their exports in order to meet the threat of nuclear proliferation, which often arises from the resale or diversion to third parties of 'special' nuclear materials, i.e., those which are fissile. In this context the stated intention of South Africa to expand exports, and its obvious willingness to sell large quantities of uranium to France, West Germany, Iran and other countries

147

which are meeting increasing restrictions on the part of American and other suppliers, must be of major concern. A number of countries have openly voiced their priority objective as being to break through their current dependence on the United States for supplies of uranium – particularly in enriched form.

The planned construction in South Africa of a huge commercial enrichment plant, whose costs will probably be well above the American and other plants, will mean that to find a market for the output South Africa will have an additional and very powerful incentive to sell uranium to countries unwilling or unable to observe the restrictions imposed by the present suppliers. In short, South Africa already is, and may increasingly become, the means by which a wide range of near-nuclear states can pursue their ambitions in spite of the best efforts of those attempting rather belatedly to give teeth to the Nuclear Non-Proliferation Treaty.

AN INTERNATIONAL URANIUM CARTEL

South Africa has the ambition of perhaps becoming the major supplier for the world uranium market. It is currently estimated that reasonably assured reserves of uranium in South Africa recoverable at \$15 per pound or less represent about 20% of the world's known uranium reserves. If the whole of this amount were mined, it would be sufficient to power 50 light water reactors (LWRs) of 1,000 MW each, for about 30 years.[79] Currently South Africa is the third largest producer of uranium in the world (after the US and Canada), and there is discussion in South Africa about making it the world's largest producer in the next ten years.[80]

For the present the South Africans are building up huge stockpiles of the metal, presumably in readiness for the commercial enrichment plant which they hope will greatly increase the value of the end product. By 1973 South Africa had already stockpiled 20,000 tons of uranium oxide, and Dr Koornhof, the Minister of Mines, announced that over the next 30 years another 100,000 tons would be stored. One way of doing this is for the gold mines to stockpile part or all of their slimes, after extraction of the gold; the pyrites in the slimes oxidises to sulphuric acid, which dissolves the uranium. This makes subsequent extraction of the uranium easier and cheaper, provided enough time elapses.[81]

Such stockpiling, of course, not only provides a reserve of material for the proposed enrichment plant, but also helps to drive prices up on the world market, particularly at a time when Australia and other countries have been meeting vociferous internal opposition to uranium mining and exports on environmental and health grounds. South Africa is virtually free of such opposition, as well as from the other major obstacle to exploitation of uranium deposits – the land claims of the indigenous people, which is a serious issue in Australia (with the Aborigines) and Canada (with the Indians and Eskimos). In South Africa, the problem exists to a much greater degree than in Australia, Canada and the United States; however, the system of land tenure, the legal structure and the detailed measures for repressing the claims of the Africans mean that the Government and the mining companies can dispose of the country's minerals – and those of occupied Namibia – without any effective opposition.

In addition to this, South Africa's marketing structure for uranium is highly centralised around NUFCOR, which is the largest single uranium-supply company in the world.[82] NUFCOR also has powerful allies among the multinational corporations, the most obvious example being Rio Tinto-Zinc. Quite apart from the massive joint venture between RTZ and the South African Government's Industrial Development Corporation (IDC), RTZ and NUFCOR have been the leading proponents and major beneficiaries· of a recently exposed international cartel to raise uranium prices and exclude certain buyers from the market. Another participant in the cartel in its own right was Rössing Uranium Ltd.[83] RTZ documents now in the hands of the US Justice Department show that the company's main motive in setting up the cartel was to ensure the profitability of the high-cost Rössing mine.[84]

The cartel was launched at a meeting in Johannesburg in June 1972. The agreement drawn up there called for the participants not to offer uranium at less than minimum prices set for each year. No uranium vendor was to sell below $7.50 a pound, or below the 'world market price' – which it was agreed would be the price in effect as set by the cartel and those acting in concert with it. The agreement also called for a 'leader' to be chosen to quote the agreed-upon minimum price in a given bidding situation, a runner-up to quote the minimum price plus a small premium of 8 cents per pound; and all others to quote the minimum plus 15 cents per pound, thus giving the appearance of competition while in fact eliminating it.

The basis of the agreement was that the participants would allocate future sales by country and by producer within each country, except that RTZ was to receive a separate quota distinct from national quotas because of its worldwide dominance and, presumably, because of its key role in setting up the cartel. Later, an informal agreement was set up with US companies which helped to provoke a major anti-trust investigation by the US Justice Department when the existence of the cartel became known. US producers increasingly imposed the 'world market price' clause on their customers, knowing that this price at the time of delivery would be a price wholly or in part fixed and stabilised by the cartel. Several producers, following a meeting in 1973, withdrew from the market and refused to quote until the customers acceded to the new prices and terms artificially raised by the cartel.[85] With roughly 80% of the Western world's uranium production outside the US controlled by the five original partners in the cartel who met in Johannesburg, it is understandable that uranium prices increased as rapidly as they did – an eightfold rise over five years, far more than the oil price over the same period.[86] RTZ files show that the major beneficiaries were NUFCOR of South Africa and RTZ, who in fact gained a much greater share of the market than the original cartel agreement had envisaged. Originally planned to be the third largest supplier in 1972-77, NUFCOR was the second largest (mainly as a result of internal opposition to Australian mining and exports); RTZ, planned as the fifth largest, in fact emerged as joint second with NUFCOR, as shown in Table 1.

Table 1: *Projected and actual uranium market shares for 1972-77*[87]

	Projected (%)	Actual (%)
Canada	33.5	37.3
France (Uranex)	23.75	4.5
South Africa (NUFCOR)	21.75	24.2
RTZ	4.0	24.4
Australia	17.0	7.6

In addition to benefitting greatly from the increased market price and greater share of the market offered by the cartel, RTZ and NUFCOR were able to increase the price of contracts already concluded. RTZ raised the price of Rössing uranium ordered by the British Government to about three times the original price.[88] NUFCOR renegotiated standing uranium contracts with Britain, Germany and Japan at higher prices. From the average price of $7 to $8 per pound arranged with these buyers in the first instance, the average level at the end of 1976 was about $20, and was reported to be still rising. The Japanese agreed to prices ranging from $25 to $30 per pound, and were agreeable to still futher escalation clauses.[89]

SOUTH AFRICA'S CUSTOMERS: CIVIL OR MILITARY?

Customers for South African uranium oxide, following the

151

phasing-out of orders by the Combined Development Agency for military purposes, have been electricity companies based mainly in France, West Germany, Britain and Japan. During the 1960s and early 70s, however, demand from this source, and associated price levels, were extremely depressed. There have been allegations relating this phenomenon to the fact that unpublished deals were made at cut rates, apparently without any of the international safeguards normally applied to public deals for civil purposes. Since 1965, the sales value of South Africa's uranium oxide production has not been made public.[90] Export statistics are also not published, nor are there any details on uranium production in the Department of Mines' bulletins.[91] As a result, it is impossible to establish exactly how much uranium is being exported, and at what prices. Only a 'modest' proportion of the uranium produced will be diverted to internal consumption, even by the end of the century.[92] Thus, increasing although unspecified amounts will be available for export.

The South African *Financial Mail* sought to explain the new policy of secrecy about exports since the mid-60s by saying that the uranium industry was involved in a '... tricky marketing operation, which required concealment of the figures from other interested parties.' The article referred to two gold mines with published contracts with the UK Atomic Energy Authority; the last statistics published showed the mines selling very much more than the UKAEA was actually buying.[93] There were reported to be large 'non-quota' sales in 1965, and company chairmen indicated that further business could be expected on that basis. The main buyer was said to be France, paying a 'non-quota' price of R2.86 per pound, as opposed to a 'quota' price of just over R8 per pound.[94]

France has become one of South Africa's main markets for uranium. In July 1977 it was announced that the French had signed a ten-year contract to buy 1,000 tons of uranium oxide from South Africa – the biggest uranium deal ever between these two countries. The French are to finance a uranium extraction plant at Randfontein with a loan of $103 million from the nationalised banks, and in return will obtain the mineral at $27 a pound compared to the current 'world price' of $40. Coming after attempts by the United States, Canada and Australia to restrict the availability of uranium except under the strictest safeguards, the *Guardian* commented that the contract '... appeared to be a deliberate attempt to break from American control on international nuclear developments.'[95]

The apparent willingness of South Africa to sell uranium oxide outside the international safeguards system indicates that serious risks are involved if it proposes to sell uranium in enriched form, which has far more immediate military potential both directly and by enabling a country to produce unsafeguarded plutonium from its reactors. In a detailed survey of the prospects for nuclear proliferation, it has been suggested that South Africa could be the agent of the uncontrolled spread of nuclear weapons capability; there was '... some suspicion of the intentions of the South African Government where the export of uranium is concerned.'[96] The sale of enriched uranium without international safeguards has also been pinpointed by a joint US-Soviet study as one of the major dangers to the Nuclear Non-Proliferation Treaty.[97] There have been allegations, too, that South Africa provided the uranium used by Taiwan in its nuclear weapons programme.[98]

It is significant that South Africa is already offering a wider range of services than any other uranium supplier. With production of tetrafluoride and, most recently, hexafluoride in South Africa (following a co-production arrangement with Britain), NUFCOR has clinched the world's first orders for uranium processed to the threshold of enrichment. Contracts have been signed on this basis with West Germany, Switzerland and Japan.[99]

South African representatives have made it quite clear that they intend to market their uranium oxide strictly in accordance with their 'national interest' and have boasted that they can bargain over price and supplies for political as well as economic ends. Dr Louw Alberts, Vice-President of the Atomic Energy Board, said in 1974 that, with one-quarter of the free world's uranium, South Africa is in a strong bargaining position in terms of the world's energy crisis. Having developed 'outstanding techniques' for low-grade extraction and enrichment, the Republic's bargaining power was '. . . equal to that of any Arab country with a lot of oil'.[100]

This was no idle boast: in fact the uranium cartel organised largely by NUFCOR and RTZ timed the increases in their uranium prices very accurately to coincide with the 1973 OPEC oil embargo, according to *Forbes* magazine.[101] A NUFCOR representative was subsequently quoted as saying: 'The uranium market is off the launching pad.'[102]

There has also been much enthusiastic talk about the vast profits to be gained from marketing the whole of South Africa's production in enriched form – regardless of long-term contracts which specify natural uranium, as in the case of Rössing. Prime Minister Vorster, in his announcement of South Africa's new enrichment process, said, '. . . it is obvious that South Africa, as one of the largest uranium producing countries of the world, will consider it in its own interest to market uranium in the enriched form.'[103] The President of the Government's Council for Scientific and Industrial Research, Dr S. Meiring Naudé, added that if South Africa sold its uranium in enriched form, the income from uranium exports could be doubled and the life of some gold mines lengthened as a result.[104] Dr Roux, head of the AEB, has written that the South African Government originally decided to develop a commercial enrichment process '. . . in order to make it possible finally for the country to market her huge uranium resources in the most refined forms.'[105]

As discussed in Chapter 5, there is room for considerable doubt as to whether the venture will be as profitable as expected; in fact, with the price of enriched uranium falling relative to that of natural uranium, because of policies of the US Atomic Energy Commission and its successors,[106] and with the possibility of complications and rising costs making the whole project a major economic catastrophe, the South Africans could be disappointed in their ambition of making a small fortune from the enrichment plant. This, however, would be a dangerous situation in terms of the possibility of unsafeguarded (and unpublished) sales of enriched uranium: in order to recoup the losses on the plant, South Africa would have stronger motivation to conclude the maximum number of sales at the highest prices, with the major selling point – as has happened in the case of reactor exports by French and West German-based companies – being the absence of stringent safeguards such as the US is increasingly imposing.

South African press reports have begun to refer to the possibility of South Africa deliberately withholding raw uranium from world markets in order to market it later in enriched form.[107] It is of interest to note the major exception to this, in that the Rössing mine has been drawing up long-term contracts even while South African production is being stockpiled.[108] This is clearly a political move to commit foreign governments to continued South African control over the Rössing mine for several years to come, in opposition to international law and the independence movement within Namibia.

For South Africa, uranium marketing strategies are already an integral part of its foreign policy. One can only speculate, at this stage, how it will use the powerful new weapon which large amounts of enriched uranium will provide in forming new and stronger alliances with other 'near-nuclear' countries.

It has already been made clear that compliance with even the spirit of non-proliferation, let alone the letter of the NPT, is far from guaranteed by South Africa. On the contrary, it will be used as a powerful weapon of blackmail to ensure that South Africa participates in the real decision-making. The former Secretary of the US Treasury, Mr William Simon, was almost certainly reflecting South African thinking when he warned a high-level business seminar:

'. . . South Africa has traditionally maintained a responsible attitude towards the marketing of its uranium. Although she is expected to emerge in the 'eighties as a major supplier of enriched uranium, South Africa's exclusion from international discussion on these critical issues could force a change of policy.'[109]

5

South Africa's nuclear industry

South Africa's research and development in the nuclear sphere is far advanced compared to the programme of any other African country. The gulf between South Africa and other African countries which could be the target of her missiles and fighter-bombers was summed up by Kurt R. S. von Schirnding, South Africa's representative on the Board of the International Atomic Energy Agency in Vienna. In an angry speech protesting against the motion not to re-elect South Africa to represent the African continent on the IAEA Board, he made the point that

> 'South Africa is the third largest supplier of uranium in the Western world. South Africa has a large nuclear research establishment with a 20 MW reactor. Attention is being given to the construction of a large uranium hexafluoride plant. South Africa is building two new nuclear power reactors, both under the IAEA safeguards, it is operating a small enrichment plant and will be building a large commercial enrichment plant. No other country in Africa has any comparable production of source material and, as for nuclear facilities, there were only three small research reactors in three other African countries.'[1]

157

Phase One: building a research team

South Africa's nuclear research centre at Pelindaba, near Pretoria, is probably the only establishment in the whole country where, among the 900 employees, there is not a single black face to be seen – even among the domestic staff, who are almost always black elsewhere in the country.[2] Other lowly jobs such as cleaning have been highly mechanised to eliminate the need for black employees; again, this is extremely rare in a country where households can dispense with vacuum cleaners, preferring the abundant cheap labour provided by black women. Even the armed forces, police and government offices employ large numbers of blacks – Africans, Asians and Coloureds.

Pelindaba, whose operations are covered in a thick blanket of official secrecy by special legislation and security measures, clearly carries a status and a degree of priority of a different order from any other project in South Africa. According to the Editor of *Science*, its main reactor building is a piece of strikingly sculptured architecture, '. . . in contrast to the surplus-Army-hut style that prevails at many such establishments around the world,' as the magazine described it.[3]

Yet South Africa does not waste its nuclear research funds on prestige or humanitarian projects, or on basic research – of which Pelindaba does very little as Mr J. P. B. Hugo, Deputy Director-General there, has freely admitted.[4] This is true of the whole of South Africa's scientific establishment; research and development spending as a proportion of Gross National Product is low, on a par with that of Spain, Portugal and Greece. Technologically developed countries will normally devote up to eight times as much money to 'R and D', in proportion to their national budgets, as South Africa does.[5] The obvious solution, given the need to develop a strong industrial and military base for the maintenance of the 'South African way of life', is to use other countries' expertise. Connections with foreign science and technology are therefore vigorously encouraged by the Government and by large corporations, and are generously subsidised when necessary.[6] Policies have been adopted aimed at bringing all available scientific and technological talents into the service of the South African Government and the economy. One important means is to encourage immigration of foreign scientists, especially from Britain, West Germany and other countries of Western Europe; another is to send South African students to carefully selected courses overseas for advanced study, related to specific South African problems.

In 1966, for example, an editorial in the *South African Industrial Chemist* complained that there were no skilled personnel available in South Africa for the nuclear pro-gramme.[7] With the recruitment of foreign scientists and the return of almost one hundred South Africans trained over-seas, the situation changed rapidly. Dr T. E. W. Schumann, deputy-President of the Atomic Energy Board (AEB), who had himself been trained originally at Göttingen in West Germany, stated as early as 1964 that a well-trained corps of nuclear scientists had already been established in South Africa, and that a further 83 South Africans were currently studying in Europe (mainly France), the United States and Britain.[8] It was reported in 1971 that there were by then over one hundred fully-trained nuclear scientists at Pelindaba, most of them from the United States and Britain.[9]

Dr A. J. A. Roux, President of the AEB and a key figure in this field, recently laid particular emphasis on the role of the United States in launching atomic research in South Africa:

> 'We can ascribe our degree of advancement today in large measure to the training and assistance so willingly provided by the United States of America during the early years of our nuclear programme when several of the Western world's nuclear nations co-operated in initiating our scientists and engineers into nuclear science.'[10]

Working alongside the foreign-trained South Africans are

some of those who helped to train them. Many of the staff at the Government's Council for Scientific and Industrial Research (CSIR), which has top-secret divisions for defence and nuclear research, are immigrants from other countries, a large number of whom have had considerable experience in highly sensitive areas before coming to South Africa.[11] One of these is Dr Adolf Engelter, whose visit to his home country, West Germany, has been described above. At the Electricity Supply Commission (ESCOM), which is responsible for the nuclear power programme, the nuclear staff is headed by John Colley, formerly with the UK Atomic Energy Authority, and most of the staff is made up of British personnel.[12] Many of the scientists at Pelindaba are also of non-South African origin. Among the more prominent scientists now operating in South Africa is Douglas Torr, who formerly worked on classified nuclear research at Harwell in Britain.[13] There are persistent reports from informed sources inside South Africa that the enrichment plant at Valindaba has large numbers of Israelis with experience in their own country's nuclear weapons programme.

In addition to individuals who contribute to South Africa's nuclear programme, some Western governments also extend assistance on a formal level, including not only training for South Africans but also the supply of equipment and special nuclear materials, as well as a range of joint research projects. Dr Roux has summed up the situation:

> 'Co-operation agreements have been negotiated with many countries including the USA, England and France, and a considerable number of impotant projects have been tackled in collaboration with prominent overseas organisations.'[14]

The basic installation on which the South African nuclear programme was built was a research reactor, which went into operation in 1965. The South African Fundamental Atomic Research Reactor (SAFARI-1), supplied by the United States under its 'Atoms for Peace' programme, has been basic to

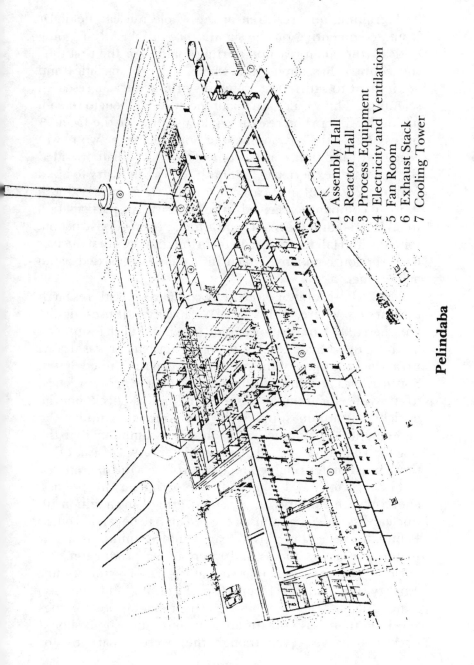

1 Assembly Hall
2 Reactor Hall
3 Process Equipment
4 Electricity and Ventilation
5 Fan Room
6 Exhaust Stack
7 Cooling Tower

Pelindaba

both training and research in the whole nuclear field. Dr Roux, commenting on the significance of SAFARI-1 'going critical' (the chain reaction starting), said: 'For the first time the country has almost the whole range of facilities and equipment to satisfy the ambitions of our nuclear research scientists.' This would be a major incentive to them to remain in South Africa. The reactor would also provide the facilities necessary for the development of South Africa's own power reactor. In addition, construction of SAFARI-1 had provided the necessary experience for South African industry to build major sections of nuclear reactors of their own.[15] South African companies received large contracts for participation in SAFARI-1 and the main contractors, Roberts Construction, joined with Hall Longmore to form the country's first nuclear construction company, which in itself was regarded as an achievement.[16]

SAFARI-1 is quite powerful by comparison with research reactors elsewhere, with a 20 MW capacity.[17] It uses highly-enriched (90%), or weapons-grade uranium, which is supplied by the United States.[18] A safeguards agreement was drawn up at the beginning of the project between the United States, South Africa and the International Atomic Energy Agency, to provide for regular inspections and a check on the enriched uranium supplied by the Americans under the agreement.[19] This was based on the assumption that the reactor would be unavailable for military use, *provided* that the United States retained total control over the operations.

The reactor is designed for experiments in physics, chemistry, metallurgy and engineering, in addition to nuclear fission.[20] The aims of the research to be carried out at Pelindaba were stated as research into uranium as a raw material, reactor studies and radioisotope production.[21] A curious feature here is that special legislation – the Atomic Energy and Nuclear Installation (Licensing and Security) Amendment Act – was passed, which among other things specified that all isotope production and transactions should be strictly secret, even though they were apparently for medical and industrial use only. The buyers were said to want secrecy regarding not only their identity, but also the quantity of material bought.[22]

In addition to SAFARI-1, South Africa's basic equipment includes a critical assembly, Pelinduna-Zero or SAFARI-2, which is also at Pelindaba. This, according to American Government sources, '. . . is in effect a reactor which one can experiment with'.[23] Heavy water and 2% enriched uranium are leased from the Americans for use in this, which was apparently built by the South Africans as part of their efforts to design their own nuclear power reactor. There are no safeguards on Pelinduna-Zero. Also at Pelindaba are a cobalt-60 source and an IBM 360/40 digital computer.[24]

Another centre conducting significant nuclear research is the Southern Universities Institute at Fauré, between Cape Town and Stellenbosch. The Institute started operating in mid-1964, when a 3 Mev Van de Graaff accelerator was commissioned there; it plays an important part in the research work. The Institute has a small permanent staff and is principally concerned with research and training. The CSIR in Pretoria also operates a cyclotron yielding 3 Mev protons, 16 Mev deuterons and 32 Mev alpha particles.[25]

A variety of other scientific institutions, including the Universities of Pretoria, Witwatersrand and Stellenbosch, conduct research in nuclear physics using a 60 Mev cyclotron, a sub-critical assembly and various accelerators.[26] The research is concentrated in universities catering to the Afrikaans-speaking rather than English-speaking community.

The nuclear power programme, and a surprise deal with the French

Ever since the beginning of the nuclear research programme, South African representatives have claimed that the objective is to build nuclear power plants for the production of electricity. Research was carried out to develop a reactor design which the United States had been working on but had abandoned. This used natural uranium, with heavy water as a moderator and sodium as a coolant.[27] The plan was abandoned after a certain amount of work, probably because the development of enrichment techniques had become much more important for South Africa and it was not feasible to have two major research projects going together, competing for funds and scarce personnel.

Like many other countries, South Africa in the early 1960s saw nuclear power as a panacea for its energy problems – which are a serious weakness in this country without its own oil and subject to constant threats of an economic boycott. Professor Zeemann, of the University of Stellenbosch, rashly predicted in 1964 that within ten years nuclear power would have replaced traditional power sources.[28] In fact, no electricity has been generated by nuclear power to date, and more rational views have started to prevail over the planning of future possibilities. It is recognised that although nuclear power for electricity is generally desirable, it would be hopelessly uneconomic anywhere in South Africa with the possible exception of a coastal site. Even there its advantage is not primarily economic; rather it is the prospect of conserving scarce fresh-water supplies. Since South Africa's rainfall is only about 20% of that of the United States,[29] this creates major problems for all South African industry, especially power generation. The 'nuclear park' planned for the Cape will use sea-water for cooling.

There have been several plans to build nuclear power facilities in the Cape, but in 1973 these were shelved in favour of expanding coal-fired capacity and building transmission lines from the stations on the Rand coalfields, on the grounds that escalating capital costs of nuclear reactors, and their excessive capacity for South Africa's immediate needs, made them an irrational choice.[30] At the end of 1973, however, the Arab states in the Organisation of Petroleum Exporting Countries (OPEC) announced a total embargo on oil sales to South Africa, which although only partially effective (because of non-compliance by Iran) produced apprehension among South African planners. It was then decided to go ahead with the shelved nuclear plans, largely to conserve South Africa's coal reserves for domestic manufacturing and for export. Two reactors of 1,000 MW each were therefore proposed, at least one of them at Koeberg, a site at Duinefontein, 40 km. north of Cape Town on the coast. The first was scheduled for operation in 1982, and the second a year or two later. The Koeberg site offered the possibility of further expansion to between 10,000 and 72,000 MW of generating capacity, in a 'nuclear park' concept which had been discussed earlier in terms of a site in the Durban area.[31]

At the end of May 1976 a surprise announcement revealed that South Africa was finally awarding the contract for its first nuclear power reactor, Koeberg A, worth about $1 billion, to a French consortium of Framatome, Alsthom and Spie Batignolles.[32] The contract was signed in August.[33] By September work had begun on preparing the site.[34] There are to be two pressurised water reactors (PWRs) at the Koeberg station; they are twin turbine generators capable of a peak output of 922,000KW of electricity each. Koeberg A is due to start up in November 1982 and Koeberg B on the same site a year later.[35]

The French contractors will be responsible for most of the work, South African companies participating in operations worth about 30% of the total costs.[36] The South African representative of the Framatome consortium announced in September that they were ready to start training South Africans in France to operate the plant, and that this would last for about a year.[37]

The reactors are to be fuelled by lightly-enriched (about 3%) uranium, the enrichment service apparently to be provided initially by the United States, with whom South Africa has a firm contract for enrichment services for both the reactors planned, on which preliminary payments totalling $4 million have already been made.[38] It is not at all certain what South Africa's intentions are with regard to the

165

question of uranium enrichment and fuel fabrication, since it has announced that that would be done by a Franco-Belgian consortium of Péchiney-Ugine-Kuhlmann, Creusot-Loire and Westinghouse.[39] It would be comparatively simple for South Africa to resell its existing contract for American enrichment. It is possible that the Franco-Belgian consortium has significantly less stringent safeguards than the Americans impose against diversion of material for military purposes.

It is worth considering the whole basis on which the French consortium won the contract, against all expectations and despite having tendered an apparently uncompetitive bid. General Electric, the head of the leading consortium, had in fact already received a letter of intent from ESCOM, and clearly understood that it was the leading contender for the contract and that Framatome had dropped out of the competition some time earlier.[40] The French claims that they won the reactor order because of their competitive prices were '. . . treated with derision by US observers,' according to the authoritative *Nucleonics Week*.[41]

The French companies admitted that they had 'snatched' the contract from the Americans at the very last minute.[42] The mystery is compounded when it is realised that one of the most important specifications for the reactor was dropped in order to accommodate the French. ESCOM had specified that Koeberg must be identical to an *overseas* light water reactor to be completed at least three years in advance, thus providing both a guarantee as to Koeberg's performance, and adequate operating experience (since reactors of this size are still to some extent experimental).[43] However, Framatome has none of the experience, nor the earlier orders necessary to meet the specifications; the South African order is its first 'turnkey' export contract for a pressurised water reactor[44] – in other words, one where it is responsible for the entire construction programme right up to the starting of the reactor. In a deal of this scale, which is the biggest contract negotiated by ESCOM and one of the biggest ever in South Africa,[45] there must have been some critical factor which outweighed all questions of the contractors' experience and competitiveness.

One possible reason for the switch was that ESCOM and the South African Government had requested all bidders to advise them whether the supply of reactor components and fuel could be guaranteed. The American response, while positive as regards '. . . a long-standing US policy of serving as a reliable supplier by fulfilling US supply undertakings', also added the proviso that nuclear export licences would have to be issued by the Nuclear Regulatory Commission (NRC), an independent regulatory agency.[46] An NRC decision, although it would almost certainly have been positive, was still pending at the time the letter of intent was withdrawn from General Electric.

Another reason which was given prominence at the time of the surprise switch to France was that South Africa was 'furious' about the opposition that had built up in the United States and the Netherlands because of South Africa's racial policies, and fears that it might divert plutonium from the reactors to make nuclear bombs.[47] This explanation, while flattering to the groups attempting to stop the deals, is inconsistent with South Africa's tactics in international affairs generally, which are to deal with key people in governments and private enterprise, usually behind closed doors, and ignore protests from opponents. It would be most uncharacteristic for the South African Government to change its policy merely as a result of public criticism, either internal or external.

Another factor in France's favour was that the credit terms offered by the banking consortium, headed by Crédit Lyonnais, incorporate interest rates which are lower than usual as well as a favourable repayment period.[48] It seems most unlikely, however, that this would outweigh the extra costs involved in buying the reactor from the French instead of the Americans. American credit was already available with the decision taken in advance by the Export-Import Bank.

There were also many rumours and allegations that the

French won the contract through promises to go easy on safe-guards against military uses; this was officially denied by the French Government. However, the French have a bad reputation for unconcern about nuclear safeguards. No safe-guards agreement had been worked out at the time the deal was announced,[49] although after much controversy one has subsequently been drawn up.

Another official denial concerned persistent rumours that the French had clinched the deal over the reactor as part of their negotiations with South Africa about selling them a licence to build the very latest *Mirage* jets, aircraft developed as part of the *force de frappe* to carry France's own nuclear weapons.[50] Informed sources close to the French Government have revealed that Framatome was awarded the contract at the precise moment that Thomson-CSF agreed to provide the latest and most sophisticated ship-based missiles to South Africa. The sources add that Vorster and his more hawkish allies in the Cabinet put great pressure on ESCOM to accept the French bid in spite of its being uncompetitive. Even though ESCOM was advised on the evaluation of tenders by an Anglo-American team (Associated Nuclear Services of London, and Gilbert Associates Inc. of Reading, Pennsylvania), it was obliged ultimately to abandon technical and economic criteria and follow Vorster's orders.[51]

The close link between the valuable reactor contract and major arms deals with the French illuminates another puzzle, namely why France was prematurely named as the contractor as far back as 1974. There was some confusion: in March of that year the Chairman of ESCOM announced that it would buy an American reactor of the boiling water (BWR) or pressurised water (PWR) type, using enriched uranium as fuel; invitations to tender for the station were invited from nine corporations, for a contract to be signed in 1975.[52] However, in June it was reported that South Africa had requested a reactor from France as part of a deal involving the supply of unsafeguarded South African uranium to France, which was testing nuclear weapons at about that time, and also the sale of further French arms to South Africa.[53] After some hesitation, it was announced in July 1974 that Framatome would supply the reactor.[54] Advertisements were placed in the French press for 'technical staff interested in operating a nuclear plant.'[55] Since then, however, the deal regarding the reactor sale apparently failed to materialise. A high-level French Government mission which visited South Africa in April 1975 was reportedly charged with ensuring that the contract for the reactor was finalised.[56] However, it was unsuccessful: a new tendering process was put in motion, and ESCOM almost got to the point of ordering the reactor from the lowest bidder. When the question of the Thomson-CSF missiles came up for a decision in 1976, the French presumably made it a condition that, this time, the deal must go through for the reactor to be ordered from Framatome – and Vorster found a way to make ESCOM, which is after all a State corporation, toe the line.

The French have a habit of following up massive arms deals with South Africa by announcing that there will never be any more. In May 1974, at about the time of the uranium-arms-reactor deal already mentioned, the new President, Giscard d'Estaing, announced 'the liberal mission of [France's] diplomacy' which would mean a halt to all arms sales to repressive regimes as a means of '. . . supporting throughout the world the cause of freedom and right of peoples – and I mean peoples – to determine their own future.'[57] A specific pledge to halt all arms sales to South Africa accompanied the subsequent missile-reactor deal in 1976. It may therefore be ominous rather than reassuring that the French Government announced in November 1976 that it would not sell any more nuclear power stations to South Africa after fulfilling existing contracts. The South African press found that informed business circles in Paris saw the announcement as very much 'political window dressing' in line with usual French practices, '. . . just as the official statements on arms embargoes which are very proper but which do not appear to be all that realistic.'[58]

South Africa's own motivation in pursuit of massive nuclear power stations, the largest and most expensive currently available, is not as simple in economic terms as it might appear at first sight. The original 1969 estimate of the cost of a reactor was about R60 million. By 1972 it had reached R100 million,[59] and since then it had jumped to a fantastic R900 million for the latest and biggest model.[60] With the choice of the French consortium, together with continually escalating costs, it is now expected to reach R1,000 million.[61] Although the external financing is covered by the French banks this does not solve the problem of the heavy local costs, and the necessity for continual refinancing so that the debt can be paid off over a much longer period than the ten years which the French credits will cover – and all this at a time of growing economic difficulties in South Africa as well as increasing difficulties in raising further loans on overseas capital markets.

Because South Africa is in a period of prolonged credit restraint,[62] the foreign bankers who have provided massive loans for ESCOM and other State corporations in the past are increasingly reluctant to increase their exposure there still further because of the political as well as economic uncertainties. Mr Len te Goen, ESCOM's finance manager, found in a visit to Europe in mid-1976 that further private placements and public capital issues were out of the question for the time being in West Germany, Switzerland, Belgium and France – the best sources of capital so far – as well as American banks in London.[63] For this reason the decision by the 'hawks' in Washington to engineer a rise in the gold price, which is discussed in Chapter 8, would be critical in determining whether the South African Government could go ahead with Koeberg, quite apart from the rest of the billion-dollar nuclear programme which it has lined up. The fact that the hidden American subsidy has been justified in terms of South Africa's strategic importance to the Western Alliance, together with the probability that the subsidy will be used for a nuclear programme that cannot be justified in economic terms, is itself a strong indication of the military motivation behind the nuclear installations.

IS IT ECONOMIC? IS IT NEEDED?

ESCOM's General Manager, Jan Smith, has admitted that the repayment period for the French credits is 'horrifyingly short term for ESCOM', with an increased burden coming on completion of the reactor as interest payments fall due.[64] In addition to these worries, ESCOM's immediate capital commitments for the reactor are extremely high, and cannot be met from the currently low electricity rates prevailing in South Africa – which are a substantial contribution to the competitiveness of South African industry, and an incentive to foreign investment. The days of cheap electricity are therefore numbered. As Smith says, 'We're still ridiculously cheap by world standards, probably half the unit price of most Western countries.' Yet only a year previously it had been a third, and with more heavy increases – largely, if not exclusively, to meet the costs of Koeberg[65] – South Africa's advantage in this field will probably dwindle substantially. This is a serious matter in a country which already relies more heavily that most on electricity as a power source.[66]

There is room for serious doubt as to whether nuclear power can ever be competitive with electricity from conventional sources, if all the costs – including eventual decommissioning of the reactor, disposal of radioactive waste, and other major items – are included.[67] This becomes even clearer in South African conditions, where there is a very high level of coal production at relatively cheap prices, and some hydro-electric capacity. South Africa will also be purchasing large quantities of power from the Cabora Bassa hydro-electric project, which is coming into full operation. There is to be a new coal-fired power station in the Eastern Transvaal, with a capacity of 3,600 MW (almost four times the size of Koeberg A), and the proposed Tugela hydro-electric

171

scheme could provide more than twice the power of Koeberg.[68] Yet Koeberg A alone is almost large enough to feed the major industrial centre at Johannesburg.[69]

It is true that ESCOM has been planning for the proposed enrichment plant, which would probably require some 2,400 MW of electricity. In addition, it has been assuming an increase in electricity demand from industry and domestic consumers of around 9% a year.[70] This is the level of increase which many other countries, including Britain and the United States, had also expected before rapidly-rising electricity prices starting in about 1973 caused demand to remain stationary or even to fall. The construction programme already under way, based on earlier estimates, led to enormous over-capacity (about 40% in Britain). Thus it could well be that ESCOM, in deciding to go ahead with Koeberg – no doubt at the urging of the Government – has bought itself a white elephant and moreover has helped, at great expense, to accelerate the rate of increase in electricity prices which could make the reactor increasingly redundant. As if that were not a sufficient problem, nuclear reactors cannot be regulated to meet fluctuations in demand, and additional heavy costs would be incurred if the elaborate pumping system for storing excess electricity were built to overcome that.[71]

There are eminent people in South Africa who foresee quite clearly that Koeberg could be an expensive and disastrous mistake from the economic point of view. Extrapolating from experience with the oil-from-coal plant (SASOL 2) and other large projects in recent years, it is obvious that not only will completion of the reactor take considerably longer than promised, but that costs will escalate with the delay. Mr S. Kuebler, the South African representative of the French consortium, has said that the project would cost at least R1,500 million if for any reason it is delayed;[72] and he is unlikely to be overestimating.

The motivation behind the order for Koeberg, then, can hardly be economic. The real reason for going ahead with it, after the decision earlier in the 1970s to abandon the idea, must be found in the accumulated evidence. It could be partly a question of providing an incentive to France for its arms sales. It could be justified in terms of South Africa's ambition to be self-sufficient in its own energy sources (particularly for military purposes) now that it plans to enrich its own uranium.[73] According to sources close to the radical West German monthly *Dritte Welt* (Third World) there is even a possibility that the power stations will not be supplied at all. The publicity around the order and delivery, it says, was conceived as a cover-up for large supplies of technical equipment destined for the uranium enrichment plant at Valindaba. Whether or not this is the case, there is no doubt that its operation would be of secondary importance to the uranium enrichment process.

Dr Roux, President of the AEB, has justified the construction of Koeberg by saying that it is needed to keep up the standards of scientific research, specifically on uranium enrichment, which is

> '. . . a long and difficult road, and is a development which, in the near future, can be carried out meaningfully only if our engineers and scientists can gain experience with the behaviour of nuclear fuel in power reactors under operating conditions, and are given the opportunity to test their own nuclear fuel under practical conditions in power reactors. Without its own nuclear programme, South Africa cannot give to its researchers the necessary stimulus to pursue this direction of development and to bring it to practical results which can be fruitfully applied.'[74]

There remains perhaps the most powerful motive of all: the use of the Koeberg reactor to produce spent fuel from which plutonium, the basic ingredient of hydrogen- or H-bombs, is made. Here it is not a question of capability – all reactors produce spent fuel containing plutonium, and the construction of a plutonium reprocessing plant is relatively easy, certainly for South Africa with its industrial and technological capacity.

173

It is worth noting that South Africa had already clearly indicated that it wanted few or ineffective safeguards on Koeberg. According to the Johannesburg *Star*, the final signing of the contract, two months after the deal had been announced, was postponed by the French: French sources said that there was a difference of opinion concerning who should have control over the reactor fuel.[75] The French side were perhaps more conscious than they usually are of their responsibilities in this area because of allegations both internally and by African governments that Koeberg would be used to produce nuclear weapons.

The safeguards agreement which was eventually signed is more stringent than it would have been had there not been such a controversy. It would be more convincing if South Africa were ever to accede to the non-proliferation treaty. However, the loopholes of the treaty, as described in Appendix 1 below, are such that South Africa cannot be prevented from carrying out the secret reprocessing of plutonium for military purposes.

The secret enrichment process

The key installation providing South Africa with the capability for building atomic weapons is the pilot enrichment plant, supplied under a West German 'licence'. The process of enrichment increases the proportion of the fissile isotope U-235 which occurs naturally in U-238 to about 3% for reactor fuel or about 90% for nuclear weapons. The South African enrichment plant is totally sealed off to outsiders. It is significant as the only plant operating outside the existing nuclear powers which is not subject to regular inspections under the international safeguards system. It is this which is the critical issue in the mounting debate about South Africa's rapidly developing nuclear programme.

A HISTORY OF ENRICHMENT EFFORTS

South Africa, it would appear, has had not just one 'secret' process 'unique in concept', but two – with a quick switch in the middle, perhaps in the hope that nobody would notice the change. Prime Minister Vorster made the original announcement about the top-secret venture on 20 July 1970, adding that the technology had been developed to a stage where a large-scale plant could be competitive with existing plants in France, Britain and the United States. The nature of the process was to be kept secret because of the enormous cost of developing it, although 'South Africa does not intend to withhold the considerable advantages inherent in this development from the world community', and would be prepared to collaborate 'with any non-Communist country' subject to an agreement safeguarding South Africa's own interests.[76] The claim was an ambitious one, since at that time in the Western world only the French, British and Americans had enrichment plants; South Africa, in proposing to build a pilot plant, would be the fourth.

The statement was at that stage treated with scepticism, given the amount of money and skilled labour required to develop a truly viable new enrichment process. Although perhaps a dozen theoretically possible processes are known, the crucial question was whether they could be made economically competitive in order to justify the very large investment.[77] It is interesting that the only foreigner who unequivocally endorsed the South Africans' claim was Dr Heinz Schimmelbusch, a key figure in West Germany's atomic industry. He said that he was absolutely convinced that the South Africans were not bluffing about their new secret process.[78]

The nature of the process which the South Africans were announcing is in fact fairly easy to deduce. South African nuclear scientists had always shown great interest in the ion exchange process. This was developed in the United States but discontinued several years ago because of technical

175

difficulties, and because it is very slow, so that at the current stage of development it would take years for a plant to produce enriched uranium.[79] Following Mr Vorster's announcement of the new process, some reports appeared which confirmed that the process was indeed based on ion exchange. The *Wall Street Journal*, responding to what were probably official Government leaks, reported that American scientists believed they had discovered the secret: the South Africans had made ion exchange *work*.

The scientific principle involved is uncomplicated and in common use. It involves a chemical process similar to that which was also developed, up to a certain point, by the Japanese: washing uranium in a solution of sulphuric acid and sodium fluoride through a synthetic resin. The solution moves down through the resin in sinking layers and the uranium in the lower layer is slightly richer in U-235.[80] This process is somewhat similar to that already employed in South Africa for extracting uranium from low-grade ore. It has been repeatedly stated that the uranium-mining companies played a crucial part in the development of the enrichment process.[81] Dr Roux, who in addition to being head of the Atomic Energy Board (AEB) is also Chairman of the Uranium Enrichment Corporation (UCOR), announced that mining-houses would be invited to participate in the first enrichment plant.[82]

The development of the ion exchange technique into a commercial process would, according to an American scientist who researched the area, '. . . require a great deal of research into the chemistry involved in the field.' There is abundant evidence that the South Africans did put a lot of work into this, and specifically into the development of low-grade ore extraction processes. Some scientists saw it as significant for South Africa that the ion exchange process would involve a large inventory of partially separated uranium, since it is '. . . a very slow process with a great many stages.'[83] Dr Roux announced in 1971:

'We have already produced enriched uranium by the
South African process – naturally not to a high degree,
owing to the limitation on the number of stages which can
be used until the pilot plant becomes operative.'[84]

The necessary inventory would have been feasible for South
Africa, with its large uranium stockpiles and a policy of
building them up still further for enrichment purposes.
Nuclear Industry commented that this was what Mr Vorster
had had in mind when he included in his original statement
the comment that the process was:

'. . . presently developed to the stage where it is estimated
that under South African conditions a large-scale plant
can be competitive with existing plants in the West.'[85]

Not quite, however. The South Africans apparently over-
reached themselves in attempting to develop ion exchange
after the Americans and others had tried and failed. It may
have been the time involved which proved a greater problem
than anticipated, or the large number of stages required – a
major problem if there was to be any question of 90% enrichment for weapons purposes. By the end of 1972, the *Financial
Times* was reporting that, following South Africa's decision
not to promote their new process at the Nuclex '72
conference and exhibition in Basel, experts were becoming
very sceptical about South Africa's ability to follow through
from research to commercial viability.[86] And with growing
confidence in the centrifuge method of enrichment, developed
by the 'Troika' of West Germany, Britain and the Netherlands
and presented in detail at Nuclex, South Africa's claims to
have a unique alternative to the basic gaseous diffusion
method no longer caused the same excitement, and the
foreign partners needed to implement South Africa's ambitions
had not appeared. Sir John Hill, Chairman of the British
Atomic Energy Authority and a frequent visitor to South
Africa, turned down the idea of partnership with South
Africa '. . . while we remained in league with our Tripartite

partners.'[87]

Dr Roux admitted in mid-1972 that UCOR's budget had been cut and the construction of the proposed enrichment plant delayed.[88] It seems that the South Africans then attempted to climb on other countries' enrichment bandwagons. Tentative approaches were reported to have been made to the West Germans, British and Dutch for participation in their 'Troika' for development of the centrifuge process.[89] The subsequent approach to the West Germans has been described in Part I.

A 'UNIQUE' PROCESS?

It is at this time that we begin to hear more about South Africa's supposedly 'unique' new process, 'a phenomenal achievement of world dimensions',[90] bearing an uncanny resemblance to another West German process, the 'jet-nozzle' system. There is some irony in the fact that the South Africans continue to make claims about 'their' process, as Roux did in 1970:

> 'It is an entirely new principle. And we have thought it out and worked it out ourselves – every calculation and every little step in the process. . . It is all the work of South Africans.'[91]

In April 1975, the Prime Minister announced the successful completion of a pilot plant with the statement:

> 'That this achievement could be reached without any assistance from foreign countries inspires enormous confidence for the future scientific and technological development of our country. The trust placed by South Africa in its nuclear scientists has been completely vindicated. . .'[92]

Considering that the pilot plant had been built with considerable technical assistance as well as essential components from West Germany, the credibility of this claim rested on the cover-up of South Africa's reliance on West Germany. The Act setting up the Uranium Enrichment Corporation in 1970 gave the Government the right to withhold from the public and Parliament any information about the corporation and its activities which could be considered 'contrary to public interest.'[93]

Interestingly, Professor Becker's jet-nozzle process is itself something of a second-hand system. Becker named the process 'jet-nozzle' as a pseudonym for what is really a centrifugal separation process, in order to avoid problems with the United States Government over the highly classified centrifuge technique. The correct name for the process, which is what Dr Roux used in announcing the 'South African' technique at a meeting in Paris, is 'stationary walled centrifuge'.[94] As it happens, South Africa's chief nuclear scientist and General Manager of UCOR, Dr W. L. Grant, who is an aerodynamicist and credited with the development of South Africa's process, also has research experience in this area and has himself taken out patents on jet-nozzle principles.[95] Dr Grant was one of the many South Africans trained by the United States in the nuclear field;[96] it would seem that his expertise in this area is a direct result of his American training, supplemented by subsequent contacts with the West Germans.

The South African process, as used in the pilot plant, has been developed beyond the work done for the installation at Karlsruhe, the West German nuclear research centre run by the Federal Government. It appears that the theoretical work for the modifications was carried out at Karlsruhe, and put into operation for the practical testing at Valindaba, the enrichment site adjoining Pelindaba. A trade publication commented in 1974:

> 'Recently published pictures of mechanical components being manufactured for the plant seem to confirm that the system was based on the German nozzle process coupled with some undisclosed innovation.'[97]

179

From Dr Roux's guarded exposé in Paris in 1975 of the South African operations – what *Der Spiegel* described as '. . . two sedate gentlemen from South Africa' performing 'a kind of nuclear energy striptease'[98] – certain clues can be picked up as to the kind of modifications made. Roux claimed that the novelty of the UCOR process lay partly in a new cascade technique called the 'helikon', in which a single axial-flow compressor can simultaneously transmit several gas streams of different enrichment without too much mixing. In addition, the efficiency of the separative element also enables hydrogen – which is used to dilute the heavy uranium hexa-fluoride gas being processed so that much higher speeds of centrifugation can be attained – to be separated cleanly from the 'hex' at the end of the process. The main impact of these refinements is that a process which is extremely demanding in terms of energy, because of the heavy gas pumping requirements and the losses through turbulence in a complex cascade, has been made considerably more efficient – although still demanding very large amounts of power.[99]

The power requirements of the process make South Africa a particularly suitable place for developing it, according to Roux, because of the relatively low cost of electricity there. In addition, it has the advantage of operating at low temperatures and pressures, and is '. . . perhaps the least exacting of enrich-ment processes so far developed.'[100] This has obviously been advantageous in the drive to create a nuclear capability in South Africa's industrial sector, where as Dr Roux emphasised, a total of 234 companies were involved somehow in the construction, and there were more than 100 subcontracts – 'There is hardly anyone in South Africa not involved in one way or another.'[101] Roux also stated that Valindaba's role '. . . in helping South African industry into sophisticated manufacture cannot be over-emphasised.'[102] As he has explained elsewhere:

'During the Second World War and the years which followed, South African industry went through a

180

transition stage from what was, in the main, a repair industry to a fully-fledged manufacturing industry. For the past decade, and even longer, it has been obvious that a new stimulus was necessary for the transition of our industry from conventional to sophisticàted manufacture. The type of manufacture demanded of South African industry for the construction of the pilot plant has given this impetus to a large extent, because of the extremely severe demands and strict specifications laid down for the manufacture of components.'[103]

Among the specific benefits involved was the fact that UCOR's orders were large enough to enable many South African firms to purchase 'numerically controlled machines'; in some cases UCOR actually paid for these machines. This greatly contributed to the capacity for producing more accurate and less expensive components for other customers within South Africa.[104] Closely related to this, and very important in its own right, was the size of the payments involved – running into millions of Rands for several contractors – at a time of dwindling orders and overall recession in the South African economy.

All of this, of course, has resulted from the contributions made by the West German Government and industry, working in close collaboration with South Africa to build up local capacity as described by Dr Roux. Details of this collaboration are discussed in Part I. We could just note in passing, perhaps, another of those statements to which the South Africans are so partial, and which can most charitably be described as tongue-in-cheek:

'Referring once again to the pilot plant, it is worthy of mention that 90 per cent of it will be of South African manufacture, and that the foreign content has been purchased through normal channels and was in no way crucial for the completion of the project. If use had not been made of certain imported components, it would have meant only a slight delay. . .'[105]

On 7 April 1975, the announcement that the pilot plant

had been put into operation was made by no less a person than the Prime Minister, at the opening of a session of the South African House of Assembly. He told the MPs:

'The new South African process has therefore not only been proved in practice but it also enables us to proceed with confidence with the erection of a large scale plant with a view to the marketing of a large portion of our uranium supplies in enriched form for commercial purposes.'[106]

The statement was apparently considered so important that the Minister of Mines, Dr Pieter Koornhof, was by-passed in its preparation.[107] Perhaps because the process was then in the critical testing stage, the Atomic Energy Board and Ministry of Mines refused to answer any detailed questions on the plant.[108] In early 1976, however, a Government official was reported as describing operations there as 'very successful'.[109]

Also described as 'very successful as a scientific study' were the joint operations with Steag, the West German licensee of the Becker process. Although our documents show the relationship to have been a contractual one involving West German technology, close collaboration, and supply of the basic components from West Germany, it was officially described as a 'feasibility study.' In April 1975 — at the same time as Vorster announced successful completion of the pilot plant — representatives of both Governments said that South Africa would be continuing on its own.[110] Steag was to be left out of full participation in constructing the commercial plant, although it still has an important role to play in supplying components.

Foreign participation of some kind remains vital to the project. The sheer size of a commercial enrichment plant, based on a new process as yet unproven on an industrial scale, make the possible losses considerable. Since the South African Government apparently intends the South African stake in the proposed international enrichment company to be held by a local company in which the mining houses and industrial interests would have a share alongside the Government, spreading the risk might also be necessary in order to persuade them to participate.[111] In addition, each of the partners in the international company would be expected to provide finance, markets and raw material in proportion to its stake.[112] South Africa would find it extremely difficult to provide enough of any of these elements by itself.

In his announcement to the European Nuclear Conference in Paris, Dr Roux stated:

> 'South Africa has always been willing to share its knowledge of uranium enrichment with friendly countries and in particular to undertake the establishment of a large-scale enrichment plant in South Africa as an international venture.'[113]

He added that negotiations with interested foreign parties were now under way, and in some cases were at an advanced stage; it was expected that a conclusion would be reached during the next few months.[114] It is not known exactly which countries were being referred to, but it seems likely that in addition to West Germany, France and Iran were interested in participation. At the end of 1975, it was reported that Iran was considering financial participation in a South African plant; it would also be a useful customer for the enrichment services.[115] However, South Africa's ambitions to launch a major international consortium do not seem to have been particularly successful, since at the Salzburg conference of the International Atomic Energy Agency, two years after Paris, the South Africans seemed to be starting all over again: there was another announcement that UCOR was now ready to discuss contracts for enrichment services with other countries.[116]

Meanwhile, South Africa is proceeding with its own plans. At the end of 1976, UCOR was reported to have completed construction of the pilot plant, only one section of which was in operation at the time of Vorster's announcement in 1975.

Further sections have now been brought into operation. In December 1977 Dr Roux announced that the pilot enrichment plant had been operating during the previous year; however, he declined to say what was being done with the enriched uranium produced by it.[117] UCOR is also far advanced with the engineering of miniature prototype modules for a large-scale plant. The plant would eventually have some 90 to 120 such modules, if built for a capacity of 5,000 tons a year.[118]

At the same time, considerable investment has been put into the manufacture of uranium hexafluoride ('hex') from the uranium oxide produced by the treatment plants at the mines — and, as noted in Chapter 4, already processed into uranium tetrafluoride by NUFCOR. There is already a pilot 'hex' plant at Pelindaba providing the feedstock for the pilot enrichment plant nearby at Valindaba, and in 1975 the Minister of Mines announced completion of a 'commercial module' with a capacity of 200 tons a year, which could form the basic unit of a large-scale 'hex' plant.[119] The Minister, Dr Koornhof, told the House of Assembly on 11 June 1975:

> 'Sir, it is with pride, therefore, that I am able to announce that a plant for the production of uranium hexafluoride has been successfully commissioned at Pelindaba so that for the first time in South Africa this compound has been produced on an industrial scale, which is very, very important. Uranium hexafluoride is a vitally important material in the production of enriched uranium for nuclear power processes. It is the only compound of uranium which is gaseous at temperatures and pressures around normal and is the feed material for all enrichment processes used today, including the enrichment process developed by UCOR.[120]

He went on to give the usual fulsome praise to '...many years of intensive effort by a team led by Dr R.P. Colborn under the direction of Dr R.E. Robinson and Mr H. James as part of the activities of the extraction and metallurgy division of the Atomic Energy Board...'[121] However, as usual, significant external assistance appears to have been given. Dr

Roux disclosed in 1970 that research for the 'hex' process had been carried out by South African scientists and industrialists in collaboration with 'overseas interests'.[122] There have been allegations of close collaboration by the British Atomic Energy Authority, which previously converted South Africa's uranium oxide to 'hex'.[123] The British contribution to South Africa's 'hex' production has been greatly increased also by the opening of the Rössing mine. At the time Rössing was started, the *Rand Daily Mail's* Mining Editor noted:

> 'The significance of the Rössing project is that its opening will certainly lead to the establishment of a uranium hexafluoride industry in South Africa. Uranium hexafluoride is processed to make enriched fuel for nuclear reactors, but none is made in South Africa because production is uneconomic below several thousand tons a year.'[124]

The stage is now set for construction of a commercial enrichment plant. The decision to proceed with it, taken in 1975, was described in the AEB's annual report as the single most important event in South Africa's nuclear energy programme for that year.[125] The plant was to have a minimum capacity of 5,000 tons a year of enriched uranium, with a potential for expansion of up to 10,000 tons. It was expected to come into operation by 1984, and to reach full production by 1986. Development costs for the project were estimated at a staggering R910 million (at October 1974 prices), of which R130 million was for further research and development, and operation of a commercial prototype unit.[126] The plant would consist of two main sections: the first to convert uranium oxide into 'hex', and the second to enrich it.[127]

South Africa was obviously in a hurry to launch its enrichment plant, which it was promoting as '. . .a serious competitor in the technological struggle for the next large-scale plant in the free world to conquer the unsatisfied market in the early 1980s', according to the AEB's publication

185

Nuclear Active.[128] The President of the Institute of Mining and Metallurgy, Mr P. Rensburg, has stressed the paramount importance to the country as a whole of bringing its own enrichment plant into production before other countries are too far committed on their own rival programmes.[129]

Thus, the decision to proceed was made under considerable pressure, and the enormous risks involved were accepted in the alleged attempt to be ahead of the competition. UCOR's process has never been tested as an *industrial* process, as opposed to an experimental one. There are always problems in scaling up a sophisticated process; even the expansion of nuclear reactors from 5-600 MW to almost 1,000 MW has introduced a whole set of problems, for example. The prototype module currently being constructed, the 'Mini-Z', which is one-tenth of the size of the enrichment modules for a commercial plant, already differs significantly from those in the pilot plant in terms of design, construction and operating conditions.[130] The implications of using an untried technology, not only for possible technical problems with the commercial plant, but also for the cost estimates, are serious. Even though the pilot plant was eventually a technical success, problems encountered during construction which had not been foreseen raised the final cost of it considerably, from the original estimate of R13 million to R38 million by 1973 and well over the R50 million mark by 1975.[131]

The original size, 5,000 tons a year, is extremely large — about the same size as total American capacity prior to a recent upgrading, and more than the total Soviet enrichment capacity, which American Government sources estimate at about 3 million tons. Koornhof said in 1975 that the final size would be determined '. . .when the additional development work will have progressed sufficiently.' It would also depend on South Africa's success in obtaining contracts for the finished product.[132]

Similarly, there was no guarantee that enough uranium oxide could be obtained by the South Africans to justify the plant's enormous capacity. In spite of being one of the biggest uranium producers in the world, South Africa could not by itself provide enough raw material for this plant, even when the recent increases in production are taken into account. The unexpected technical problems at Rössing and the political as well as economic risks of opening up the other low-grade deposits of uranium in Namibia make forecasts of future growth somewhat difficult. The 5,000-ton minimum output of the plant would require 20,000 tons of uranium oxide a year, compared to current production of about 3,000 tons a year.[133] Even the perhaps over-optimistic projections of the Chamber of Mines show a total of no more than 15,000 tons by 1985.[134] This left the following possibilities for the South Africans: to insist that all uranium deliveries be in terms of the much more expensive enriched form, even perhaps for pre-existing long-term contracts; and/or to rely on overseas partners to supply their own uranium oxide.[135] It was this latter requirement which forced Steag of West Germany out of the intended full-scale participation in the plant. And the prospect of South Africa unilaterally changing the terms of long-term contracts (for which legislation has already been passed) was hardly conducive to good relations with its customers, who may want either natural uranium for CANDU or Magnox reactors or other purposes, or prefer to enrich the oxide in their own or other countries' facilities. Many of the customers already have enrichment contracts with the United States.

Undeterred by these risks and uncertainties, however, the South Africans were — at least in public — full of enthusiasm for the prospects of their proposed plant. Dr Koornhof boasted that it could ultimately be even more valuable to the country than the discovery of diamonds.[136] He also claimed that South Africa could enrich uranium about 30-40% cheaper than the United States and others.[137] This was subsequently modified to 'up to 25% cheaper' by Dr Roux.[138] It was also claimed by Roux that the capital cost of the plant would be less than 65% of that of a gaseous diffusion facility of the same capacity.[139] Overall, Koornhof predicted that

187

with the value added being over 200%, the enrichment plant would bring in a minimum profit of R250 million a year and possibly much more, thus helping considerably to ease the chronic balance of payments deficit.[140] The pro-Government *Transvaler* exuded confidence that the first plant would be 'completely paid off after 10 years' and even '. . .if we may dream a bit about the future, it seems that uranium may soon make South Africa sought after in the world.'[141] Or as the *Financial Gazette* expressed it:

> 'Members of the Nuclear Club of Five, and many nations hostile to South Africa, are slowly digesting (to them) an unpalatable fact: the Republic is going to have its yellow cake and eat it too during the world competition for the enriched uranium fuel market in the Eighties.'[142]

It almost seems unkind to question the assumptions on which these claims were based. However, it can by no means be assumed that South Africa will have an adequate market for large-scale enrichment work. Estimated nuclear power-station capacity has been very drastically cut every year, and programmes continue to be cancelled, particularly in the United States which has the world's biggest nuclear industry. In 1976 and 1977 not one new reactor order was placed there. Public opposition to reactors on health and environmental grounds is mounting rapidly in countries such as West Germany, France and Britain — all important customers for South Africa. The uncertainty surrounding this whole area is deterring many governments and multinational companies which would have just as good prospects as South Africa, if not better, from launching an enrichment process. An authoritative West German study shows a margin of uncertainty of 55% in the most recent estimates of nuclear power capacity in the mid-1980s, when South Africa's plant was due to come on stream.[143]

The United States has already contracted for the equivalent of about 280 gigawatts[144] of enriched uranium by 1985, estimated to be the year of peak demand; but the effective demand from the world's power stations is likely to be only about 110-120 gigawatts.[145] In addition to this a considerable new enrichment capacity is already being built in America as well as in France and other countries. It is not surprising, therefore, that the American offer to share its own enrichment technology with some other countries (which did not include South Africa) was turned down by each one; even American corporations have proved unenthusiastic about the invitation extended to them to participate in a commercial venture.[146] Nor is South Africa unique in working on a new process; a number of these are being actively investigated, including a new laser enrichment system which offers the possibility of commercial-sized plants costing only $80 million.[147]

With the prospect of a saturated market in the mid-80s and beyond, it is obvious that the price which can be asked for enriched uranium would be unfavourable. Already the United States concludes enrichment contracts at less than an economic rate; and this could well become even more marked, as a strategy to counter the threat of nuclear weapons proliferation which the construction of enrichment plants involves. As a study by the American Congress's Office of Technology Assessment put it: 'The U.S., by guaranteeing enrichment at a low fee or at cost might slow down the spread of advanced enrichment technologies.'[148]

The question of South Africa's accuracy in calculating the costs of its process therefore becomes crucial. As already mentioned, the South African record in this respect is one of massive cost overruns and delays, which can only become more serious since inflation did not begin to bite in the country until comparatively recently. The rising cost of the enormous power requirements — a factor which Dr Roux has admitted to be 'deleterious'[149] — could be critical here. Energy consumption is the main determinant of running costs, and is already much higher for the South African process than for the European centrifuge project; it is comparable with the energy-hungry diffusion process, even after modifications which reduced considerably the energy requirements of the nozzle principle.[150] In terms of capital costs, too, South African claims (which again relate to the diffusion process, which is more for military use than economic, and which is heavily subsidised by those countries involved) may be unrealistic. The South African pilot plant costs three times as much as the centrifuge one.[151]

American Government sources inverviewed by the authors have stated flatly that estimated capital costs for a 5,000-ton plant of just over $1 billion should more accurately be seen as around $4 billion — excluding research and development costs. They saw this as much too big a venture for a relatively small economy such as South Africa's, with an industrial base so much weaker than that of the other countries which have so far taken the plunge into enrichment. They predicted that South Africa would find it impossible with their process, and 'barely possible' with any other process, to compete with American prices for enrichment work; that they could be left with a half-built plant for which they could not get advance orders.

Clearly, South Africa would find it difficult, to say the least, to raise the necessary capital, particularly in view of the other large projects (including the Koeberg reactor) already under-way or committed. Dr Roux has in fact admitted that this is one of the factors which motivated the attempt to obtain foreign partners for the project, the others being the provision of guaranteed markets and the ensuring of the existence of the very large industrial capacity required.[152]

So South Africa would have been launching itself into what Dr Roux has described as 'a stupendous undertaking'[153] without adequate backing from foreign interests which could help to assure its success and spread the risk of failure. Despite rumoured opposition to the project within top government circles, Dr Roux announced in 1975 that the plant would go ahead as planned whether or not the negotiations with 'foreign partners' were successful.[154] Or as the more extravagant Dr Koornhof boasted in the House of

Assembly: '...we are able in South Africa to build a large-scale uranium enrichment plant without any foreign partner and without having to depend on a foreign country in any way.'[155]

At the end of 1977, the South Africans suddenly changed their tune — although in rather muddled fashion. An official Government announcement was made that UCOR had awarded a R500 million (£305 m.) contract to the South African company Murray and Roberts to build the commercial plant at Valindaba, with work starting in early 1978 and the final size to be determined only at the end of that year. The *Financial Times*, in reporting this, commented:

> 'In purely commercial terms the decision cannot be justified today. Commenting on the short-term market, the chief executive of a rival enrichment supplier said yesterday that "the situation for the rest of us is sufficiently dismal to put anyone off".'[156]

The announcement was followed, however, by a flat denial from a Murray and Roberts spokesman that they had received any such contract.[157]

This was then followed by a completely inconsistent new announcement by the Minister of Mines, Fanie Botha, that the existing pilot enrichment plant would be converted into a 'relatively small' production facility. The Government would decide later whether to extend it further into a commercial plant, 'depending on economic circumstances, development plans for South Africa's nuclear energy programme, and the demand for enriched uranium on the world market...'

Some broad hints were offered that the plant was now seen more in terms of closing South Africa's nuclear fuel cycle, freeing it from safeguards: 'The plant will have the capacity to meet South Africa's needs during the course of time,' and 'there are strong indications, to say the least, which put a question on the free availability of nuclear fuel for South Africa.' This is apparently a reference to American delays in

191

supplying enriched uranium to South Africa because of their fears of nuclear proliferation. The Koeberg station would require some 200-300 tons of enriched uranium a year, and South African is committed to self-sufficiency in nuclear fuel, according to Botha, 'particularly when it concerns the strategically important commodity of electric power.'[158]

This report attributes the sudden switch of plans to the soaring costs of electricity, steel and other items, and the inability of South Africans to find support on foreign capital markets. This, as we have shown above, is a very real factor in limiting the Government's nuclear ambitions. There is some doubt, however, as to whether the confused and conflicting official announcements reflect the real truth about this new plant of smaller capacity. Researchers of the Anti-Apartheid Bewegung in Bonn, using their own sources, have told the authors that a separate commercial plant is in fact being built, not at Valindaba but at the SASOL 2 (oil-from-coal) plant, near Evander/Trichardt, 140 km. east of Johannesburg. This project is going ahead in unprecedented secrecy, with deliveries under two quite separate code-names, appropriately enough 'white' and 'black'; the enrichment plant, the researchers say, is being built at breakneck speed in order to be ready in 1981-2, and many West German concerns are working flat out to produce the vital compressors and other equipment. Financing is stated to be by a consortium led by the French bank Crédit Lyonnais, under cover this time of the Koeberg project.

These reports are given credibility by the obvious interest of the South Africans in maintaining secrecy over their whole nuclear programme; political opposition to *apartheid* is increasingly focussing on investment and trade, and would certainly be directed at any bank or company participating openly in the enrichment plant. The South Africans also have an interest in completing their project as quickly as possible, in order to have a *fait accompli* with which to confront the Americans and others who are trying to pressurise them into accepting safeguards as a condition of help in constructing nuclear installations.

Civil Or Military?

The crucial question is whether the South Africans' nuclear programme is intended purely for 'peaceful' or commercial purposes, as they claim, or whether it is concerned with the production of nuclear weapons. South Africa has the technical capacity to produce such weapons, and the Soviet and French Governments have stated categorically that is *is* doing so and that a weapons test was being prepared in mid-1977 before Western pressures halted it.

Both highly-enriched uranium and plutonium are already being produced: this would only be for military purposes. The almost simultaneous completion in the 1980s of the Koeberg nuclear power station and the commercial enrichment plant will give a very large production capacity for nuclear weapons. At present, the major potential for weapons construction is the pilot enrichment plant, together with a small, secret plutonium plant for reprocessing.

South Africa has long been interested in so-called 'peaceful nuclear explosions' (PNEs), which are identical with nuclear bomb explosions, and there have already been unexplained earth tremors in the Namib Desert, an ideal testing area.

The South Africans themselves, while throwing out hints as to their military intentions, have categorically denied reports about their nuclear weapons programme. These denials lack credibility, however — and they are a vital part of the military strategy, allowing little or no warning or ability to retaliate to countries suddenly threatened with a nuclear attack in the event of a new regional upheaval in southern Africa.

The motivation behind South Africa's nuclear weapons programme is not hard to understand. The bomb bestows enormous political as well as military power on its possessor; and in addition, it could mean forcing the United States into

193

an open military alliance with South Africa, bringing it under the American nuclear umbrella as has already happened in the case of South Korea. The effect of such an alliance would be to increase sharply polarisation of the African continent along East-West divisions, and escalate tensions in the area as African states seek a nuclear capability or guarantee from one of the other nuclear powers, probably the Soviet Union or China.

SOUTH AFRICA'S TECHNICAL CAPABILITY

Non-proliferation experts consider South Africa to be one of the most prominent instances of a 'near-nuclear' or 'Nth' country, in terms of both the capability and the motivation to obtain its own nuclear weapons. A detailed report prepared by the United States Arms Control and Disarmament Agency (ACDA) places South Africa in the same category as several industrialised western countries (including West Germany) and Taiwan, as having '...full access to the special nuclear material needed for weapons' as well as a broad base of personnel skilled in nuclear technology. South Africa is judged to be ahead of other near-nuclear countries such as Israel, South Korea, Iran and Brazil in these respects.[159] The American Congress's Office of Technology Assessment has set out the basic requirements necessary if a country is to construct its own nuclear weapons; it also makes it clear that South Africa, among others, fulfills all those requirements:

> 'Many nations are capable of designing and constructing nuclear explosives which could be confidently expected, even without nuclear testing, to have predictable and reliable yields up to 10 and 20 kilotons TNT equivalent (using U_{235}, U_{233} or weapons-grade plutonium) or in the kiloton range (using reactor-grade plutonium).
> A minimal national program is an effort to produce, *without nuclear testing*, a weapon which is *very confidently* expected to have a *substantial yield*. This will call for a group of perhaps a dozen well trained and very confident

194

persons with experience in several fields of science and engineering. They would need the support of a modest, already established, scientific, technical and organisational infrastructure. If these requirements are met and the programme is properly executed, the objective might be obtained approximately *two years* after the start of the program, at a cost of a few tens of millions of dollars.'[160] [Emphasis in the original.]

Myron Kratzer, a senior official in the State Department dealing with science and technology, commented flatly: 'South Africa has the technical ability to take steps towards making a bomb...'[161] A 'well-informed American Government source' is quoted by the *Washington Post* as saying that the South Africans could produce a bomb in a matter of months if they concentrated their funds and personnel on it.[162] West German military experts have also been quoted, in *Der Spiegel*, as being in no doubt that South Africa is already in possession of atomic weapons as well as having the means to carry them to their targets.[163] They were proved right in August 1977 when the Soviet and American reconnaissance satellites detected South Africa's nuclear weapons testing site. A recent delegation to the IAEA in Vienna was told that nothing could be done to stop South Africa's nuclear weapons programme.[164]

M. Raymond Barre, Prime Minister of France, confirmed in February 1977 that South Africa '...already has a nuclear military capability' and, defending the French sale of power reactors to South Africa, said that this would 'add nothing to it'.[165] This official statement echoed a private admission by an unnamed senior French official to this effect, two months previously.[166] Finally, the Soviet leader, Mr Leonid Brezhnev, announced South Africa's nuclear weapons plans, as repeated in the official statement by TASS:

> 'According to information reaching here, work is now nearing completion in the South African Republic for the creation of the nuclear weapon and preparations are being held for carrying out tests.'[167]

There have been indications that the Soviets have used agents on the ground in South Africa as well as satellite reconnaissance to get hold of this information; in fact it was reported by a Chinese source that a mysterious exchange of agents involving West Germans held by the Soviets and a single Soviet agent, Victor Loginov, was connected with their priority concern with South Africa's nuclear plans; Loginov had in fact been caught in South Africa.[168] How he came into the hands of the Germans for exchange against their own people is something of a mystery, but presumably yet another aspect of the close co-operation between the two countries over nuclear, military and other issues.

The Soviet intelligence work and satellite reconnaissance were crucial, since only that country had a real interest in exposing South Africa's nuclear weapons plans. The information, once released at the highest level, was then confirmed by French sources once again. The Foreign Minister, Louis de Guiringaud, responded to the Soviet alert by saying that they had themselves received 'more precise indications which have increased our concern' and that he could confirm that the South Africans were making preparations for a nuclear explosion.[169]

It would perhaps be illuminating to consider the various items of nuclear equipment available to South Africa's nuclear research teams, in terms of the requirements for a nuclear weapons programme as summarised by the Office of Technology Assessment team:

> 'The material for a nuclear weapon might be plutonium, or uranium with a high concentration of either one or two uranium isotopes — U_{233} or U_{235}. Using either form of uranium or weapons-grade plutonium it is possible to design low technology devices that would produce explosive yields reliably up to the equivalent of 10 or 20 kilotons of TNT. With reactor-grade plutonium it is possible to design low technology devices with probable yields which are 3 to 10 times lower than those mentioned above (depending on the design); thus yields in the

kiloton range could be accomplished.
Thus militarily useful weapons with *reliable* nuclear yields in the kiloton range can be constructed with reactor-grade plutonium, using low technology.'

This report sees commercial-grade plutonium-240, as distinct from weapons-grade plutonium-239, as '...not desirable in nuclear weapons because of its high rate of spontaneous fission.'[170] However, a small group of British second-year students in physics and chemistry successfully designed a workable H-bomb, using only published information, based on commercial-grade plutonium; this should therefore be added to the list. As *The Guardian*, which sponsored the experiment, commented: '...how much more easily could the job be done even by a small government-backed establishment?'[171] It was subsequently announced that the Americans, too, had demonstrated the use of commercial-grade plutonium in a secret test in the Nevada desert, confirming theoretical studies in nuclear weapons laboratories which had concluded that countries could built nuclear weapons from plutonium stockpiled ostensibly for use as reactor fuel.[172]

The original research reactor known as SAFARI-1 could be useful in a military programme, if its fuel came from a country which was not as concerned as the Americans are about stringent safeguards. During South Africa's negotiations with the Americans in 1967 over the renewal of the contract for highly-enriched uranium supplies, it was reported that French officials had offered to supply them with the material if they did not wish to renew the American agreement; *Newsweek* commented that '...the French have hinted that enough fissionable material might be supplied for military purposes as well.'[173] Two subsequent reports from London indicated that France was in fact providing South Africa with enriched uranium.[174] The Congressional report already cited reminds us that a substantial potential for diversion to military uses exists in terms of 'research reactors...that are fuelled with highly-enriched uranium' and also in terms of 'the critical assembly in several countries that use plutonium.'[175] South Africa of course has one of the research reactors, capable of producing enough plutonium for a bomb every three or four years.[176] It also has several critical assemblies, one of which, Pelinduna-Zero, is of indigenous manufacture and therefore not subject to safeguards. The plutonium from critical assemblies 'is essentially uncontaminated by fission products and is of high quality for use in weapons.'[177]

The plutonium that will be produced by the Koeberg reactor is also of interest here; the *Washington Post* estimates that the two proposed plants '...would produce as by-products every year a total of about 1,000 pounds of plutonium, enough to make 100 atomic bombs, each with a force equal to the bomb that destroyed Nagasaki in Japan in 1945.'[178] A similar conclusion was reached by the French newspaper *Le Figaro*.[179] Koeberg could make all the difference between an A-bomb and a plutonium-bomb programme of the Nagasaki type, and would considerably increase the number of nuclear bombs that could be produced in a relatively short space of time.

If South Africa planned to use plutonium rather than enriched uranium as the basis of its nuclear weapons, it would need a reprocessing plant to recover plutonium from the spent fuel of either SAFARI-1 or Koeberg. Although it has never been publicly announced, informed sources in French Government circles as well as a knowledgable British observer, Patrick Keatley, are quite sure that a 'small working model' processing plant has been built in South Africa.[180] This is not very surprising in view of the statement by the Congressional study that '... a small reprocessing plant capable of separating enough plutonium for several bombs per year is well within the capability of most Nth countries, even if an economical [sic] commercial plant is not.'[181] Once a stockpile of separated plutonium has been built up, it could be made into bombs by almost any determined government in a matter of days or weeks.[182] Dr Albert Wohlstetter, an authority on proliferation, has named South Africa as one of the countries which — once it had a reprocessing plant — would be able to make bombs 'within a few days or even hours'. Controls on plutonium are so loose, he added, 'as to make it perfectly legal to accumulate stocks of plutonium in a form days or hours from insertion in a nuclear explosive.'[183] Patrick Keatley states that South Africa has in fact already accumulated some two or three kilogrammes of plutonium derived from SAFARI-1.[184]

Perhaps the major obstacle to the separation of plutonium from spent fuel is the extreme danger involved for the personnel handling this hazardous material; as the American 'father of the H-bomb', Theodore Taylor, has put it:

> 'Some use the phrase "self-protection" in the sense that it [plutonium] is very heavy, radioactive, and difficult to handle without extreme danger. I think that is not a terribly strong inhibition against a country doing what is necessary to extract it . . .'[185]

The inhibition becomes weaker in a country like South Africa, where there is high unemployment among all races and little choice of jobs, especially for black people. The operators of a small 'table-top' reprocessing plant, unlike those at most other nuclear installations, would not have to be familiar with nuclear physics and could carry out their work without being aware of its nature. South Africa could dispense with some safety measures – which constitute much of the cost of reprocessing. Workers suffering from radiation-induced illness might not realise the cause, especially as the symptoms can take some time to appear. In any case, legislation prohibits discussion of nuclear matters.[186] Claiming compensation could be very difficult. It appears that in India, people doing this kind of work were not aware of the dangers involved; and the US Government is refusing to take responsibility for such illnesses among former soldiers exposed to nuclear weapons tests.

Availability of spent fuel in large amounts will probably

have to await the commissioning of the Koeberg reactor. Here the significance of ordering it from the French, and particularly of switching the contract for enrichment services to non-American companies, becomes apparent — as does the relative unimportance for the Government (if not for the management of ESCOM) of whether the French bid is competitive or not. It appears that nowhere in the original Franco-South African reactor deal is there mention of where the spent fuel (which will actually be owned by the South Africans) is to be stored, nor of which country, if any, is to reprocess it for extraction of plutonium. It was only after the Soviet revelations that the French claimed that all spent fuel would be returned to France for reprocessing.[187]

Perhaps an even simpler approach might be the use of some of South Africa's large reserves of thorium, another fissionable material. After passage through a reactor it produces uranium-233 which, as already observed, is a fissile material readily usable in nuclear weapons.

If the Government is not treating Koeberg strictly as a commercial venture, which seems to be the case judging by the switch to the French, then it could manipulate the reactor's operation in order to obtain more of the weapons-grade plutonium which is easier to use than the commercial grade. By sacrificing about half the power, and producing about a quarter as much plutonium per kilo of fuel, it would obtain this weapons-grade plutonium with 7% or less of Pu_{240} instead of the roughly 20% in material derived from normal operation of the reactor.[188]

This mode of operation would probably reveal the fact that the South African Government was using the reactor for military purposes; other countries could also learn of South African military intentions from IAEA inspections showing that some of the plutonium had been diverted from the normal channels. However, not only is the inspection and safeguards system very far from foolproof, but South Africa could at any time refuse access to IAEA inspectors. Moreover, by that time South Africa would have the full complement of nuclear installations necessary for quite large-scale production of nuclear weapons, and any attempt at sanctions is unlikely to have much effect. If this took place in a regional emergency or war situation, any threat of IAEA measures would be insignificant in relation to the security which South Africa would feel in deploying a substantial number of nuclear weapons.[189]

The combination of the Koeberg reactor and a commercial as well as pilot processing plant is very significant in terms of any military ambitions. For one thing, operation of the reactor to produce more weapons-grade plutonium approximately triples the fuelling requirements; an abundant supply of locally-enriched uranium then becomes important, particularly if external supplies are cut off in an attempt to enforce IAEA safeguards. In fact, it could be no accident that the commercial enrichment plant and the reactor are due to come on stream within a year or two of each other in the 1980s. As observed by Robert Alvarez, an American environmentalist:

> '. . .if these projects occur at the same time, let's say around 1985, if operation begins on the two 1,000-megawatt plants plus operation of the enrichment facility, there is no stopping South Africa from that point on. Curtailment of nuclear materials is meaningless because they have their own uranium and will have a sufficient technological base to ignore all treaties.'[190]

Once that stage is reached, of course, South Africa would also have the capacity to offer highly-enriched uranium or plutonium to other friendly 'near-nuclear' countries. This could be the most important reason for close relations between South Africa and West Germany, Israel, Iran and Brazil in the nuclear field.

Until that time, the immediate South African threat to international non-proliferation policies derives mainly from its pilot enrichment plant, which is the only unsafeguarded enrichment plant outside the existing nuclear-weapons states, the United States, Soviet Union, China, France and Britain. The few countries which have made the vastly expensive investment in gaseous diffusion — which is the process to which South Africans generally compare their own — have all done so primarily to produce weapons-grade uranium. Enrichment for nuclear fuel comes second.[191] South Africa's pilot plant is the cause of considerable concern among non-proliferation experts and officials who admit that it is one of the major dangers to the attempts to limit nuclear proliferation. As Fred Iklé, then Director of ACDA, put it: 'Their own enriched uranium is not under international safeguards and the South African Government can do with it what it wants . . .'[192]

While South Africans have said that they would consider placing the commercial plant under international safeguards (although with reservations about preserving secrecy that call even this vague commitment into question),[193] this only serves to highlight the fact that at no time has any move been made to discuss safeguards for the pilot plant. This alone is a strong indication that the plant is being used for military purposes. Political and military analysts have been quoted by the *New York Times* as believing that this is in fact the case.[194] Highly-enriched uranium is needed not only for atomic bombs, but it is also the only known trigger for hydrogen bombs.

American Government sources interviewed by the authors point out that in order to produce weapons-grade enriched uranium, a jet-nozzle plant would probably have to be specially designed for that purpose. They dismiss claims made at various times by the West German Government that '. . .the German separation nozzle process for uranium enrichment is not suitable for the manufacture of material usable for weapons . . .'[195] and state that it certainly can be used for weapons-grade enrichment — although with substantial loss of throughput as highly-enriched uranium would require large numbers of stages, which involve halting the whole plant in order to feed the batches of material back

202

through it. In addition, the amount of separative work units (SWUs) needed for a given volume of material goes up at each stage, so that the amount of highly-enriched uranium finally produced is a small fraction of the uranium hexafluoride which is initially fed into the plant. This is only a problem, however, if there is a shortage of the raw material — not a problem for South Africa — or if the plant is intended to run on an economic basis. It has never even been suggested that the pilot plant at Valindaba has any commercial basis. The plant's capacity is a closely-guarded secret which, because of unfamiliarity with the exact details of its operation, non-proliferation experts have been unable to discover.[196] It is not known, therefore, exactly how much weapons-grade uranium can be produced each year at Valindaba. One factor that could make the output greater than a plant of similar size based on a different process is that, as Dr Roux has said:

> 'The South African process has a separation factor between stages which is considerably higher than that of the gaseous diffusion process of the USA. This means that the South African process requires many less stages than the gaseous diffusion process for the same degree of enrichment.'[197]

The big question then is whether the South Africans designed the pilot plant from the beginning with a view to enriching uranium to weapons-grade levels. *The Washington Post* quotes American sources as saying that the plant can do this;[198] *Der Spiegel* quotes a Director of the IAEA, Mr Rometsch, as saying the same thing.[199] Dr Roux made a point of noting the military capability of the plant while also claiming that his Government would refrain from using that capability:

> '...if a country wishes to make nuclear weapons, an enrichment plant will provide the concentrated fission material if the country possesses the necessary natural uranium to process in the plant.'[200]

Curiously enough the announcement by the West German Ministry of Science and Technology, while announcing the 'feasibility study' arrangement between Steag and the South African AEB, mentioned that the resulting product would have a military potential as well as supplying energy for peaceful purposes.[201] Prime Minister Vorster told an interviewer in 1976 that South Africa 'can enrich uranium and we have the capability of mounting a nuclear defence'.[202]

The question of whether the proposed commercial enrichment plant could or would be used to produce weapons-grade uranium is somewhat less clear. As pointed out by *Newsweek*, there would be plenty of enrichment capacity left over to make dozens of weapons if the South Africans do not sell it all overseas.[203] However, recent discussions in the Nuclear Suppliers' Group, according to an American Government source, have resulted in a decision that certain supplies — including the vital compressors — should be withheld from an unsafeguarded enrichment plant; thus, the South Africans may be forced to accept some kind of safeguards agreement in order to get it built, although there are experts still worried that the South Africans could get around that. The smuggling out of compressors from West Germany, as noted in Part I, is relevant to this.

In any case, safeguards agreements outside the context of the Non-Proliferation Treaty can be of widely varying degres of effectiveness. South Africa has stipulated that the inspections would not be allowed to interfere with the plant's operation or to risk disclosure of confidential proprietary details of the enrichment process[204] (which would include the vital fact of whether it was designed to produce weapons-grade uranium). The IAEA inspectors, even given the best safeguards agreement, are overworked. And a commercial enrichment facility, according to the Congressional study, is particularly difficult to safeguard:

'Nations possessing a commercial facility . . .could covertly dedicate a portion of it to weapons-grade enrichment, use the same technology to construct another facility for weapons-grade production, or abrogate the safeguards and overtly convert some or all of the plant to highly-enriched uranium. Covert diversion from an enrichment plant would be difficult to detect, even with sophisticated safeguards.'[205]

We have obtained a copy of a confidential study by the International Atomic Energy Agency which goes even further than the Congressional study in highlighting the problems encountered in the attempt to monitor any diversion of fissile material from civil to military use. The study, *Special Safeguards Implementation Report*, was produced in June 1977 and is the first published account of the IAEA's overall safeguard operations.

A number of the more serious verification problems apply to the South African case, although the report is careful not to mention any country by name. 'Significant and continuous improvement' is called for in systems of accounting and controlling nuclear material operated by States, this being:

> '. . .particularly necessary in States that have fuel cycles containing fuel conversion and fabrication facilities, in States that operate on-load fuelled reactors, and in respect of certain large critical assemblies and research reactors.'[206]

The latter two cases are particularly relevant to South Africa. The IAEA estimates that in three cases of research reactors fuelled with highly-enriched uranium, like SAFARI-1, 'the technical objective was not obtained'; in other words, safeguards were not adequately applied.[207]

Particular difficulty is noted in the case of States which have not signed the NPT and therefore have only some of their facilities inspected; the possibility of moving fissile material temporarily to the inspected installation to cover up diversion is specifically mentioned.[208] Inspectors are heavily dependent on whether States choose to facilitate their work:

'In other words, the actual effectiveness of safeguards is closely related to the effectiveness of national systems of accounting and control of nuclear material and the extent to which States' legislation and operators' practices enable the Agency to carry out the planned verification activities in a timely fashion.'[209]

In the category of twenty-seven countries (which clearly includes South Africa) which have only research installations under safeguards, two are classed as 'needing improvement' in the measurement system, two in the system of records and reports, and one each in physical inventory procedures, stratification of inventory lists, evaluation of unmeasured inventory and unmeasured losses and provisions to ensure correct operation of accounting and control procedures.[210] Altogether five of the States in this category show 'deficiencies in accounting and control'.[211] There is no indication which countries are meant; the most important point here is that countries can perform very inadequately in this respect, making adequate inspection impossible, *without incurring any penalties.*

The proposed commercial enrichment plant, even if it were safeguarded, would be in the category of 'bulk-handling facilities' over which the IAEA expresses particular concern. Out of eleven such facilities inspected in 1976, at no less than five 'the objectives could not be fully attained' for a variety of reasons.[212]

The serious inadequacies of the international safeguards system as revealed here, and which is further elaborated in Appendix 1, are all the more sobering since, as the IAEA itself estimates, a nuclear weapon could be made from diverted material, depending on its chemical composition and physical form, in ten days to six months — which could be considerably less than the usual interval between inspections:

'Thus diversion, if it had occurred, would not necessarily have been detected within the critical time period (i.e.

before the explosive was made).'[213]

The safeguards sytem does not pretend to prevent States diverting fissile material for nuclear weapons; it is now apparent that it cannot even provide early warning of a weapon being made, before it is ready to be used. South Africa, in short, has very little to fear from any safeguards, even if it does agree to an inspection system for its Koeberg reactors and its commercial enrichment plant. 'Safeguards' would make little or no difference if the Government decides to divert fissile material for weapons production.

NUCLEAR EXPLOSIVES TESTING

Well before the Soviet announcement of a weapons testing site in the Kalahari Desert there had been indications that South Africa had carried out nuclear explosions, some under the pretext (used also by India, with as little credibility) that they were 'peaceful'. The American State Department has revealed that:

> 'A literature study of the peaceful application of nuclear explosions and some model experiments have been conducted [by South Africa] with information derived mainly from the US Plowshare program.'[214]

'Plowshare' is a programme which was terminated by the Americans when they came to the conclusion that there can be no distinction made between nuclear explosions for civil and military purposes, and that they are of no particular value for civil purposes in any case, while presenting unacceptable radiation and other hazards.

In 1969 the AEB reported that it was conducting research in the use of nuclear explosions for earth-moving.[215] In 1970 it was reported that the Government was 'keeping abreast of the latest developments in the use of nuclear explosives in civil engineering projects.'[216] The French said in 1977 that the atomic explosion planned by the South Africans would be 'peaceful'.[217]

The idea of a 'peaceful' nuclear explosion is a thin pretext for nuclear weapons tests. Here again, the Americans have shown South Africa the way. In 1958, under cover of International Geophysical Year (in which they collaborated quite closely with the South Africans) the United States conducted secret atomic weapons tests in an area of the South Atlantic, some distance from South Africa, called the Cape Town Anomaly, where the ionised layer comes relatively close to the earth. The tests, which were revealed by the *New York Times* a year later, were intended to examine the effect on the ionosphere of nuclear explosions. Monitoring was conducted by sending in bombers to take samples of the atmosphere in the surrounding area.[218] Because of the geographical location of South Africa, it was important to the Americans to involve the South African Government in this and other monitoring activities, including satellite observation. For many years, the National Aeronautics and Space Administration (NASA) has had an observatory in South Africa run jointly with the CSIR. After repeated attacks in the Congress on the *apartheid* conditions operating at the observatory, NASA finally agreed in 1974 to phase it out, ostensibly because technical innovations had rendered it obsolete.[219]

In 1967 Martin Walker, a young British journalist working for the well-connected Afrikaner publisher Otto Krause, discovered that a somewhat similar exercise was going on from South Africa itself. Douglas Torr, a nuclear scientist straight from the British nuclear research establishment at Harwell, was working on a joint project of the CSIR and the University of the Witwatersrand, ostensibly on meteorological research, in the area of the Anomaly. Douglas Torr, it appeared, was using three Hastings bombers of the South African Air Force for his research . A check on what he had been doing at Harwell indicated that it had been highly

The nuclear-industrial complex in West Germany

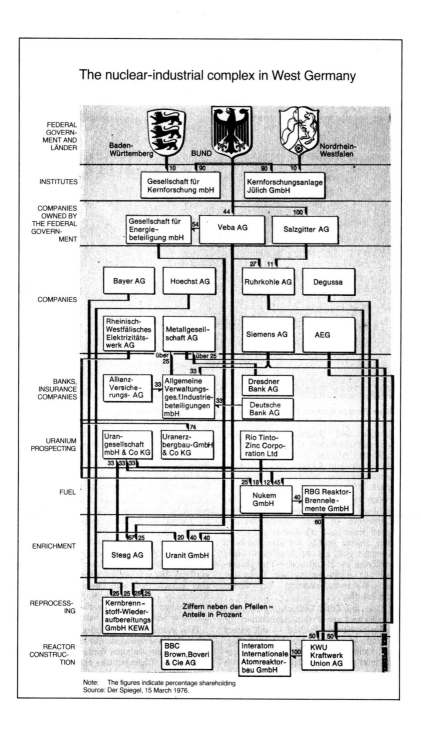

Note: The figures indicate percentage shareholding
Source: Der Spiegel, 15 March 1976.

SA Embassy, Bonn (*Dritte Welt*)

SA Foreign Minister H. Muller and Ambassador
D. B. Sole   (*Dritte Welt*)

A. J. A. Roux (*Camera Press*)

Georg Leber (*Camera Press*)

Pelindaba/Valindaba (*Peter Davis*)

Pelindaba (*Peter Davis*)

Tank containing separation nozzle tubes (*The Society for Nuclear Research, Karlsruhe*)

Separation tube containing nozzles being inserted into tank
(*The Society for Nuclear Research, Karlsruhe*)

Inside Pelindaba

confidential. The CSIR building where he was based was heavily guarded and had no windows; it was reputed in the local area to be for top-secret nuclear research.[220]

Any testing of nuclear explosives in that area would probably have to be carefully concealed; the Anomaly is situated between South Africa and Antarctica. The Antarctic Treaty negotiated in 1958-59, which declared the Antarctic a nuclear-free zone, includes South Africa among its signatories — something which the Americans were at the time very proud of achieving. An ideal opportunity to cover up any testing, however, could be furnished not only by the fact that the Anomaly area is already highly radioactive, and avoided by shipping for that reason, but also by the series of French tests going on in the Pacific in the early 1970s. South Africa was itself monitoring radiation from the French tests in collaboration with Harwell in Britain, and tied in with the American monitoring network.[221] Thus it would have access to detailed information on radiation emissions from there. (Atmospheric· testing provides no opportunity for seismic detection.) As soon as Martin Walker started investigating the activities around the Anomaly, he was ordered by Krause to drop the story.[222]

Informed sources close to the South African nuclear programme have told us, however, that atmospheric testing is not feasible at this point: of much greater concern is the likelihood of underground testing. They point out in this connection that there have been a number of officially reported 'tremors' in the northern Cape and the Namib Desert in recent years, of which three were small underground tests (one being in the third quarter of 1976). The Namib is not in an earthquake zone, although the northern Cape is, and there is apparently no seismic equipment sophisticated or near enough to the Namib to distinguish clearly a natural from a small man-made tremor.

Another factor making the Namib an excellent underground testing area is the fact that much of it is totally deserted and under heavy permanent guard to keep out any intruders by security forces of an Anglo-American Corporation subsidiary, Consolidated Diamond Mines. This is to cut off all access by unauthorised personnel to the diamond areas along the southern coastal area, named the 'Skeleton Coast' by early explorers for its bleakness. There are numerous ways of covering up an underground explosion, which are listed in Appendix 7.

OFFICIAL SECRECY, AND CONFLICTING STATEMENTS

Despite a South African Government official, questioned about the visit of General Rall from West Germany, who said: 'We have nothing to hide at Pelindaba', the whole operation is in fact shrouded in secrecy. The very name Pelindaba is a contraction of a Zulu expression (the name of an ironsmelting works in the vicinity in pre-colonial days) which means 'we don't talk about this any more'. Valindaba, the site of the pilot enrichment plant, means 'we don't talk about this at all.'[223] Dr Roux has translated this, rather dramatically, into modern terms:

> 'To those in South Africa — some of them in responsible positions at that — who expound great theories as to what South Africa should do with this development [i.e. enrichment], a word of friendly advice: say nothing — and if the temptation to speak becomes very great, take counsel with yourselves. This delicate matter can be left in the hands of the Government with the utmost confidence.'[224]

Dr Carel de Wet, then Minister of Mines, spelt out right at the beginning of the enrichment venture in 1970 the importance of secrecy, not only among those concerned with the project but also among those in the press who might come across any information. Congratulating everybody on the degree of secrecy which had been achieved, he added:

210

'This in itself goes a long way in proving the self-control
and loyalty of the personnel at Pelindaba — not only
those intimately concerned with research and development, but also, and probably especially, those not directly
concerned, who were aware of the fact that a project was
being undertaken at Pelindaba which could not be talked
about and who did not try to pry further into the matter,
much less discuss secret work being undertaken at
Pelindaba. It is perhaps fitting that in the interests of
South Africa and especially the communication media,
the attention of everybody is drawn to the strict
stipulations of the Atomic Energy Act regarding the
question of secrecy.'[225]

Government officials have taken considerable trouble to
maintain ambiguity as to whether South Africa has a nuclear
weapons programme, throwing out hints from time to time,
and refusing to either confirm or deny the speculation from
abroad. When some statement is made that seems to confirm
the speculation, it is usually quickly denied.

The earliest report that South Africa was capable of
producing its own atomic bomb seems to have been in 1960,
when Dr Roux said it could do so if it was '...prepared to
isolate the best brains in the country and give them all the
funds they needed.'[226] He repeated this in a broadcast in 1962,
adding: 'It is my sincere hope that we shall never be called
upon to engage in this activity.'[227] In 1965 Dr W.L. Grant,
speaking for the AEB, said:

'On several occasions the Director-General has indicated
that South African scientists, in common with those from
most developed countries, do have the technical ability to
develop nuclear weapons . . .'

He denied, however, that any military research was being
done in that area.[228]

The politicians, however, were more outspoken. Inaugura-
ting SAFARI-1, Prime Minister Verwoerd said: 'It is the duty
of South Africa not only to consider the military uses of the
material but also to do all in its power to direct its uses for

211

peaceful purposes.'[229] Growing pressure was reported for the Government to start work on weapons manufacture. Dr A. Visser, a member of the AEB's board, advocated production of a bomb for 'prestige purposes'. 'Money is not problem. The capital for such a bomb is available.' He proposed using nuclear weapons against the 'loud-mouthed Afro-Asiatic states', and he had strong support from many whites for his suggestion. In December 1968 General H.J. Martin, the Army Chief of Staff, was quoted as saying that the Republic was ready to make its own nuclear weapons, and this was linked with the work of missile development then in progress, resulting in *Cactus/Crotale (see* Chapter 7).[231]

Speculation and denials continued, but with a distinct change of emphasis in the early 1970s, when the question of nuclear proliferation became a serious one for several countries. Soon after the explosion of a nuclear device in 1974 by India, which maintained that is was for peaceful purposes, Dr Louw Alberts, vice-president of the AEB, said that South Africa had the technology and resources to produce a bomb, although its programme was for peaceful purposes, and added: 'Our nuclear programme is more advanced than that of India.'[232] The statement was made against a background of increasing activity by the liberation movements in Rhodesia and South Africa.[233] He also pointed out:

> 'Any third-year student in physics has the know-how to make the atom bomb and a fourth-year student can do it better. Obviously, any nuclear research body here or elsewhere will be able to do it even better than a fourth-year student.'[234]

What seemed rather like a semi-official announcement that South Africa had in fact 'done better' came recently from Prime Minister Vorster. During an interview with *Newsweek* in early 1976 in which he was asked whether South Africa's defences include a nuclear capability, he said:

'We are only interested in the peaceful applications of nuclear power. But we can enrich uranium, and we have the capability. And we did not sign the nuclear non-proliferation treaty.'[235]

This caused quite a stir in diplomatic circles, and the American State Department instructed its Embassy in Pretoria to seek clarification. They were assured, according to the State Department sources of the authors, that this phrasing resulted from a contraction of a number of things said at different points of the interview, and had not actually been said in quite that way.

It can hardly be coincidence, however, that virtually the same formula appeared in a reply given by the then Minister of the Interior, Connie Mulder, to a similar question in an interview by *Bild-Zeitung* of Hamburg in June:

'I should like to reply in this way: we have made it clear that we possess uranium and we have given the green light for building up a big uranium industry. We have not signed the Non-Proliferation Treaty...'[236]

When the Soviets announced in August 1977 that the South Africans were ready to test a nuclear device, and high-level approaches had been made by Western governments to South Africa, Foreign Minister 'Pik' Botha denied all the allegations, by the French as well as the Soviets, as 'wholly and totally unfounded'.[237]

On 23 August President Carter announced at one of his regular press conferences that:

'In response to our own direct inquiry and that of other nations, South Africa has informed us that they do not intend to develop nuclear explosive devices for any purpose, either peaceful or as a weapon, that the Kalahari test site which has been in question is not designed for use to test nuclear explosives and that no nuclear explosive test will be taken in South Africa now or in the future.'

This, on the face of it, is as categorical a denial as one could imagine – even covering the peaceful nuclear explosions which South Africa is known to have been working on. According to the State Department, the assurance had been repeated in a letter from Vorster to Carter on 13 October. On the 17th, however, in an interview with ABC-TV, Vorster denied having given such an assurance.[238] Owen Horwood, South Africa's Finance Minister, has also announced that his country reserved the right to use its nuclear potential for other than peaceful purposes to suit its own needs. South Africa, he said, had given an assurance that its nuclear programme was aimed at peaceful purposes but that if it decided to use its nuclear potential in any other way it would do so according to its own needs: 'I, for one, reject absolutely and entirely that anyone should tell us what we should do.'[239] Thus the inconsistencies and contradictions in government statements continue, apparently deliberately as a means of maintaining ambiguity about South Africa's real intentions. One critical issue is becoming less ambiguous, however: South Africa's increasing reluctance to sign the Nuclear Non-Proliferation Treaty.

South Africa's position with regard to the NPT has shifted markedly since it started acquiring an enrichment capability. The Treaty is of far more than theoretical importance, since any country which signs and ratifies it is thereby undertaking to allow inspection of all its nuclear installations, regardless of whether they have been subjected to a specific safeguards agreement with the IAEA. The NPT also prohibits the manufacture or acquisition of nuclear explosives of any kind. Parties to the NPT undertake not to supply nuclear equipment and materials to non-treaty countries other than the existing nuclear powers, except under safeguards. As Mr Botha himself put it in 1968:

> '... all nuclear materials produced locally or imported and all nuclear facilities, whether self erected or erected with outside assistance, would be subject to safeguards and their international inspection.'[240]

214

When the General Assembly of the United Nations voted on the draft Treaty in 1968, it was reported that '. . . the treaty sponsors took particular satisfaction in the favourable vote of South Africa, a major producer of uranium and the only Southern African country with a real potential for nuclear industrial development.'[241] South Africa's representative claimed during the debate that his country had many years ago decided:

> '. . . to do absolutely nothing in the context of uranium sales to foreign buyers which might conceivably contribute to an addition to the ranks of the nuclear-weapon States . . . And, so far as our own atomic energy programme is concerned, this programme, as we have so often stated . . . is devoted to peaceful purposes only.'[242]

He also said that South Africa would vote for the resolution as 'an earnest of our good will and our wholehearted support for the objective of non-proliferation of nuclear weapons.'[243]

When the NPT was opened for signature later in the year, however, South Africa expressed apprehension that the applicability of safeguards through inspection might be 'economically harmful' to it as a major uranium-producing country. In July 1970 Vorster said that South Africa would accede only if there were a satisfactory agreement on the safeguards and inspection issue, in order to preserve secrecy and efficiency in its nuclear operations, and South Africa would not be limited in its pursuit of nuclear energy research. In October 1973 Foreign Minister Muller told the Assembly that his Government would negotiate an agreement with the IAEA for the application of safeguards to its enriched uranium production.[244] No move to follow that up has apparently been made. Renewed pressure on South Africa to sign the Treaty resulted from the Western diplomatic offensive to stop South Africa's intended weapons test in August 1977; if anything, South African resistance to the idea was even stronger than before.

Although the safeguards system has been accepted by

several other countries with uranium mines, such as Canada, Czechoslovakia, Denmark, Madagascar, Morocco and Sweden, South Africa shows little real interest in doing so, and stresses repeatedly that the safeguards arrangements are not acceptable.[245] Some countries which at various times have raised equally strong misgivings about the Treaty have finally signed, although not yet all have ratified it. Japan's signature in 1970, after long hesitation, was apparently in order to take part in the negotiation of the safeguards system. South Africa, as a non-signatory, cannot participate in these negotiations.[246] This raises doubts about the sincerity of its expressions of concern to influence the negotiation of agreed safeguards procedures. South Africa stood out also as one of only four states which in 1972 voted against a General Assembly resolution calling for an international commitment to forego the use of both conventional and nuclear weapons.[247] Since that time its objections to the NPT have become increasingly uncompromising, in spite of representations from a variety of countries.

, There is a clear contradiction between South Africa's increasing intransigence in the face of demands that it sign the NPT and allow inspection of its unsafeguarded facilities, and protestations that the Government's motives are entirely non-military. However, as one diplomat based in South Africa has pointed out: 'It's like devaluation. You go on swearing you'll never do it, until you do.'[248]

An issue of such extreme sensitivity would certainly have the status of 'deniability' – a principle widely used by government which means simply that they will deny any reports about their most secret and delicate operations, regardless of whether the reports are true or false.

There are good reasons for South Africa to continue denying its military intentions as regards nuclear power. Firstly, an open admission that the programme is for military purposes could easily mean cancellation of the deals with the French and West German Governments for the supply of power reactors and enrichment capability respectively –

which are of vital importance to the nuclear weapons programme.

Secondly, such an admission would almost certainly provoke frantic attempts by the African states most directly threatened to obtain nuclear weapons of their own, or to offer nuclear missile bases to the Soviets or Chinese. Whichever way it happened, South Africa could lose the initiative, and its own weapons could become vulnerable to a first strike by the other side. The cost of providing invulnerable delivery systems and the appropriate command and control system would be prohibitive. The temptation to strike first in a confrontation would be increased by the fear that the opponent may be about to do that, for identical reasons; this would greatly exacerbate the insecurity and tension already present in the southern African region, and would be an enormous burden on South African resources.

Thus it is much better, from the South African point of view, to preserve at least an element of ambiguity about its military plans for nuclear power. This gives it the 'time' for which South African politicians are always pleading, to gain a decisive military lead, based on a flourishing 'commercial' programme, over any rivals. As Sverre Lodgaard of the International Peace Research Institute, Oslo, observed, the best way to obtain a nuclear weapons capacity these days is indeed by the 'civilian route':

> 'It may be simpler then to go most of the way under a civilian flag, trying to leave the outside world in uncertainty with shorter warning time, and with reduced possibilities for designing and enacting sanctions. ... civilian utilization of nuclear energy may enable a state to produce nuclear bombs without ever disclosing it, thereby reducing the warning time to zero. ... Experience from civilian utilizations of course facilitates the construction and operation of clandestine facilities for bomb production.'[249]

One State Department source has told the authors that it would be impossible for the United States to continue its nuclear co-operation with South Africa if the latter were to explode a nuclear bomb. No doubt this was one of the threats used to dissuade the South Africans from proceeding with a test. A further consideration is the massive amount of money required to finance the Koeberg reactors and the commercial enrichment plant, money which could dry up if the programme was too obviously military. Meanwhile, the Kalahari test site has not been dismantled. *The Guardian* quotes an American official in touch with intelligence analysis as saying that South Africa could explode a bomb there at any time.[250]

Africa

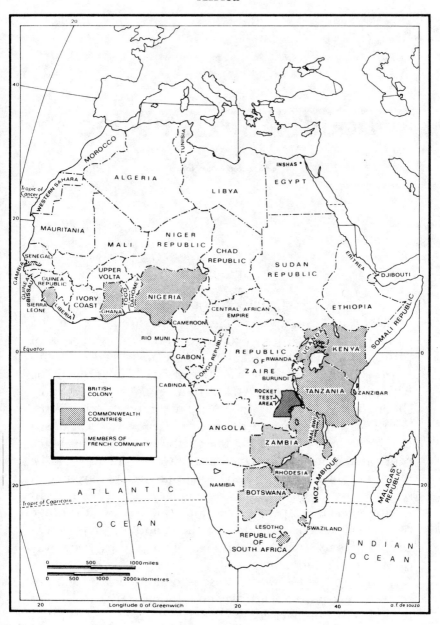

BRITISH
COLONY

COMMONWEALTH
COUNTRIES

MEMBERS OF
FRENCH COMMUNITY

a. f. de souza

6
South Africa's nuclear strategy

When South Africa finally shows its hand and announces that it has a nuclear weapons programme, what motivation and strategy should be taken into account? This is a topic of profound disagreement among those interested in South Africa. Some argue that nuclear weapons will not prevent the white minority in South Africa from being overwhelmed by sheer force of numbers from within. Others say that if the South African régime were to use a nuclear weapon against a neighbouring African state, the resulting world-wide abhorrence that would be directed at the Government would be insupportable. Neither of these, nor any related argument, seems to us to outweigh the advantages to South Africa of having nuclear weapons, of continually expanding the range of delivery systems available for them, and of being seen as prepared to use them if the survival of the 'South African way of life' is called into question.

There are many powerful reasons for non-nuclear countries to seek to acquire nuclear weapons. Some of the main ones have been summarised by the US Congress's Office of Technology Assessment as follows:

220

South Africa's nuclear strategy

'1) *Deterrence.* Several states on every list of potential new nuclear weapons states (Nth countries) have reason to fear direct attack or long-term deterioration of their security *vis-à-vis* neighbors or regional adversaries. 2) *Increased International Status.* As a symbol of modernity and technological competence, nuclear weapons are often viewed as a source of status, prestige and respect. Aside from its symbolic significance, a nuclear weapon capability will augment national military and political power in real terms. 3) *Domestic Political Requirements.* For many of the same reasons that nuclear weapons can enhance a nation's international reputation, they also may bolster a government's domestic political support. The Indian detonation may have been motivated in large part by such considerations. 4) *Increased Strategic Autonomy.* Even if it is already protected by an alliance, a nation may feel it has more options to pursue national objectives as a nuclear state, than as a non-nuclear state. France is an example of this reasoning. 5) *Strategic Hedge Against Military and Political Uncertainty.* Uncertainty about the reliability of allies and the intentions and capabilities of adversaries may make nuclear weapons attractive. 6) *Possession of "A Weapon of Last Resort".* Nuclear weapons may be perceived by a state like Israel as offering an ultimate guarantee against extinction. 7) *Leverage Over the Industrialized Countries.* Certain developing countries may conclude that acquiring nuclear weapons is a means of compelling the developed nations to take more serious account of the interests of the less developed.'[1]

All of these factors are applicable to the South African minority government.

Sources close to the South African Government have told us that the ignominious defeat of the South African column in Angola, left without the expected American support, was a serious shock to the military planners and the South African Government. In particular, weapons such as the 122mm. ballistic missile, known as the Stalin Organ, had a devastating effect on South African troops who for the first time were faced with real, as opposed to imaginary, 'communists'.[2] Another shock to the inner circles of government came from

221

an opinion poll of young Afrikaners who, asked if they would stay and fight if the South African way of life were threatened, showed little enthusiasm for such a sacrifice.

Even with a defence budget far beyond that of most countries of a similar size, a budget which increases rapidly every year, and with the decision to extend conscription from one year to two,[3] South African policy-makers feel increasingly vulnerable in the face of refusals by the major Western countries to make an overt commitment to their survival.

The sheer destructive power of nuclear weapons might therefore seem to some to be the logical answer to the South Africans' dilemma. William Epstein, an expert in nuclear non-proliferation, has summed it up:

> 'The essence of the nuclear arms race is power – military, political, and economic. ... [some states] see nuclear weapons as promoting their security, enhancing their prestige, augmenting their influence, and improving their economic conditions.'[4]

This is echoed by *Die Volksblad*, a newspaper very close to the Government.

> 'The world respects the country that is militarily powerful ... in addition [a powerful defence force] is a useful instrument for the creation of good relations outside the country. Friendly approaches are usually made towards a country with a powerful military fist, because such a country can be a valuable market or a seller of armament and strategic material.'[5]

Applying this concept specifically to nuclear weapons *Die Beeld* observed:

> 'Mr Vorster has not yet said categorically that South Africa will never make an atom bomb. In view of this fact, people will have to look at us in a new light. South Africa now becomes an altogether different proposition if you want to tackle it. This bargaining power can be used in various fields in the difficult years that lie ahead.

America, for example, would have to revise its strategy toward us.'[6]

The special reference to the United States is revealing, and we will return to that later. Since 1969 and 1970, when these comments were made, the South African press has been remarkably quiet about the value of nuclear weapons as the ultimate deterrent. There has been no public discussion of a nuclear option. Planning on this – as indeed on all sensitive issues – is confined to an extremely small inner circle who make the key decisions. In any case, it is unlikely that there would be any serious opposition to the development of nuclear weapons; there has for decades been unanimous support among the whites for the Government's defence and foreign policy strategies. The conflict between 'verligtes' and 'verkramptes' (soft- and hard-liners), which has been much exaggerated, has never involved any questioning of the desirability of maximising South Africa's military strength, nor of South Africa's attitude to the NPT. It is not particularly surprising to see Visser's call in 1965 for a nuclear bomb to be aimed at 'loud-mouthed Afro-Asiatic' states repeated in 1977 by Dr Christian Barnard: 'If there was international intervention, whichever country that comes from, against South African whites, those countries should not forget that we have the atomic bomb.'[7] As the African National Congress of South Africa puts it: 'Driven to despair, the regime will not hesitate to launch mass murder on the African continent.'[8]

It is difficult and probably unprofitable to speculate about the reaction in the international community to a threat by South Africa to use nuclear weapons, or to their actual use against one of the neighbouring African countries. Even a hint that South Africa might use nuclear weapons would have repercussions far beyond the continent of Africa, of course. Chronic instability in southern Africa has already become a subject of international concern, and Dr Kissinger's frantic shuttle diplomacy there was a signal that the United States had realised its potential as a trigger for more generalised conflict. The area is by no means devoid of a military commitment by the nuclear-weapons powers: two examples are the French military bases off the coast of East Africa, and close Soviet ties, including a mutual defence 'Friendship Treaty' with Mozambique. There are always submarines armed with nuclear warheads from at least four of the five nuclear powers which are or could quickly come within range of southern Africa.

An appeal for support from African countries threatened with devastation by a nuclear attack or conventional bombing could be directed at either or both the Soviet Union and China, and there could even be competition between them as to which will offer the toughest response to a threatened South African attack. And as far as the United States is concerned, the reaction to any hint of 'communist' intervention in a crisis of this kind could depend on who is in control of the White House at the time. A conservative American President would be more likely to back the South African whites. And the Cuban missile crisis of 1961 has shown that an isolated incident can lead rapidly to nuclear confrontation.

African states are already reacting to growing evidence that South Africa is planning a nuclear-weapons capability. An international symposium, 'Perspectives in Afro-German Relations', held in Bonn in October 1975, was dominated by angry African reactions to West Germany's nuclear co-operation with South Africa which had just been exposed by the African National Congress in its publication, 'The Nuclear Conspiracy'. African delegates protested furiously that the German provision of nuclear technology to the South Africans could mean that their continent is on the verge of a nuclear arms race. The Nigerian delegate warned that economically more advanced African states might also try to acquire nuclear technology. His concern was echoed by diplomats from Tanzania, Sudan, Ghana and others.[9] Concern was also voiced by African representatives at South Africa's announcement of its own commercial enrichment plant, and the offers from American, German and French companies to build a nuclear reactor in South Africa.[10] Already Nigeria has ordered its own nuclear power capacity. As Mason Willrich, a professor of Law and long-standing expert on proliferation, put it:

> '. . . a strong South African nuclear power program . . . is likely to increase general nuclear incentives elsewhere in Africa. These incentives may be channeled in civilian directions initially, but some military undertones may also develop.'[11]

Or, as the American State Department has admitted, it may not be simply a matter of the relatively slow build-up of a national nuclear capability:

> 'A nuclear weapons capability on the part of South Africa would no doubt be seen as a security threat by neighbouring countries . . . were South Africa to develop such a capability, neighbouring countries could feel sufficiently threatened to seek greater security assistance from outside powers hostile to South Africa.'[12]

In other words, it would greatly increase the Soviet (or possibly Chinese) military involvement in the whole of Africa, something which is already a major preoccupation with American military planners.

It is precisely this polarisation, and the splitting-up of Africa into rival East-West camps, that South Africa could be aiming at with its nuclear programme; the remark by *Die Beeld* already noted could be the key to the whole strategy. In this, South Africa is not alone. Non-proliferation experts are in fact most worried about the category of countries in which South Africa is pre-eminent, which are in the western camp and have major conflicts with their neighbours yet, either because they seem expendable in terms of their location or because their extremely repressive policies make them unacceptable as allies, are not part of any formal alliance with the United States and cannot count on its military or nuclear

225

'umbrella' to protect them against external threats. As Epstein puts it:

> 'Those non-nuclear countries that are not under the nuclear umbrella of any of the nuclear powers and have no alternative means of ensuring their security feel that they may ultimately have to rely on nuclear weapons and in the meantime are developing nuclear weapon options.'[13]

The State Department's concern is reported to focus on the potential 'outlaw club' of countries long taken for granted as allies of the Western powers, but now seen as a 'potential league of the desperate'.[14] They include Israel, Brazil, Iran, Taiwan and South Korea (which received a shock analagous to that of South Africa over Angola, when the Americans abandoned Vietnam). It is generally assumed that, as US Treasury Secretary Michael Blumenthal commented:

> '. . . we are more likely to see nuclear weapons used in the future by a new or secret member of the nuclear club than by one of the two great nuclear superpowers.'[15]

SOUTH KOREA FIRST: SOUTH AFRICA NEXT?

Although barely noticed in most countries, it will not have escaped the notice of South Africa that South Korea has managed to force the Americans to extend the nuclear umbrella over their heads, precisely by threatening to deploy nuclear weapons: despite having signed the NPT, President Park Chung Hee warned that the country might have to develop its own arsenal of nuclear weapons if the American nuclear umbrella failed.[16] However, South Korea has now become one of the worst offenders in the abuse of human rights, about which President Carter seems so concerned. Strictly, it should also be one of the first countries affected by Carter's declaration that Americans are '. . . now free of that inordinate fear of Communism which once led us to embrace any dictator who joined us in our fear.'[17] It seems only a

226

matter of time before the American troops which have been in South Korea since the war there will be withdrawn. So the South Korean Government has continued working on nuclear weapons production. In May 1977 President Carter announced that South Korea would be protected by American military might – with the implication that it would be used 'if necessary' – after the withdrawal of the troops.[18]

The International Institute for Strategic Studies, in a recent analysis, points out that increasing Western, and particularly American, concern over nuclear proliferation:

> '. . . enhances the risk that for a number of countries [and it cites South Africa together with Israel, Pakistan, Iran and Turkey] a minimal nuclear option might become attractive, not for reasons of military security but as a lever to bargain for American concessions in non-nuclear fields, particularly over the supply of conventional arms.'[19]

The logic of current non-proliferation arguments actually points in the same direction as the most right-wing or pro-*apartheid* position: an alliance with South Africa. Mason Willrich suggests that 'In view of the emerging situation, the United States might benefit from building an interdependent relationship with South Africa in the nuclear power field.'[20] William Epstein's best suggestions for countering a country's nuclear ambitions are similar:

> 'The disincentives are largely potential, ranging from effective security guarantees through adequate supplies of conventional armaments to assurances concerning future supplies of fissile materials.'[21]

Such a response would encourage South African whites in their determination to make no compromise with the black majority within their own borders, and to make no concessions to international opinion and law with regard to Namibia and Rhodesia. They would be further encouraged to adopt a policy of 'pre-emptive strikes' against neighbouring countries suspected of helping the internal resistance to *apartheid*, secure in the knowledge that the Western alliance, with all its military might, lay behind them. This would give the independent African states no alternative to a close alliance with the Soviets or Chinese. For South Africa, its apparently uneconomic investment in a nuclear capability could reap enormous dividends.

7
Delivery Systems

According to Dr Frank Barnaby, the Director of SIPRI,

'. . . a new nuclear weapon power should today have little difficulty in producing an atomic bomb with a yield-to-weight ratio (a measure of the efficiency of a bomb) of about 20,000. For comparison, a yield-to-weight ratio of a conventional bomb is about 0.5. Such a weapon would weigh about 1,000 kg. A warhead with these characteristics could be transported by many delivery systems some of which are already in the arsenal of many near-nuclear countries.'[1]

South Africa not only has her missile industry, she also has an Air Force capable of carrying the bomb. The French Mirage III Fighter/Bomber has a maximum weapon-load of 4,000 kg, sufficient to carry four bombs. South Africa has ninety-five Mirage IIIs in its Air Force.

The SIPRI Year Book also pointed out that most civilian airlines have aircraft, the Boeing 707 for example, which are more sophisticated than the B-29 bomber which dropped the atomic bombs on Hiroshima and Nagasaki. These could be provided with the avionics needed to convert them into effective long-range bombers capable of delivering even very crude (and therefore heavy) atomic bombs.

The South African order of four Airbus aircraft jointly produced by West Germany and France, equipped with special, huge fuel-tanks used for refuelling jet fighters in the air, indicates that South Africa is well versed in converting civilian aircraft for military purposes.

This chapter looks at the development and growth of the South African Air Force and rocket systems. One of the most serious constraints on the nuclear ambitions of relatively small powers has been the availability of delivery systems with sufficient range and accuracy; but increasing numbers of countries are acquiring such systems in the form of missiles, fighter bombers or both. South Africa is well ahead of the field in this regard, and has a range of options. Most recently the acquisition of French-built Mirage F-1s has provided one of the most sophisticated airborne delivery systems available. These are complemented by ship-borne missiles provided by the Israelis and the French.

Not all missiles or aircraft of a particular range would be earmarked for nuclear weapons. They have other uses such as carrying conventional bombs or poison gas or other forms of chemical and biological warfare, in which South Africa has been interested.

South Africa has Canberra 12 bombers, easily convertible to carrying nuclear weapons, which have a 4,000-mile range. They were supplied by Britain in the early 1960s. There is also a squadron of Buccaneer light bombers, with medium-range capability, also supplied by Britain.[2] The Mirages, as we have already indicated, form the vanguard of its Air Force.[3] These are currently being replaced with 48 of the latest and deadliest of the Mirage range, the F-1 fighters. The initial batch of 16, delivered in 1974, are of the F1CZ all-weather interceptor type similar to those currently being delivered by the French Air Force in connection with France's own *force de frappe*. These were to be followed by 48 Mirage

F1AZ ground-attack fighters with Aida II radar range-finders.[4] The number has now been increased to 100.[5] The Mirage F-1 is being assembled under licence from Dassault, the manufacturer, by the South African Government's Atlas Aircraft Corporation. The first was scheduled for 1977-78.[6] 16 were received in 1975 directly from France, and another 32 are to be assembled as the 'Atlas Mirage'.[7]

The French *force de frappe* is based on a type of rocket-bomb carried by a supersonic bomber to within 'stand-off' launching distance of the target and then released to seek that target, driven at very high speed by its own rocket motors.[8] All the Mirages are equipped with French missiles, notably the AS-20 and AS-30 air-to-surface missile and the Matra R 530 air-to-air missile. The Mirage F-1 will carry the Matra 550 Magic short-range air-to-air missiles.[9] Jet fuel for the Mirages is to be supplied by France.[10] The arrangement of the aircraft in the South African Air Force is shown in Table 2. The four A-300 Airbus tanker-transport planes, mentioned earlier, have also been supplied by France to support the Mirage force, for delivery in 1977.[11]

Table 2: *The arrangement of Mirage aircraft in the South African Air Force*[12]

Squadron	Aircraft
3 Ground-attack fighter (FGA)	16 Mirage III EZ 14 Mirage IIIDZ
1 FGA	15 F-86 (being replaced by Mirage F1AZ)
1 Fighter/reconn.	27 Mirage III CZ/BZ/RZ
1 Interceptor	16 Mirage F1CZ
(32 Mirage F1AZ on order)	

The South African Government was probably given high priority as a customer for the Mirages, since Dassault has a long waiting-list for them.[13] As early as 1971, *Die Burger* had described the South African deal with Dassault for Mirage III and F-1 jets as a clear indication that 'South Africa and France have virtually entered into a military partnership'.[14] The announcement of the first deliveries of the Mirage F-1, in 1974, was made at about the same time as the reported deal for the French Government to supply South Africa with a nuclear reactor in part-exchange for South African and Namibian uranium.[15] The Government-controlled news-paper *Die Transvaler* has intimated that the uranium deal included the supply of French weapons despite prior commitments by the incoming Government led by Giscard D'Estaing to stop the sales of arms to repressive régimes.[16] In January 1977, soon after the signing of the final contract for Framatome reactors, President Giscard told an African audience that 'France will not sell any more arms to South Africa'. This was a few days after the news that the Atlas Aircraft Corporation was to assemble 100 Mirage F-1s, instead of the original 32.[17]

The South African guided missile capability is based on French as well as West German deliveries, licences and technical assistance. In April 1963 it was announced that a new National Institute for Rocket Research would be set up near Pretoria to develop guided missiles, with its own rocket-testing grounds. This was allegedly to give South Africa '. . . a foothold in space and weather research . . .'[18] – although there is no evidence of any particular South African contribution in these two fields. Dr Roux announced that South Africa had been forced by events in Africa to enter the missile field.[19] In the same year, Defence Minister Fouché told the South African House of Assembly: 'We need rockets. They will have to take the place of many other weapons which are fast becoming obsolete.'[20] In 1964, the new Institute for Rocket Research announced the development of a ground-to-air missile that could carry nuclear weapons or poison gas.[21] This, it turns out, is in fact the French-built Crotale missile, known in South Africa as the Cactus. It has a range of

13 km. and carries a 15 kg. warhead.[22] It has been reported as 'probable' that the Cactus missile could be developed for a rocket-propulsion system suitable for nuclear weapons delivery.[23] Backed by South African financing and participation, the French are now marketing the Crotale abroad, and have attempted to sell it to the US and other NATO countries as well as to Libya and elsewhere. South Africa currently has 18 Cactus/Crotale missiles in its Army.[24]

It was no doubt in connection with Crotale/Cactus that the head of the French concern Sodetag, Mr Maurice Belpomme, paid a visit to South Africa in 1968, much of it cloaked in secrecy. Sodetag is closely involved in the design, testing and building of nuclear warheads in France and has played a major role in the Sahara and Pacific atomic tests. It has also commissioned the building of missile test ranges and several satellite tracking stations – among them the Paardefontein French tracking station near Pretoria. Mr Belpomme said during his visit that he would have consultations with the South African Atomic Energy Board, the Munitions Productions Board, the Provincial Administration and the Council for Scientific and Industrial Research. He declined to say anything about discussions with defence authorities or his interest in South African missile development. 'You will understand that is strictly classified', he said.[25]

On 9 October 1968, Defence Minister Botha announced that a missile base was to be established on the coast of Zululand, about 150 miles north of Durban, in co-operation with an unnamed European company. The base and test site was to be on a 10-mile wide strip of land stretching from Cape Vidal in the south to Ochre Hill in the north, taking in a parallel belt of the ocean and also incorporating part of St Lucia Lake.[26] This area, although geographically a part of Zululand, has never been available for the KwaZulu 'homeland' or Bantustan – the collection of dozens of scattered pieces of land, all overcrowded and eroded, which the four million-strong Zulus are supposed to consider their territory.[27] Botha also said that the rocket-testing site was 'of great strategic value in that personnel could be stationed there permanently.' The base would serve scientific and industrial research in armament production organisations as well as the Army, Navy and Air Force.

On 17 December 1968 the first rocket was successfully fired from the new rocket-launching range at the base in St Lucia Bay. This was the anniversary of Blood River, the battle where the Boers had finally defeated and massacred the Zulu armies during the Great Trek. A restricted list of about 50 people were present at the scene.[28] In 1971 the Defence Department announced that the South African missile, using an automatic heat-seeking device, had been fired successfully from a Mirage fighter.[29]

In addition to the St Lucia Bay launching range – which is not far from the South African border with Mozambique – there is an operational military base much further to the north, Katima Mulilo, at the tip of the Caprivi strip of Namibia: this could also be used as a rocket-launching base. The strip juts out into Zambia, and is a thousand miles closer to the heart of Africa, within easy reach of Luanda, Lusaka and other African capitals, than are South Africa's own borders. The occupied territory of Namibia – where, according to the original League of Nations mandate, all military activity is banned – also has a new base called 'Drumpel', or 'Threshold', which could provide for a range of military operations directed mainly at internal opposition within Namibia, and operations against the new MPLA Government in Angola. The base is at Grootfontein, not far from the Namibian border with Angola.[30]

It has also been alleged that guided missile testing is carried out at the rocket research centre and ionosphere station near Tsumeb in Namibia, not far from Grootfontein. The station was established by the West German Institute for Aeronomics, Lindau am Harz, an establishment which is reported to have its projects financed by the West German Ministry of Defence.[31] There are two West German companies engaged in rocket production there, according to SIPRI: Waffen und

Luftrüstung AG (Armaments and Aerial Munitions Co.), and Herman Oberth-Gesellschaft of Bremen.[32] SIPRI also describes the Tsumeb centre as a direct result of the initial contacts between West Germany and South Africa over German participation in South Africa's nuclear research programme.[33]

The station at Tsumeb is run on the South African side by Dr Theo Schumann, an expert on 'meteorological rockets'. He also, however, has a seat on the Atomic Energy Board.[34] The South African Institute for Rocket Research is reported to carry out its own rocket-testing at the Tsumeb site.[35] Since 1973, there has also been a special centre set up to work on development up to the production stage of missiles and their warheads, propellants and propulsion systems. This is known as the Propulsion Division of the National Institute for Defence Research.[36] Overall, it is quite clear that the work on rockets and guided missiles, although frequently justified in terms of space or meteorological research, is very closely linked with South Africa's military establishment; and moreover that French and West German participation is by overtly military agencies and companies, with close links to their respective governments.

There have been some rumours that the performance of South Africa's Cactus missiles has been disappointing. It may be for this reason that South Africa bought British-built Tigercat missiles in a secret deal with Jordan in 1974, as an addition or back-up system for Cactus.[37] The Army currently has 54 Tigercat missiles.[38]

The newest developments seem to revolve around ship- or submarine-based missiles. South Africa has taken delivery of three Daphne-class submarines from France in the last few years.[39] An order for four more was announced at about the same time as the 1974 nuclear deal outlined above. It is not known exactly what, if any, guided missiles the submarines will carry. However, sources close to the French Government indicate that Thomson-CSF signed an agreement with South Africa for the delivery of some extremely sophisticated missiles, which they are currently developing for France's own ships and submarines. Some of the missiles, it appears, have hollow warheads suitable for poison gas or other substances. The deal was signed at the same time as Framatome won the contract for the Koeberg reactors (see Chapter 5).

Informed sources in Washington indicate that the development of the Franco-South African Cactus missile was carried out in parallel with the Franco-Israeli Gabriel, a seaborne missile with a range of 22 km and carrying a warhead of 160 kg – very much larger than the Cactus warhead.[40] There is now a new missile, the Gabriel II, with a greatly extended range, which is currently used on Israeli patrol-boats. Following Prime Minister Vorster's controversial visit to Israel in 1976, when overt contacts between the two countries were greatly expanded, it was announced that the IAI Gabriel II would be built under licence in South Africa[41] for use with six 'Ramata'-class patrol boats.[42] South Africa already has Gabriel mark I rockets, delivered by Israel in December 1974 to equip seven new ships acquired by the South African Navy at that time. In addition, Israel has been building two long-range gun boats, also equipped with Gabriel II missiles, and training 50 South African naval personnel to operate them.[43]

South Africa has also obtained two type A69 frigates recently from France; four corvettes are also under construction for them at the Bazan shipyard at Cartagena in Spain, using components supplied by Blohm and Voss of Hamburg. The corvettes are being equipped with Exocet rockets, using components from Messerschmitt-Bolkow of Munich.[44] It also appears, according to observers in Washington, that the United States is approaching a major policy decision on naval co-operation with South Africa.

It is no longer difficult to obtain nuclear weapons delivery systems, nor is it particularly expensive. As mentioned earlier, a Boeing 707 aircraft can easily be converted for this purpose.[45] For South Africa, therefore, the main question is not whether it has a nuclear delivery system, but to what point it can increase its range and accuracy. South Africa's acquisition of the latest missile-equipped Mirages and sub-marine- and ship-based missiles provides it not only with scope for reaching any target it might wish to threaten within Africa, but also with a considerable range beyond the shores of Africa.[46]

The combination, in the hands of the white minority in South Africa, of nuclear bombs *and* the latest delivery systems represents a grave threat to the whole of Africa. While a number of international observers tend to dismiss the possibility of the South Africans ever using nuclear weapons as an irrational move that would ultimately rebound on the whites, many of the politicians and military leaders of the Republic clearly see the issue in far simpler terms.

PART THREE
OTHER COUNTRIES

8

The United States of America: policy conflicts

Although the focus of this book is on the part played by West Germany in South Africa's nuclear programme, the role of the United States has been critical in the collaboration. In addition, America's own links with South Africa are of great significance in themselves.

The United States gave South Africa the possibility of setting up its nuclear research development capability in the first place, by developing a uranium mining and processing industry for the American post-war nuclear arms race. America then provided the research reactor, SAFARI-1, and trained large numbers of South Africans in nuclear technology across a wide spectrum – collaboration which is continuing to this day. America still delivers fuel for SAFARI-1 and, linked with the Kissinger initiative in southern Africa, has agreed to sell and provide credits and fuel for the proposed Koeberg power reactor. Two computers, vital elements in the enrichment plant, have also been exported to South Africa from the United States.

Attitudes within the Administration and Congress are extremely varied, ranging from acute distrust of South

238

Africa's intentions as regards an allegedly peaceful nuclear programme to warm friendship and co-operation with South Africa as a valuable nuclear partner. It is quite possible that, while the new Carter Administration takes a relatively tough stand against South Africa, highly secret collaboration in the development of South African nuclear weapons is under way. Understandings reached between Kissinger and Vorster, and American concern about containing 'communism' in Africa since the Angolan war, probably still have some weight in the thinking of some American defence-planners. This possibility is borne out by the top-secret change of policy towards raising the price of gold in order to help South Africa militarily.

In this chapter we also examine the opposing forces operating in Washington with regard to American policy on South Africa, and we note in particular the power of the South African propaganda system in Congress and the Administration, as well as the vested interests of the powerful American corporations in the nuclear field in circumventing the controls on exports of sensitive technology and hardware.

US assistance to South Africa's nuclear industry

Dr A. J. A. Roux, President of the South African Atomic Energy Board, recently paid a rare public tribute to the American role in developing South Africa's nuclear technology: 'We can ascribe our degree of advancement today in large measure to the training and assistance so willingly provided by the United States of America during the early years of our nuclear programme...' He noted that SAFARI-1 'is of American design' and that '... much of the nuclear equipment installed at Pelindaba is of American origin, while even our nuclear philosophy, although unmistakeably our own, owes much to the thinking of [American] nuclear scientists.'[1]

The basis for this American commitment to building up a

239

nuclear industry in South Africa was laid almost immediately after the Second World War, when the US started a crash programme of building up its own nuclear weapons stockpile. Apart from taking South Africa's total uranium output in the early years just after the war, the US, together with Britain, made a major contribution to perfecting South African techniques of extraction and processing; and they financed the entire cost of the programme. Altogether, the Americans bought approximately 43,000 tons of uranium for their nuclear weapons programme, at a cost of $1,000 million.[2]

The American and British nuclear authorities approached the South African Government, shortly after the end of the war, about the possibilities of extracting uranium from the Witwatersrand gold mines. A joint American-British team reported back that '. . . the Rand may be one of the largest low-grade uranium fields in the world.'[3] A joint research programme was set up immediately with South Africa; at the same time large quantities of uranium were being discovered elsewhere in South Africa. On 23 November 1950 the Combined Development Agency, a joint US and British uranium procurement organisation, entered into an agreement with the South African Atomic Energy Board to provide for large-scale uranium production on four mines (later extended to cover 27 mines). The first uranium processing plant in South Africa was opened in 1952. By 1956 the capital costs of building the uranium oxide production plants, as well as plants to produce the sulphuric acid used in the extraction process, had reached R66m.[4] All the capital requirements were met by loans raised in the US and Britain, as part of the 1950 agreement. The CDA was also, at that time, the sole customer for South African uranium oxide. The price was fixed for a ten-year period at a level above that applying to American and Canadian ores, making it possible for South Africa to repay the loans within that time.[5]

The CDA also built an additional electric power station in South Africa for its own operations, and the construction of 17 uranium extraction plants formed a major part of the country's industrial activity since building began in January, 1951. Between then and 1957 R140 million was invested in these and in 9 ancillary sulphuric acid plants.[6] This investment in South Africa's uranium mining and processing industry was of major importance in the establishment of a large industrial base, and in the expansion and consolidation of the gold mines, both of which have been fundamental to the development of South Africa's economic base as we know it today.

The shipments of the CDA continued on a large scale from 1953 to the mid-1960s when the stockpile requirement fell off and other markets for uranium oxide became available to South Africa. In the 1960s alone, the CDA purchases in South Africa amounted to about 25,000 tons of uranium oxide, representing 'an important part of US requirements' according to an ERDA representative, Nelson Sievering. He added, 'Although the purchasing program was concluded in 1967 and formal cooperation on uranium production has terminated, we still have periodic exchanges of visits by United States and South African experts in this area on an informal basis.'[7]

A proposal was made in 1965 for a multi-million dollar barter deal between the US, South Africa and what was then the Congo (now Zaire), involving surplus US agricultural products, South African uranium and Congolese industrial diamonds. The proposal, which was put forward initially by South Africa, was dropped when the US discovered that '. . . both the uranium and the diamonds are controlled by monopolies and . . . the barter contractors had to be ones that would be acceptable to South Africa and the Congo,' according to the US statement announcing withdrawal. The monopolies in question were the Anglo-American Corporation of South Africa, which controls gold and uranium production, and its associate company De Beers, which controls the diamond market.[8]

On a more serious level, American co-operation with South Africa has shifted from building up South Africa's uranium production capacity to providing the Republic with a nuclear industry of its own. The formal basis for this extends back twenty years to the signing of a US-South African Agreement for Cooperation Concerning Civil Uses of Atomic Energy. Myron Kratzer, formerly of the State Department, has stated that '. . . while the cooperation originally provided for in this and similar agreements [with other countries] was on a research scale, eventual cooperation in nuclear power was contemplated from the outset.'[9] The Agreement was originally to expire in 1977; amendments to expand its scope and duration were made in 1962, 1967 and 1974. The 1967 amendment renewed the provision for supply of weapons-grade uranium for the research reactor for another ten years, and specifically provided for cooperation including the export of power reactors. The final amendment in 1974 provided for the export of the enriched uranium necessary for 2,000 MWe of nuclear power capacity. It also amended the original agreement to provide for its application for fifty years from the date of signature in 1957; the formal programme of co-operation therefore extends to the year 2007.[10]

The initial agreement was signed with the main objective being construction of the SAFARI-1 research reactor. The major contractor was Allis Chalmers, and the reactor was based on the Oak Ridge (Tennessee) model. Eight other US organisations were involved: the Argonne, Brookhaven and Oak Ridge National Laboratories, Massachusetts Institute of Technology, Reno Research Center, Rochester University, the University of Illinois and New York University.[11] In effect, the entire range of US research and development on nuclear power was made available to the South Africans for SAFARI-1.

Although this project was subject to safeguards on the assumption that the reactor would be unsuitable for military use – provided regular checks were made on the fissionable material provided by the US – in fact the whole of South Africa's corps of nuclear scientists and engineers came about as a result of the training and practical experience that were provided in this deal. 'Expert sources' in the US are said to feel that the operation of SAFARI-1 was an essential element in the training of South African scientists who later developed the unsafeguarded enrichment process.[12]

SAFARI-1 is designed to run on highly-enriched, or weapons-grade uranium, and this has been supplied either directly or indirectly – through the British Atomic Energy Authority – by the United States. Material supplied by the UK was actually of US origin, fabricated into SAFARI-1 fuel elements at Harwell. Its subsequent transfer from the UK to South Africa was authorised by the United States.[13] A complete list of all shipments of highly enriched uranium for SAFARI-1 up to April 1975, is given in Table 3. Table 4, showing re-exports of spent fuel, indicates the role played by third parties such as Britain and France in US dealings with South Africa.

Table 3: *US-supplied uranium for SAFARI-1 research reactor*[14]

Date	Kgs. U	Kgs. U-235	Fabricator/Shipper
Feb 65	4.315	3.874	Babcock & Wilcox
May 65	3.446	3.096	Babcock & Wilcox
Oct 67	2.526	2.273	UKAEA, Harwell, UK
Nov 67	8.299	7.469	,, ,, ,,
Dec 67	2.541	2.286	,, ,, ,,
Dec 69	7.558	6.802	,, ,, ,,
June 70	5.767	5.190	,, ,, ,,
Dec 70	.975	.878	,, ,, ,,
July 71	13.334	12.001	,, ,, ,,
Sept 71	.663	.596	,, ,, ,,
May 72	.971	.874	,, ,, ,,
Dec 72	4.641	4.177	,, ,, ,,
June 73	2.139	1.925	,, ,, ,,
Aug 73	3.167	2.850	,, ,, ,,
Apr 73	.300	.270	,, ,, ,,
Oct 73	3.089	2.780	,, ,, ,,
Jan 74	5.217	4.696	,, ,, ,,

Apr 74	3.691	3.322	,,	,, ,,
May 74	3.092	2.780	,,	,, ,,
June 74	3.530	3.177	,,	,, ,,
Nov 74	4.299	4.006	US Nuclear, Inc.	
Dec 74	3.437	3.203	,,	,, ,,
Mar 75	3.444	3.209	,,	,, ,,
Apr 75	4.879	4.552	,,	,, ,,
TOTAL	95.320	86.288		

Less Returns	(34.251)	(28.154)
Less Burnup thru 12/31/73	(11.473)	(13.897)
Less Burnup 12/31/73 to 4/30/75 at 0.333 kg/mo.	(5.330)	(6.460)

Inventory	44.263 kg	37.777 kg

Table 4: *Re-export by South Africa of US-supplied fuel used in SAFARI-1*[15]

Shipments to UK

Kgs. U	Kgs. U-235	
3.039	2.465	
.019	.015	
11.055	9.183	
.421	.320	
(.213)	(.220)*	
4.989	4.134	
.457	.327	
.034	.041	
3.306	2.711	
6.293	5.233	
29.400	24.209	At least 11.300 kgs of this amount has been returned to the US

Shipments to France:

4.851	3.945	Returned to the US 7/30/74.	
34.251	28.154	TOTAL RETURNS	34.251

* Adjustment to reflect the difference between computed and measured burn-up.

Summarising the situation in June, 1976, Nelson Sievering of the US Energy Research and Development Administration (ERDA) said that a total of about 104 kilogrammes of the highly-enriched uranium had been provided to South Africa in the form of fuel elements, in which the uranium is alloyed with aluminium. Of this total, some 22 kilogrammes had been returned to the US, and about 13 kg. transferred to Britain for reprocessing with American consent and purchased by Britain. An estimated 21 kg. had been burned up during the operation of the reactor; about 5 kg. were in the reactor core; about 20 kg. were in irradiated fuel elements in the cooling tanks. The remainder was in fuel elements ready to be loaded into the reactor. Annual consumption of fuel by SAFARI-1 in its normal operation is about 13 kg.[16]

The amount of highly-enriched uranium already shipped to South Africa by 1975 would be enough for seven nuclear weapons, according to disarmament experts. With the 28 lbs (12.7 kg) still owed under the current contract, a further two weapons could be made. Despite strenuous objections, the licence authorising the shipments has been amended several times to extend the expiry date,[17] although court action by the Congressional Black Caucus and others has held up deliveries since 1975.

A major expansion of American involvement in South Africa's nuclear industry was proposed in 1976, in the form of two General Electric boiling-water reactors for the Koeberg site. GE, in partnership with Brown Boveri of Switzerland and Benucon of the Netherlands, had submitted a bid when South Africa's Electricity Supply Commission (ESCOM) sent out invitations to tender, in 1974.[18] ESCOM had requested G E and the other bidders to advise on whether they could assure the regular supply of reactor components and fuel if the contract were awarded to them. The response – obviously prepared in consultation with friendly agencies of the US Government – was that nuclear export licences were now issued by the Nuclear Regulatory Commission, an independent regulatory agency, but that there was a long-standing US policy of 'serving as a reliable supplier' by fulfilling supply contracts undertaken by US companies.[19]

G E, which had apparently by that time received a letter of intent to purchase from ESCOM,[20] filed an application on 10 May 1976, requesting an export licence for two boiling-water reactors. The value of the exports was stated in the application from G E as $200 million, and included the reactor coolant pressure boundary, instrumentation, fuel-handling equipment, spare components, equipment and tools, liquid control-system and clean-up filtering-systems.[21] Sources inside the NRC said at that point that the licences would probably be approved.[22]

G E had already received a 'preliminary commitment' from the Export-Import Bank to guarantee financing for this project, issued in 1974. The preliminary commitment determines whether the Bank is likely to provide financing support for a potential export sale, and sets forth the terms and conditions on which it would be willing to consider an application for a direct loan or a financial guarantee. The total amount of private financing to be guaranteed by the Bank was to be approximately $256.5 million, covering the enriched fuel as well as the reactor itself. Repayment would be made in 24 semi-annual instalments beginning no later than 1983 and 1984, for Units 1 and 2 respectively.[23]

When opposition to the deal surfaced in the Netherlands and the US, ESCOM withdrew its letter of intent to G E and switched to the Framatome consortium.[24] US companies may have had the last laugh, however. Framatome markets a 925

MWe pressurised water reactor designed by Westinghouse; licence fees are paid for it to Westinghouse, which also has a 15% stake in Framatome. This reduced shareholding – down from 45% – was part of a deal with the French Government to obtain uranium for its outstanding contracts in the US.[25]

In August 1974 ESCOM concluded a long-term, fixed-commitment contract for enrichment services for 1.4 million pounds of uranium between 1981 and 1984 with the US Atomic Energy Commission; this followed the latest amendment to the 1957 Agreement for Cooperation which was made in 1974 specifically to provide for US participation in the Koeberg project. The South African contract was evidently a last-minute affair, since according to State Department sources it was the last one before the deadline set by the AEC for the conclusion of enrichment contracts. ESCOM has already made two advance payments required under the terms of the contract totalling more than $2 million.[26] The total cost of enrichment and fabrication of the initial fuel cores is estimated at about $60 million. This, like the proposed reactor sales, is covered by the Export-Import Bank's preliminary commitment to guarantee private financing. Repayment is set in 10 half-yearly instalments beginning in 1983.[27]

The South African announcement of the contract with Framatome indicated that actual fuel fabrication would be in Franco-Belgian hands.[28] However, the contract for US enrichment services still stands, and a GE application for an export licence, filed on 4 May 1976, is still pending at the time of writing, in the face of a legal challenge from members of Congress and others.[29] American and South African sources have been quoted as saying that this commitment to provide for South Africa's uranium enrichment requirements up to 1984 carries South Africa up to the time when its own commercial enrichment plant is scheduled to be in operation.[30]

The contract for large-scale enrichment of uranium for Koeberg follows a wide range of supplies of special nuclear materials to South Africa over the years. Table 5 shows the

247

deliveries of low-enriched uranium in the form of fuel elements manufactured by Nukem in West Germany from American uranium hexafluoride, provided for the South African-built critical assembly in 1966-67. Table 6 shows the export licences for special nuclear materials which were approved for sale to South Africa.

Table 5: *US-supplied fuel for the critical assembly, National Research Centre, Pelindaba*[31]

Shipment No.	Date	Kgs. U	Kgs. U-235
1	Aug 66	64.984	1.294
2	Sept 66	65.586	1.307
3	Sept 66	66.682	1.328
4	Sept 66	67.179	1.343
5	Nov 66	67.989	1.355
'6	Nov 66	67.863	1.352
7	Dec 66	67.941	1.353
8	Dec 66	67.989	1.355
9	Dec 66	59.371	1.187
10	Dec 66	4.416	.088
Adjustment	–	.088	.002
11	Mar 67	6.675	.133
TOTAL		606.763	12.097
Less: Transfers Out		(605.532)	(12.062)*
Less: Consumption		(.928)	(.012)
BALANCE:		.303	.023**

* Transferred to UK for reprocessing. 602.522 Kgs U and 11.954 Kgs U-235 were returned to US on September 8, 1971.
** Purchased by South Africa as unaccounted losses.

In terms of hardware, one of the most important US exports for the South African nuclear establishment was the two industrial-process FOX 1 computers and spare parts sold by the Foxboro Corporation for the enrichment plant and licensed for export by the US Department of Commerce. The Foxboro catalogue describes the equipment as 'the most powerful sensor-based computer system available'. Modifications were made at South Africa's request to increase the memory capacity from 32,000 bits to 48,000 bits, and the computers were specially prepared for installation in the pilot enrichment plant. A team of engineers from Foxboro carried out the installation. The company also sold the South Africans the standard computer-operating software.[33] The computers are a crucial element in the operation of an enrichment plant, and it is quite possible that South Africa would not have been able to find suitable computers if the US Government had not authorised this sale.

Perhaps even more important than the hardware and special nuclear materials, however, is the training and continued programme of technical assistance which the US has provided for the South Africans, and without which the Republic of South Africa would have no nuclear capability at all. Between 1955 and 1974 the US Atomic Energy Commission trained 88 South Africans in various aspects of nuclear technology.[34] In addition, there have been frequent visits in both directions for training and orientation purposes, with several US scientists training South Africans in their own installations. A staff member of the Oak Ridge National Laboratory, Tom Cole, was loaned to Pelindaba as a consultant for the SAFARI-1 reactor.[35] Among the special training opportunities made available to the South Africans were classes in plutonium reprocessing — which were finally closed to foreigners only in 1972. Government documents show that probably over a dozen South Africans attended these classes.[36]

In addition to these official, intergovernmental exchanges, a variety of unofficial contracts are officially permitted.

Table 6: *US export licences for sales of special nuclear materials to South Africa*[32]

Licensee	Licence No.	Material	Maximum Enrichment %	Purpose or Use	Date Licence Issued
SPECIAL NUCLEAR MATERIALS					
1. Major Quantities					
US Nuclear, Inc. Oak Ridge, Tenn.	XSNM-508	12.5Kg uranium 11.7Kg U-235	93.3	Fuel, SAFARI-1	01/22/74
	XSNM-508 Amend. 1	12.5Kg uranium 11.7Kg U-235	93.3	Fuel, SAFARI-1	10/02/74
Note: Only 4.3Kg U containing 4Kg U-235 were shipped under XSNM-508 and its amendment prior to expiration of the Licence on 1/1/75					
US Nuclear, Inc. Oak Ridge, Tenn.	XSNM-690	20.8Kg uranium 19.3Kg U-235	93.3	Fuel, SAFARI-1	01/06/75
	XSNM-690 Amend. 1	20.8Kg uranium 19.3Kg U-235	93.3	To extend exp. date to 8/1/76	01/30/76
2. Minor Quantities					
Gulf Oil Corp. San Diego, Calif.	XSNM-193	1.26 grams uranium-235	93	Fission Chamber neutron monitor	05/04/71
Gulf Oil Corp. San Diego, Calif.	XSNM-257	1.26 grams uranium-235	93	Fission Chamber neutron monitor	01/14/72
Gulf Oil Corp. San Diego, Calif.	XSNM-342	2.558 grams uranium-235	93	Two fission chambers	09/22/72
US Steel Corp.	XSNM-583	1.76mg plutonium-238 20mCi iron-55 3mCi cadmium-109		Fluorescent Analyzer	09/09/74

Licensee	Licence No.	Material	Maximum Enrichment %	Purpose or Use	Date Licence Issued
US Steel Corp.	XSNM-714	1.76mg plutonium-238 20mCi iron-55 3mCi cadmium-109		Fluorescent Analyzer	04/16/75
Texas Nuclear Corp.	XSNM-795	.703gm plutonium-238		Fluorescent Analyzer	07/30/75
Gulf General Atomic	XSNM-83	1.25 gram uranium-235		Fission Chamber neutron monitor	04/09/70
SOURCE MATERIAL					
Kerr-McGee	STE-8145	96.7 lbs. thorium		Chemicals	10/29/73
	STE-8145 Amend. 1	add: 96.7 lbs. thorium			
Zirconium Corp. of America	STE-7998	3 lbs. thorium		High temp. Metallurgical application	11/30/71
Picker International Corp.	SME-7363	14.5 lbs. depleted uranium 1710 Ci cobalt 60		Teletherapy unit w/shielding	05/23/66
BY-PRODUCT MATERIALS					
Beckman Instruments	XB-4-110	carbon 14, cesium 137, chlorine 36, strontium 90		Instruments	11/30/71
Vernon Craggs	X-19-715	20mCi strontium 90		Cigarette gauge	10/07/74
Vernon Craggs	X-19-717	20mCi strontium 90		Cigarette gauge	10/21/74
UTILIZATION FACILITY					
Allis-Chalmers Mfg. Co.	XR-42 Doc. No. 50-179	ORR-Tank type, light Water cooled and moderated Materials Test Reactor		SAFARI-1 Reactor Located at Pelindaba, near Pretoria	06/14/61

Several South Africans attend US universities to study nuclear-related subjects, and in 1972, the South African Government wrote to all South African graduate students in the US, regardless of whether they were officially sponsored by the government, asking them to apply for jobs in the nuclear programme 'in all fields'. Columbia University passed on the names of their South African students to the South African Government.[37]

A number of American scientists have also been recruited. One of the most noteworthy is Dr Sverre Kongelbeck, one of the key US nuclear missile scientists and developer of the Mark II, the world's first fully-automatic guided missile launcher, currently installed on scores of American Navy warships. Kongelbeck visited South Africa in 1971 to seek employment after his retirement as chief engineer of the US Navy's main missile laboratory. He seems to have had a warm welcome, and told the local press: 'I believe I could help South Africa in the field of missiles, radar and satellites...It is God's own country. I'm not bothered about the racial situation.'[38]

In addition to training, the US has shown a particular interest in joint projects with the South Africans — which are important to the latter because they may have a training component built in, and they also contribute greatly to the standing of South Africa in the eyes of the US officials and scientists concerned with the nuclear programme. A forerunner of joint projects dates back as early as 1929, when Namibia (then South West Africa) was chosen by Dr G.C. Abbott of the Smithsonian Institution as a site for the Solar Radiation Observatory, and work continued there for some years.[39]

A later counterpart to this approach was an experiment to detect high-energy neutrinos, financed and organised by the US Atomic Energy Commission in conjunction with the University of Witwatersrand.[40] Even more significant is the collaboration over weapons testing. In 1958, the US conducted nuclear weapons tests in the Atlantic Ocean off Cape Town, and the effect on the ionosphere was measured in a co-operative project with the South Africans. This was done under the cover of International Geophysical Year, in which South Africa was a major participant [41] (see Chapter 5). South Africa also worked 'in intimate collaboration' with the British Atomic Research Institute at Harwell (obviously with the full knowledge and backing of the US) in monitoring radioactivity resulting from the series of nuclear weapons tests carried out in the Pacific Ocean in the mid-1960s by France.[42]

ATTITUDES IN WASHINGTON TO SOUTH AFRICA'S NUCLEAR AMBITIONS

The US Atomic Energy Commission (AEC) has worked with South Africa to develop its nuclear programme since 1957, and has provided twenty years of continuing help and encouragement. This arose in the first instance because of South Africa's importance as a source of uranium for the nuclear arsenal. Myron Kratzer, who dealt with nuclear technology at the State Department, explained the relationship to the Senate Foreign Relations Committee:

> 'Beginning in 1953, South Africa became an important supplier of uranium to the US for defence purposes, a role South Africa continued to play until defense requirements declined in the early 1960s. It was therefore natural that South Africa should be among the countries with which co-operation in this field would be established.'[43]

State Department officials have also indicated that this previous co-operation for defence purposes is behind the current deals, including the deliveries of enriched uranium and negotiations to sell GE reactors.[44] Since the whole basis for the relationship was orginally military, it is hardly surprising that the US defence establishment should be

253

favourably disposed towards South Africa in the nuclear field, and could explain to some extent the reluctance of US Government officials to recognise South Africa as a proliferation threat until very recently.

The good relations go far beyond the strictly military, however. One of the areas in South Africa least affected by international ostracism is the whole field of science and technology, where there are close and fruitful relations with leading research centres all over the Western world, and particularly with the United States. US Government research organisations are involved in this, including those concerned with nuclear research and development. Even where the co-operation is on an individual or private basis, US Government funds are often to be found supporting the work.[45]

Sources close to the Arms Control and Disarmament Agency of the State Department, which was set up partly as the watch-dog for nuclear proliferation, have admitted that until very recently, South Africa was held in high regard there. Warm and uncritical attitudes towards South Africa's nuclear programme were expressed by Charles Von Doren, the main author of the Nuclear Non-Proliferation Treaty (NPT) and currently a key official in the Non-Proliferation and Advanced Technology Bureau of the Arms Control and Disarmament Agency (ACDA).

James Blake, a State Department official in the African Bureau, later told the House Sub-committee on Africa:

> 'The Department of State believes that the willingness of the United States to continue to act under these bilateral cooperative agreements [with South Africa] will have a direct bearing on the receptiveness of other governments toward our views on nuclear proliferation.'[46]

Myron Kratzer enlarged on this theme in defence of the decision to finance the sale of the two GE reactors: 'There is no question in my mind that our action as a supplier in any country has an effect not only in that country itself but on our reputation elsewhere in the world.'[47] The fact that co-operation with South Africa has been condemned by the majority of other countries seems not to have entered into the calculation. This may be because the more basic reason for assistance has far more to do with US policy towards South Africa itself than concern for the US 'reputation' elsewhere. Kratzer explained the tactic as one of trying to kill by kindness:

> '. . .while I agree with you that the fact that South Africa is not a country that has ratified the treaty is unfortunate, we feel that the best prospect for bringing that about is to continue the process of co-operation and that if we can continue the dialogue — and we do have one with them — that it may lead to their becoming parties to the treaty.'[48]

The State Department's obsession with 'dialogue' under Kissinger is notable because it was substituted for all normal means of diplomatic pressure on a country whose policies create problems for the US. In fact, the concept of 'dialogue' is one which seems to have been specially introduced into the diplomatic vocabulary to cover the enormous range of assistance made available to the South African Government by the US and other Western countries, assistance which lies behind the present dilemma with regard to South Africa's position as a major threat to nuclear non-proliferation policies.

A nuclear carrot for South Africa

The Kissinger initiative in southern Africa in 1976 was readily compatible with close and friendly relations between the two countries in the nuclear field. In a series of meetings with South African and Rhodesian representatives, Henry Kissinger seems to have employed his customary 'carrot and stick' approach to solving the problems of Namibia and

Rhodesia — although the talks with South Africa involved more carrot than stick. In return for South African promises to put pressure on the Smith regime in Rhodesia, Kissinger offered a variety of concessions involving relaxation of pressure on South Africa's racist policies and measures to raise the price of gold. After the Carter administration took over, it was revealed for the first time that Kissinger had also reached an 'understanding' with South Africa over Namibia.[49] A well-informed observer has reported that the proposals made on Namibia in September 1976 at the Kissinger-Vorster meeting in Zurich included a public guarantee of South Africa's own frontiers by the United States, backing for a new Black Namibian armed force, trained, equipped and financed by the US for at least a decade, and a massive 'aid' programme.[50]

The National Security Council of the US was reported in May 1976 to be preparing urgently a series of recommendations for closer American support for South African Government policy, and easier economic relations. It was also reported at the time that a key indicator of any further softening of US policy on South Africa would be the response to the General Electric request, filed at just that moment, for financial guarantees for the export of nuclear reactors.[51] This was not a new departure for Kissinger: a major row was sparked off earlier by his attempt to gain concessions in the Middle East by offering to finance and construct nuclear reactors for both Egypt and Israel.

A number of official statements underlined the link between the Kissinger deals with South Africa and the question of nuclear co-operation. One State Department official is quoted as arguing, in justification of the Department's support for G E's application, ' . . .if you expect South Africa to play a role in helping to bring about negotiations in Rhodesia, you can't treat the South Africans as outcasts.[52] Soon afterwards Myron Kratzer told the Senate Foreign Relations Committee that the G E deal would help Washington establish a 'more intense dialogue' with South Africa over ways of achieving change in Rhodesia.[53] James Blake, another senior State Department official in charge of African affairs, reinforced the argument with references to the deal as an incentive to 'dialogue' on South Africa's part, one of Kissinger's major objectives, and added:

> '. . . at a time when South African cooperation and understanding will be important to us . . . we cannot forget the importance of certain things to South Africa itself. This transaction is obviously of importance to the South African Government.'

When pressed for clarification, he finally conceded that '. . .we would hope it would be significant enough in South Africa so that it would be a sort of piecemeal approach to our concerns [in Namibia]'.[54]

The full story of Kissinger's deals with the South Africans has never been divulged. It cannot be ruled out that alongside what South Africa's Prime Minister John Vorster called Kissinger's 'superficial platitudes' in favour of majority rule in southern Africa,[55] some secret plans were laid which might result in the continued domination by the minority. This would be fully compatible with the 'Nixon Doctrine' (which has survived Nixon) under which the United States will no longer send its own troops to defend threatened allies, as in Vietnam, but will provide all the means necessary for them to do the job themselves — which involves a massive build-up of the latest and deadliest military hardware in unstable regions.

Angola seems to have been a turning-point as regards American policy on southern Africa, and it is widely assumed that South Africa invaded Angola in the war of 1975-76 with the encouragement and promises of support from the United States.[56] The Administration indeed tried very hard to obtain funds for supporting the anti-communists in Angola, but was unexpectedly held back, mainly because of painful memories in Congress of a similar commitment in Vietnam. Behind the scenes, however, a sense of obligation to the South Africans remained among American military planners.

As the South African Foreign Minister, 'Pik' Botha, explained: 'South Africa was left in the lurch by people who created other expectations in people who wanted to support the Free World in Angola.' Mr Botha added that one advantage of the South African intervention was that it had shown Africa that South Africa could do on a large scale what she was prepared and able to do on a small scale. 'We made friends who are still our friends today despite what is said in public at international meetings.'[57]

A well-placed source in Washington has let it be known that the US has deliberately permitted South Africa to acquire a nuclear weapons capability, defending this decision by saying that the Administration in Washington would have six to eight hours' advance warning of any plans to use such weapons, and could take steps to prevent this.

On the face of it, this admission is incompatible with the official expressions of concern about nuclear proliferation which emanate from the White House and the State Department. However, the phenomenon of independent and inconsistent policies toward South African being pursued by different agencies of the American Government — or even different departments within one agency — is to be found also in connection with export policies, the arms embargo, scientific and industrial collaboration and many other important areas. The fact that the Arms Control and Disarmament Agency may be trying to stop South Africa getting its own nuclear weapons while certain sections of the military establishment are secretly encouraging these ambitions is quite credible in this context. To the extent that American intelligence and military work is carried out in collaboration with the Israelis, too, many of the government departments not directly involved could be excluded from the 'need to know' list.

Has the US deliberately encouraged South Africa's nuclear ambitions? The record of full technical co-operation, plentiful supplies of equipment and material, the training of South Africa's nuclear scientists and co-operation on highly senstive

projects related to nuclear weapons testing all indicate that
the US Government, through at least some of its agencies, has
indeed been pursuing a policy of total collaboration in the
nuclear sphere, military as well as civilian.

The realization in Washington that there can never be any
question of using US armed forces in a racial conflict in
southern Africa — because of the importance of blacks in
these forces as well as the likely provocation of riots in US
cities — is compatible with a decision to provide South Africa
with the ultimate deterrent. There are indications that top
military planners arrived at this conclusion before Kissinger
started on his 'shuttle diplomacy' in southern Africa in 1976.
With the arrival of a Carter administration heavily indebted
to black voters, the use of US forces to protect white South
Africa directly has become even more unthinkable. As
Andrew Young himself, a personal friend and political ally of
President Carter as well as US Ambassador to the United
Nations, put it:

> 'I see no situation in which we would have to come in on
> the side of the South Africans . . .You'd have civil war at
> home . . . An armed force that is 30% black isn't going to
> fight on the side of the South Africans.'[58]

The argument could well be that South Africa, without
enough whites to sustain a major war without the whole
economy grinding to a halt, and without the prospect of outside intervention, would need nuclear weapons in order to
survive. Insofar as the survival of white supremacy in South
Africa is seen in certain quarters of Washington as being in
US interests, both strategically and in terms of corporate
investment, this would appear to be the obvious solution. It is
possible, then, that charges of a major conspiracy to give
South Africa a nuclear capacity, in a deal with West Germany,
Israel and Portugal — with the endorsement of the United
States — may have at least an element of truth in them. East
German and Chinese sources have made serious and quite

detailed allegations, apparently based on intelligence reports, about joint development by these countries of nuclear, biological and chemical weapons of mass destruction, using American technology.[59] The announcement in 1977 by the Soviets, confirmed by the French and Americans, of a South African nuclear test site in the Kalahari Desert is instructive not only in terms of the information conveyed, but also of the continuing and intense interest of the Soviet intelligence agencies in South Africa's nuclear ambitions, and particularly the American connection.

On the surface, American policy towards South Africa was fundamentally changed after Jimmy Carter took over the White House. However, it is too often forgotten that many of the most important decisions are taken not in the White House or the State Department, but in the Pentagon, the Central Intelligence Agency, the Treasury and the Federal Reserve Bank. Even aside from the direct or indirect assistance provided by the Americans in terms of nuclear technology and hardware, there is the crucial question of American financing for the programme, past and future: the Koeberg nuclear power station, the huge commercial enrichment plant and the whole research and development network behind these. As already outlined in Chapter 5, the South African economy in its present state of economic recession and with the difficulty in getting overseas loans for capital expenditure could not support alone such massive expenditures. Since official American policy has been to put pressure on the minority régime, and since there has been pressure from African states for economic sanctions, any *open* subsidies would be out of the question.

It is quite clear, however, that even under President Carter there is a strong tendency to cultivate South Africa, even in its nuclear ambitions, rather than put meaningful pressure on its minority régime. A report in July 1977 cited State Department officials as saying there was a widely held view that ' . . . to isolate South Africa when it is a major exporter of uranium is very short-sighted', and that ' . . . the non-proliferation concept gets nowhere when you isolate the country with 10-20% of the world's uranium reserves.'[60] This view is reflected in official statements at the highest level: thus Andrew Young, American Ambassador to the United Nations, told an interviewer that to end nuclear co-operation '. . .would only encourage separate development of South Africa's own nuclear potential,' and that: '. . .I think by maintaining some kind of relationship we do have the possibility of influencing them to sign the nuclear non-proliferation treaty and accepting all of the safeguards that go with the International Atomic Energy Agency.'[61] Zbigniew Brzezinski, the President's national security affairs adviser, explained that the Carter Administration had a deliberate policy of refraining from putting excessive pressure on South Africa: '. . .we feel that too many sanctions, on too grand a scale, could be counterproductive.' The attempt to get South Africa to stop supporting the illegal régime in Rhodesia, he argued, was:

> '. . .one of the reasons why we feel that our response [to massive arrests and bannings of opposition groups] sought to focus on the specific events. . .rather than to go wholesale at the generic causes of the problem.'[62]

SECRET SUBSIDIES

The authors have been shown a secret document, circulating in the most exclusive international financial circles, outlining a change of American policy on the price of gold. As a result of pressure from the Pentagon, the Americans were to take steps to raise the price of gold in order to provide a discreet but vitally important subsidy to the South Africans. Both the South African Government and the gold-mining companies stand to gain immediate benefits worth millions of dollars; it is no accident that these two are in fact the major investors in the uranium and nuclear industry.

261

Since we saw this document, in the middle of 1977, the price of gold has indeed been rising steadily, helped by a number of discreet but significant moves by the United States authorities. For example, international business contracts governed by US regulations can now for the first time include a clause allowing for valuation in gold in case of sudden devaluation of the dollar — the first step in remonetising gold after years of American policy aimed at removing its monetary role. And after the price had risen to the highest level for two years, obviously helped by a filtering through of news about the new American policy, the *Economist* reported that '. . .the American treasury was not tempted to sell when the gold price reached $160, so that bogey has been quietly forgotten.'[63]

In fact, the American sales of gold reserves from Fort Knox, established under the Nixon Administration, had been a serious influence in depressing the speculative gold price and furthering the trend to its demonetisation. R.W. Johnson has argued that this was aimed quite deliberately at South Africa.[64] The policy is now being quietly reversed with the complete cessation of all American gold auctions, in spite of pressure for more sales from Congressman Henry Reuss, Chairman of the Joint Economic Committee's Subcommittee on International Exchange and Payments. A confidential review of gold prices in 1977 by the International Monetary Fund, which has been made available to the authors, comments:

> 'On the supply side, diminished expectations concerning the likelihood of further official gold sales by the United States helped to support bullion values.'[65]

Officials in the State Department, when challenged on this policy which effectively props up the ailing South African gold-mining industry, reply defensively both that there has been no real change of policy, and — at the same time — that there was a need to protect certain countries with large gold reserves, the one cited being Portugal. The argument lacks credibility as well as consistency. The United States has never shown much interest in supporting the Portuguese Government since the overthrow of the right-wing Salazar régime. Such weak and contradictory arguments in support of a radical change in Administration policy on gold — which has important repercussions on the strength of the dollar — can hardly be taken seriously.

American manipulation of the world gold price to influence South Africa is actually a long-standing although little-known practice; but ironically, under previous American Administrations ostensibly more favourably disposed towards South Africa than President Carter's, the influence was more often used to drive the price down. As South Africa is by far the biggest gold producer in the Western world, and still heavily dependent on the metal to underwrite its whole economic development, her fortunes are closely tied to the international gold price. While trying to force South Africa to put effective pressure on the illegal regime in Rhodesia, the US Treasury coolly manipulated the price of gold and the international financial market. Johnson describes how the US Government also issued false estimates of the Soviet grain harvest (closely linked to Soviet sales of their own gold), which had the effect of driving the world gold price strongly downwards.[66] It is perhaps no accident that reports were circulating in financial circles about excellent Soviet harvests in 1977, strongly supporting the gold price rise, although the Soviet leadership itself subsequently expressed dismay about agricultural output.

As if the influence of the United States Treasury alone were not enough, the Americans seem to be using the International Monetary Fund (IMF), in which they have great influence, to subsidise South Africa. The IMF has changed its rules to allow central banks to buy gold at a price several times the previous official monetary price of gold ($35 per ounce, rising to $42.22 before the two-tier system was abandoned). South Africa was allowed to effect a new 'swap' arrangement, in 1976 and again in 1977, selling large amounts of gold (the second deal involving about one quarter of official South African gold holdings) on a spot basis from its reserves for repurchase forward.[67] At the time the second deal was made, about two-thirds of South Africa's official gold reserves must have been disposed of in this way, probably yielding a total of over $1 billion. The official reason for the second swap, as put forward by Governor De Jongh of the South African Reserve Bank, was '. . .to make advance provision for any possible adverse effect on foreign capital movements of the anticipated seasonal increase during May and June in net bank credit to the government sector.'[68] The idea of mortgaging most of the country's gold reserves for such an apparently minor reason seems unconvincing. It is true that South Africa has to pay off heavy interest on its existing foreign loans, and some of the proceeds from the swap may have been used for this purpose, although this amounts to raising a loan to cover interest payments on other loans, a process which cannot continue indefinitely. Such a serious undertaking, however, seems to us to be motivated primarily by some very heavy and unusual expenditure of top priority to the South African Government. Since construction of the commercial enrichment plant is under way, involving very heavy costs without, apparently, any credit having been raised specifically for the enrichment plant, this deal may well be intended to help pay for the enrichment gamble.

Essentially, according to the confidential IMF documents:

> '. . .the swap is similar to a loan collateralized by gold, for it allowed the Reserve Bank to increase the liquidity of its reserves without adversely affecting the price of gold through an increase in supplies on the private market. In this light when the announcement of the transaction was made on May 2 [1977], the market responded positively, bidding prices up, as the swap was interpreted to mean that the amount of gold coming onto the market would probably be less than otherwise.'[69]

Thus, South Africa is allowed by the IMF (with heavy

backing from the United States) to have its cake and eat it too: the swap arrangement provides in effect an enormous loan of $1 billion, while helping to drive the gold price further up.

Still more decisions have been made which are helping South Africa to raise money in spite of its poor international credit rating. Still in relation to the international gold price, South Africa benefited directly from the reduction in the amounts of gold being put on the market by the IMF in its series of gold auctions, whose function is to help the demonetisation of gold and at the same time use IMF gold stocks to raise money for the Trust Fund for the direct benefit of developing countries. At its ninth gold auction on 4 May 1977, shortly after President Carter had taken over from Gerald Ford, only 524,800 ounces of fine gold were available, less than half the offering of 1,278,000 ounces at the previous auction, despite the fact that investor interest was considerable and bids had been received for more than twice the amount on offer, 1,316,400 ounces.[70] Obviously, with a smaller amount offered, the lowest bid accepted was considerably higher than it would otherwise have been. The winner was South Africa, the losers the developing countries which had only just over half the proceeds they could have had from this particular auction. The system of bidding had also been changed: whereas previously all purchasers paid the lowest acceptable bid, they now paid whatever amount they had named, making prices higher.

It is ironical therefore that one of the current Administration's major projects should be the establishment of a 'Witteveen Facility' in the IMF for supplementary financing of deficit countries, with direct contributions including $1.7 billion from the United States. It is interesting that South Africa is excluded from participation in the Trust Fund built up by the IMF's gold sales, which is explicitly for developing countries, but will be eligible for the Witteveen facility which is ostensibly designed to supplement it.

Another, perhaps even more important, set of decisions was taken in 1976 and 1977, and involved the provision by the

IMF of a huge loan to South Africa which now totals $464 million: this makes the Republic the biggest borrower of IMF funds after Great Britain and Mexico, and is more than twice the total loaned to all other African countries combined over the same two years. The United States, because of its important position in the Fund, provided $107 million of the total lent to South Africa.[71]

Secret minutes of the IMF's Executive Board Meeting on 21 January and 10 November 1976 indicate a number of interesting insights into American, British and IMF staff support for the South African cause. The only doubt raised by delegates from industrialised countries at the January meeting involved whether South Africa had caused its own economic problems by deliberately withholding gold from the international market to force prices up. Peter J. Bull, representing the United Kingdom, declared that the standby arrangement for South Africa would give the authorities 'some feeling of international support, which they deserved'. He 'fully supported the proposed decision'.[72]

Of particular interest is the role played by the IMF staff; a proposed decision, once presented to the Executive Board, is in fact a foregone conclusion and Directors either support it or 'go along' with the proposal. The decision to propose such massive loans to South Africa, therefore, was a staff decision. In fact not only the size of the proposed loan but also its rationale are extremely questionable and raise issues of undue American and/or South African influence on staff policy-making. A recently retired member of the IMF staff is in fact employed by the South African mission to the Fund as a 'consultant' to lobby the IMF and its sister organisation, the World Bank. Given South Africa's minimal relations with the Bank, this lobbyist, Mr Albert S. Gerstein, obviously devotes most of his time to the Fund. He is paid $15,000 a year for the following activities, as reported to the Department of Justice in Washington under the Foreign Agents Registration Act:

'...In the IMF concerning reform of the international

monetary system, including studies of prepared amendments to its Articles of Agreement . . .and in general in connection with the collaboration of South Africa as a member of the IMF.'[73]

In another statement he makes clear that one of his activities related to American policies on gold was

' . . .the preparation of correspondence with government officials to ascertain the applicable statutes and regulation in the US pertaining to the private ownership and transfer of interests in gold.'[74]

It is probable that Gerstein had contacts with the officials preparing the case for a loan to South Africa, and that this was related to the 'suspect' calculations in IMF documents which troubled Lamberto Dini, the Director representing Italy and other countries. At the Board meeting of 10 November 1976 to discuss a proposal for a loan to South Africa from the compensatory financing facility, which required evidence that the country's balance of payments problems are externally caused by circumstances beyond its control, the minutes record Dini's objection that:

' . . .so far as the [balance of trade] shortfall was concerned, the staff had departed from current practice, since it had not arrived at an estimate of the possible shortfall in the leading export commodity, namely, gold. Naturally, forecasting the future of gold prices was difficult, but if the Fund intended to allow purchases under the compensatory financing facility, it should be prepared to make forecasts of market developments in all relevant commodities, including gold.'

He also criticised 'the staff view that the shortfall in exports other than primary commodities was due largely to external circumstances, and [he] would like the staff to provide some explanation.' Even more seriously, he continued:

'The staff's calculations were the more suspect because the shortfall calculated by the staff appeared to be exactly equal to the amount — 50 per cent of quota — requested by South Africa plus the amount of calculated double compensation under the previous drawings from the Fund.'[75]

The West German Director, Eckard Pieske, went even further in stating flatly that the shortfall projected by the IMF staff appeared to have been 'tailored to the size of the requested drawing.'[76]

In the restrained and clubby atmosphere of an IMF Executive Board meeting, such accusation of staff collusion with the South Africans to produce a recommended loan of the maximum entitlement amounts to an unprecedented condemnation of IMF staff integrity and impartiality. Given the strong American influence and large numbers of American staff members at the Fund, at least some Americans, with close links to the Treasury, would have been implicated in this deal. As Johnson points out, the Fund and the Bank both

' . . .have their headquarters in Washington, where the US Treasury looms over them. It is much bigger than them, has more money and has more and better personnel . . .There is a strong tendency for IMF and Treasury staff to be virtually interchangeable, and even non-US staff tend to be selected partly on grounds of their congeniality to the US viewpoint.'[77]

It is very interesting that Thomas Leddy, the official American representative, strongly supported the proposed loan, arguing that the IMF staff had done their job well:

' . . .the existence of a balance of payments need in South Africa was clearly established . . .South Africa's export shortfall was probably due to factors largely beyond its control . . .the staff's approach (in respect of the gold question) was reasonable.' [see Document 29]

Emerging opposition

There has always been an element of doubt behind the brash salesmanship of the Atoms for Peace programme, and the concern about nuclear proliferation and its attendant dangers has recently become perhaps the major American foreign policy issue. Proliferation has been brought to prominence partly because of the concerns expressed during and after the Presidential election campaign by Jimmy Carter, himself a former nuclear scientist who had been employed on naval reactors.

The US was already the major force behind establishment of the so-called 'London Club', the first meeting of which was held secretly in the autumn of 1975 to tighten up export restrictions on nuclear supplies in an attempt to slow down or obstruct the spread of nuclear weapons capacity. The policy of the London Club is evaluated in Appendix 1. Perhaps the single most important event sparking this rather belated concern was the announcement by India of a nuclear explosion in 1974, using US and Canadian technology and materials.

In addition to secret discussions within the London Club the US Administration has been coming under increasing pressure from Congress to take a tougher line with West Germany and France over their nuclear export policies.[78] This issue constituted the most serious conflict between President Carter and West German Chancellor Helmut Schmidt at the seven-power summit in London in May 1977. A vocal group of legislators, headed by Senator Ribicoff, Congressman Les Aspin and others, has focussed increasing pressure on the Executive branch to take measures such as cutting off all supplies of enriched uranium for the reactors of countries which are more or less openly seeking a nuclear weapons potential. However, it was not until Carter took over the Presidency at the beginning of 1977 that a genuine attempt was made to formulate a decisive policy to limit

269

nuclear proliferation.

With an on-going review of policy in this area, Carter made the first of a series of announcements in April 1977, on a Bill to be introduced in Congress by the Administration itself. The proposed package of measures includes a general policy to embargo the export of enrichment and processing plants, tougher safeguards on all material and equipment supplied by the US, and a ban on the reprocessing or transfer of fuel supplied by the US without prior US approval. Certain discretionary loopholes are also included, however, which are thought to be intended to allow the President to authorise the export of reactors to Israel. At about the same time as the announcement of this Bill, Carter also stated that the US itself would defer its plutonium-based fast breeder reactor programme to 'set an example' to other countries — such as Britain and France — which seem committed to this vastly expensive and dangerous form of technology.[79]

These measures still remain unsatisfactory for a number of members of Congress who are very alarmed about nuclear proliferation. In particular, there is profound concern about the implications of West Germany's sale to Brazil of the complete nuclear fuel cycle, which provides the Brazilian Government with everything it needs to produce nuclear weapons if it should so wish. Brazil was to have bought nuclear reactors from Westinghouse of the US, and switched to Kraftwerk Union of West Germany after it became apparent that the US Government would not authorise the export of enrichment and reprocessing plants to accompany the reactors.[80]

The fears expressed by many members of Congress about proliferation are, if anything, even more deeply felt by those in the Administration with specialist knowledge of the situation. Ken Owen of the Johannesburg *Star*, a well-connected South African correspondent who was based in Washington, has noted American fears of a potential 'outlaw club' of countries, including South Africa as well as Israel, Brazil and Iran. With such additional allies as Taiwan,

270

Rhodesia and Paraguay, this 'potential league of the desperate', to quote Owen's Administration source, constitutes what Owen terms a 'nightmare' forming 'in the minds of American policymakers'.[81] Even allowing for a little exaggeration, interviews by the authors in Washington in 1977 confirmed that US officials are indeed extremely worried by South Africa's nuclear potential, and its implications for regional and world peace.

It is an anxiety which has been forming for some time: even under Kissinger a senior official of the Energy Research and Development Agency (ERDA), Mr Nelson Sievering, when asked by Congressman Charles Diggs what could usefully be discussed in a forthcoming meeting between Vorster and Kissinger, said: 'I would certainly think the whole question of South Africa's attitude and approach to the Non-Proliferation Treaty would probably be a subject that will appear high on the agenda.'[82] In fact, an official in the State Department revealed in 1977 that several approaches had been made by the Department to the South African Government over the last few years to persuade it to sign and ratify the NPT. Another commented that the South African responses, which were beginning to echo the line of Brazil and India that the Treaty is a plot by the haves to exclude the have-nots from the nuclear club, were sounding increasingly like 'principled opposition' — in other words, tantamount to a refusal to consider signing.

Particular concern is voiced over the fact that South Africa has built its own pilot plant for uranium enrichment without any safeguards at all — the only unsafeguarded enrichment plant outside the five established nuclear countries. In fact, apart from the Urenco plant at Almelo, in the Netherlands, it is the only enrichment plant of any kind outside these five countries. An analysis of photographs taken by the reconnaissance satellites shows that at any rate it is on a scale which puts it in a class of its own among such pilot installations.

It has not escaped notice in the US that while South Africa announced that it planned to place the proposed commercial-scale enrichment plant under IAEA safeguards (with provisos to protect the technology), no moves have been made to do the same with the pilot plant. In fact, ACDA officials are far from sure that they would be able to enforce the placement of safeguards on the commercial enrichment plant. Since US pressures on West Germany and others had resulted in an agreement that compressors and other key items would not be supplied for an unsafeguarded enrichment plant, the South Africans' public position in favour of safeguards there might simply be a ploy for obtaining the necessary components, and several officials of ACDA remain suspicious of their intentions.

These officials are in 'no doubt' that South Africa has the capability to produce nuclear weapons of which it has so often boasted. They consider South Africa to be 'rather high' in the list of near-nuclear countries: in fact, among the top five.[83] Considering that at least three states – Israel, Taiwan and India – are generally accepted as being already in possession of nuclear explosives, this indicates the high priority given in concerned sections of the US Administration to South Africa as a major nuclear proliferation problem.

A surprising range of actions has been undertaken in an attempt to give substance to this concern, either in response to public and Congressional pressures or by the Administration's own initiative. One official admitted that 'the Dutch saved us by about 48 hours from welshing on the deal' involving the export of GE reactors.[84] With the decision on supplying reactor fuel pending in the early months of the Carter administration, it appeared to State Department officials that some stringent condition such as demanding signature of the NPT might have to be attached, otherwise there could be no prospect of getting it through Congress in the face of an alliance between liberals, led by the Black Caucus, and the environmental lobby.

A well-placed source has revealed that new conditions have been placed on the export of highly-enriched (weapons-grade) uranium for the SAFARI-1 reactor, conditions which are not part of the original agreement with South Africa. The material now has to be returned to the US immediately after it cools down, following irradiation in the reactor. Even more important, it is now to be shipped in two batches instead of one, so that at no time would there be enough in South Africa to be diverted for a nuclear device. At the time of writing all shipments have been officially suspended since 1975 because of the legal challenge from the Congressional Black Caucus.

Perhaps the most significant action so far has been the unprecedented attempt to stop the export of anything at all going from the US for the South African enrichment plant – even where it was quite clear that the items could easily be obtained from other countries. While it is impossible, under present export-licensing regulations, to stop a wide variety of items, including some machine tools, from being exported, a special effort was made to stop anything whatever that could be regulated. In revealing this fact, an Administration official conceded that the practical effect might be little more than to delay construction of the plant somewhat; however, she added, 'for us it's a question of principle'.[85]

The US is far ahead of all the other nuclear supplier countries in this respect, and recognises that the prospects for getting concerted action in other countries at the same level as the US are extremely remote. The Americans consider, however, that some progress has been made within the 'London Club', particularly with regard to West German policies. The Americans acknowledge that South Africa built its first pilot plant with the aid of Steag, and with compressors and other major items from West Germany, but claim that there were at that time virtually no German procedures for controlling such exports. After pressure from the Americans, the Germans allegedly changed their procedures to enforce close control over the export of sensitive items, of which compressors would certainly be an important one. If this assurance is valid, then the supply of compressors ostensibly to the Matla coal plant and the smuggling of reactor parts from West Germany in the South African diplomatic bag

273

must have relied on tacit Federal Government authorisation.

A variety of issues remain to be resolved within the relevant agencies of the Administration. There is the final decision to be made as to the export of fuel for the Koeberg reactors, and whether to continue with deliveries of highly-enriched uranium for SAFARI-1. The question of Namibian uranium sales is apparently also an issue.(America is on record as opposing economic collaboration with the South African occupation in Namibia insofar as this is regulated by governments). Finally, and perhaps most important, there is the problem of monitoring developments within South Africa and West German involvement with the nuclear programme there.

Even now, virtually no public criticism is being made by the US Government of West Germany's role; however, say officials, this does not mean that there is no pressure behind the scenes. It has been recognised, after repeated heavy criticism and public pressures directed at West Germany's deal with Brazil, that this can be counter-productive and arouse a strong backlash of resentment and nationalism in the countries concerned.

Sabotage or collaboration?
Policy conflicts in Washington

In mid-1975, it was reported in South Africa that Government officials there had strong evidence of CIA personnel visits under diplomatic cover. At least 12 such visits had been identified in recent years, where the diplomatic visas had been requested by the State Department for people who could not be traced to any identifiable Government position.[86] Shortly afterwards, it was revealed that the South African Embassy in Washington had concluded that the CIA had been trying to learn the secret of South Africa's uranium

274

enrichment process, and that this was what had been attracting all the unidentified visitors.[87]

Officials dealing with questions of nuclear proliferation have admitted in interviews that there has indeed been an effort to find out the details of the South African enrichment process. All orders for components destined for the pilot plant were carefully scrutinised to see whether they shed any light on the process. The kinds of machine tools being bought were found to indicate fine machine tooling. Deliveries of components from elsewhere were also watched as closely as possible; the South African units were seen to be much larger than the standard units for the Becker process. Attempts were also made to assess the capacity of the pilot plant, using photographic and other data. No doubt satellites were used as a source of data – although their usefulness is limited in attempting an assessment of the internal workings of a new kind of plant. Apart from the military satellites which constantly monitor the earth's surface, NASA has an Earth Resources and Technology Satellite (ERTS) surveying large areas of South Africa, which could be a source of photographs, in addition to the specifically reconnaissance satellites. (See pp.279-81 for maps of ground tracks of the Big Bird and Cosmos satellites over South Africa.)

At the same time, it is hard to imagine that the US is not also receiving information from sources inside West Germany. German industry is very closely linked with US multinational corporations which have a network of investments, licensing agreements and other forms of control dating from the days of the Marshall Plan. The degree of this control was only realized in West Germany itself when, during a recent strike of workers demanding greater participation in and control of their industries, the various hidden American interests emerged as the major stumbling-block, refusing to make any concessions to the strikers' demands. The nuclear industry in West Germany is particularly closely linked with US interests, with the country's two major companies both producing reactors relying on US technology. KWU uses both GE and

Westinghouse designs although it has a licence arrangement only with GE.[88] The CIA also has a major presence in West Germany with particularly close links with the ruling Social Democratic Party (SPD).[89] It is thought in Bonn to be inconceivable that the US Government, whether through the CIA or other operators at the mammoth US Embassy in Bonn, would not have taken a close interest in the enrichment deal with South Africa, and followed the deal with close interest through the multitude of sources available to them. The authors have received confirmation from a high-level source in the State Department that in fact the Americans knew about the West German nuclear deals with South Africa from the very beginning, and deliberately refrained from interfering. The source argued that even if the Americans had wanted to step in, they could not have done so without revealing their sources of intelligence.

A key question about whether the US Government was really attempting to monitor and control the West German-backed enrichment project in South Africa, or whether elements in the US Government were themselves implicated in these deals, arises in relation to the computers supplied for the pilot plant. In spite of the fact that these computers, although not specially designed for the enrichment process, had been 'specially prepared' for it, it appears that the inter-agency system for review and control of nuclear equipment, components and material was not applied to the transaction.[90] The Department of Commerce, which issued three export licences for the computers, made a unilateral determination to the effect that no special treatment or consultation would be required, in spite of the fact that the licences were made out in the name of the South African Uranium Enrichment Corporation, and that the stated purpose was 'operation of experimental facilities and pilot plants for nuclear research and development.'[91] The Commerce Department handled the application as an 'arms-embargo' item – of rather low priority in inter-agency dealings. The CIA, Defence Department, National Security Council, ACDA and the Atomic

Energy Commission were not consulted. The State Department approved all three licence applications, and the AEC only the third, which covered a second sale of spare parts.

The Foxboro executive said that the company had received no substantive inquiries on the transaction before or after licence approval, from Commerce, State, AEC or any other Government agency. He said that there were some 'routine' communications with Commerce regarding completion of the first licence application form prior to approval. He also reported, however, that the Boston office of the CIA had enquired as to the nature of the South African facility – *after* installation of the computers was completed.[92]

Some observers have suggested that the US deliberately tried to use the sale of the Foxboro computers to find out what was happening inside the pilot enrichment plant. The Administration has acknowledged that '. . . the intelligence community was aware of the purposes of the computers *before* the export licence was requested, and reports were routinely disseminated to the intelligence components of State and AEC.'[93] Company technicians had entered the plant, although South African officials restricted them to the computer area and monitored their activities closely. It would also be theoretically possible for American intelligence to capture signals from the computer after installation as a further aid to the discovery process – although according to knowledgeable intelligence sources, this is an extremely expensive and difficult way to conduct the inquiry.

This case perhaps highlights the fact that conflicting factions within the US Government are at work on the question of South Africa's nuclear ambitions. On the one hand, arms-control and other officials have been trying to stop whatever exports of components for the enrichment plant they could, and at the same time were trying to gather information on the exact nature of the process, as well as the plant's capacity. However, the Commerce Department, well-known as a close ally of the multinationals, failed to alert the ACDA and intelligence agencies to this major export, absolutely essential

277

to the operation of the plant. Although some parts of the intelligence community were aware of the impending transaction, the CIA and NSC were never actually consulted about the licence application,[94] and it seems likely that these agencies, or the officials within them who were following the South African programme, were quite happy to allow the transaction to go forward with only a token effort at gathering information by an open approach to the company's officials at the headquarters in Massachusetts.

Therefore, while it is possible that the computer's operations are being monitored, there is no indication that this is so. Moreover, if the US Government was genuinely concerned about South Africa's enrichment process, a strong and unambiguous commitment to non-proliferation would have required a refusal to license this sale, rather than an attempt to use it to provide information which could be gathered in various other ways. It is worth noting that the export licences for the two computers and their spare parts were issued between 1971 and 1973, at the height of the Nixon/Kissinger 'dialogue' policy with South Africa, whereas concern over South Africa's enrichment plant as a threat to proliferation became an important factor in the Administration's policy only by about 1976. By the time any inter-agency mechanism had been set up for obstructing the supply of components to the pilot plant, it was probably too late.

Another major question which arises is whether or not the American reconnaissance satellites photographed the Kalahari test-site before the official alarm was sounded by the Soviets. We have obtained maps of the ground tracks of the Big Bird satellite over Africa in July and August 1977 (see pp.280 and 281), which indicate not only that the site was clearly visible from the satellite, but that its trajectory seems to be targeted on that area. Even more interesting is the fact that this same trajectory covers simultaneously not only the Kalahari test-site but also the large region of Zaire where the West Germans are testing their missile systems. Considering that these two areas contained by far the most interesting new developments

Ground tracks of the Soviet 1977-69A Cosmos 932 satellite over South Africa, July 1977*

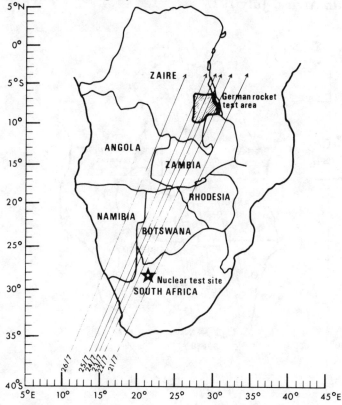

Launched 20 July 1977

Orbit lowered on 22 July to keep ground-track close to launchsite for next three days. Raised again after pass on 25 July in order to complete global coverage of mission – note increased separation of last two tracks.

Height of satellite at time of pass over Kalahari site averaged 220 km, suitable for high-resolution photography.

Site possibly located by area-survey Cosmos 922 during June 1977.

Cosmos 932 was recovered 2 August giving time to interpret photos before notifying USA.

* The date is indicated for each ground track.   Source: BH.

279

Ground tracks of the US 1977-56A Big Bird satellite over South Africa, July 1977*

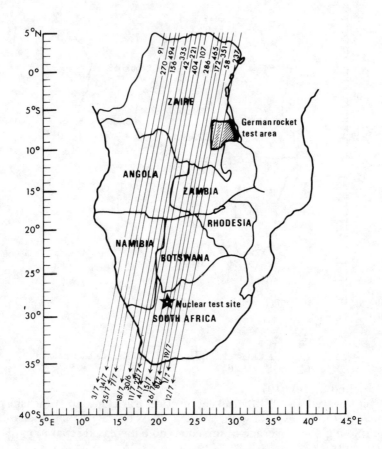

* The date and orbit number are indicated for each ground track.

Source: SIPRI

280

Ground tracks of the US 1977-56A Big Bird satellite over South Africa, August 1977*

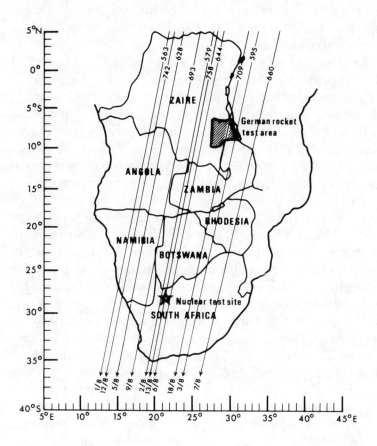

* The date and orbit number are indicated for each ground track.

Source: SIPRI

281

in the whole of Africa at that time from a military point of view, it is hardly conceivable that the satellite would pass overhead without taking multiple photographs in detail. Yet there have been a series of denials from the State Department that they were aware of the Kalahari test-site before the Soviets announced it.

The explanation seems to be similar to that in the case of the Foxboro computers: that certain parts of the Government were aware of what was happening but did not choose to disseminate the information to the agencies concerned. Information has subsequently come to light in Washington revealing the existence of the National Reconnaissance Office, which works in such secrecy that even its existence is classified. The most important source of technical intelligence gathered by the United States as a whole is that collected by the photographic and electronic reconnaissance satellites of the NRO, which is attached to the Office of the Secretary of the Air Force. The finance comes from the Pentagon but policy decisions on its allocation are made by the Executive Committee for Reconnaissance, which consists of only three people: the Assistant Secretary of Defence for Intelligence, the Director of Central Intelligence and the Assistant to the President for National Security Affairs (formerly Dr Kissinger, now Mr Brzezinski). It appears that Kissinger ordered the NRO to start military satellite coverage of Africa only in 1970, and that it was used, among other things, to compile a complete inventory of South Africa's weapons systems, including weapons type, numbers ordered or delivered, country of manufacture or licence and conditions of spare parts.[95] A copy of the 'NRO Restricted Armies Inventory South Africa' is reproduced as Document 27.

It is clear from this daily exercise that all South African territory was being very closely surveyed for any indications of military activity and equipment. A test-site for nuclear weapons would be one of the most obvious new features to be photographed and analysed. It seems probable, therefore, that some sections of the American military establishment, which received NRO reports, were aware of the Kalahari installation, while the State Department apparently was not. It is not clear whether the White House would have been informed, although informed sources in Washington insist that President Carter devotes a considerable amount of time to intelligence, and other official reports on South Africa.

The internal struggle over South Africa and the non-proliferation issue continues, compounded by the fact that policy, especially under the Carter administration, is made more by issue – such as nuclear proliferation or arms sales – than by region, particularly for Africa. Thus, many officials involved in the review of policy on South Africa or the southern African region may not be kept up to date with issues dealt with in the context of the review of policy towards non-proliferation. Inter-agency rivalry and special interests also operate in the policy split, with members of the Nuclear Regulatory Commission and ERDA controlling much of the basic information and closely tied to the nuclear-based corporations, while most of the State Department and some elements of the Defense Department are opposed to the uncontrolled spread of nuclear technology. Nor is this a simple question of left and right: some of the extreme right-wingers are adamantly opposed to nuclear proliferation, even to such favourite allies as South Africa, because of the patriotic desire to see questions of war and peace firmly under the control of the United States. Other right-wingers, however, are quite happy with the prospect of West Germany or Japan getting nuclear weapons, especially if it benefits American business. It is far from clear where the CIA, the military intelligence branches and the National Security Council stand on this issue, although indications are that at least some persons in these have been quietly encouraging South Africa in its nuclear plans.

Those opposed to this policy, apart from being kept in the dark in line with strict application of the 'need to know' principle, are strangely tentative about putting real pressure on South Africa, even while expressing great concern about its intentions and the danger they pose for the United States. Another major concern in their minds is the reaction of independent African states to nuclear deals with South Africa. James Blake told the Senate Subcommittee on Africa that 'three or four African nations', including Nigeria, had criticised the proposed sale of GE reactors in strong terms.[96]

State Department officials have seriously claimed that there is very little 'leverage' available to the US in relation to South Africa – apparently ignoring the very great importance, qualitative as well as quantitative, of US investment and official co-operation and support in the military, political and diplomatic fields, which in turn are of vital importance in promoting the confidence of the international business community in South Africa.[97] A whole range of South African interests, in international organisations as well as in bilateral deals, are fairly heavily dependent on the goodwill of the United States, so that in fact it would be far more accurate to say that it is South Africa which has almost no leverage over the United States, if the latter chose to take steps to bring pressure on the minority régime.[98]

THE CONGRESS AND PEOPLE

The split in Administration thinking and action on South Africa's nuclear programme is to a large extent a reflection of the deep division in American society and in Congress. On the one hand is the emerging alliance already mentioned, between liberals, black Americans and environmentalists; on the other, although operating much less publicly, is a powerful lobby for South Africa, including the multinational corporations which see it as an actual or potential profitable area of activity.

The lobby which is increasingly demanding a halt to all nuclear co-operation with South Africa has emerged as a serious force only since 1975, and its activities have been focussed on open confrontation in Congress and the Courts to oppose the various deals as they are announced. In April

1975, Congressman Les Aspen, a vocal environmentalist, revealed in a press statement the US Government's approval of highly-enriched uranium deliveries to South Africa.

> 'South Africa has the fear to want to build a bomb and it has the technical skill to be able to build a bomb. All it needs is weapons-grade uranium and the US government is now supplying that.
>
> What is the use of controlling the by-product when the very material we are shipping is weapons quality? . . .
>
> There is no way we can exercise effective control over the use South Africa makes of that uranium. What it comes down to is this: if they want a bomb, we have given them a bomb.'[99]

Two days after this announcement, Congressman Charles Diggs – an old antagonist of South Africa, being the longest-established black member of Congress and Chairperson of International Affairs Sub-committees dealing with Africa – introduced a Bill aimed at South Africa which would prohibit the transfer of nuclear materials and technology to countries which have not ratified the Non-Proliferation Treaty. Little has been heard of the Bill since, partly because it was referred to the Joint Committee on Atomic Energy (JCAE), whose Chairman and Vice-Chairman supported the uranium shipments to South Africa.[100] Among those supporting the Aspen-Diggs position was Senator Henry Jackson, a member of the JCAE and also a hard-line anti-Communist and supporter of ever-increasing Pentagon budgets. This unexpected alliance with the liberals on the question of nuclear proliferation is an interesting indication of the American right-wing's opposition to allowing even major allies to be out of the control of the United States.

A petition to the Nuclear Regulatory Commission was made in July 1976, to intervene in proceedings involving an amendment to a licence for the export of highly-enriched uranium for SAFARI-1. It raised questions about South Africa's refusal to sign the NPT, and its possession of enrichment technology as well as weapons capability. Among the petition's signatories were 14 members of Congress. As one NRC official said, 'When Congressmen are involved, you can't just ignore the petition.'[101] However, the licence amendment was later approved – although with the new stipulations regarding deliveries which have been noted above. This could well be seen as a direct result of Congressional pressure.

Meanwhile, Congress had been testing its powers in relation to the export of nuclear materials to India following the announcement of its atomic explosion. In 1974 a lawsuit had been brought under the National Environmental Policy Act, seeking to force the Administration to write an environmental impact statement on the issue. This would have to include the implications of the deal for nuclear non-proliferation objectives, as well as an assessment of radiation dangers. In 1975 the Senate Government Operations Committee held a series of hearings on India's nuclear programme and the US contribution to it, which attracted considerable attention in Congress and elsewhere. This led to a legal challenge to the NRC, brought in 1976 by the National Resources Defense Council (NRDC) together with the Union of Concerned Scientists and an environmental organisation, the Sierra Club, seeking to prevent the NRC from issuing export licences for enriched uranium shipments to India. The challenge was lost over a technical ruling that the plaintiffs lacked 'standing'; however, they felt that a moral victory had been achieved since a public hearing of the issues had been held.[102]

The momentum thus built up over nuclear proliferation issues, environmentalists turned their attention to South Africa when the proposed sale of the GE reactors was announced. The legal challenge was filed on 28 May 1976;[103] the next day it was announced that South Africa had cancelled the deal with America.

This action complemented hearings on the proposed GE sale by Senator Dick Clark's Senate Subcommittee on African Affairs in May. Announcing the hearings, Senator Clark –

who, like Diggs, is becoming a regular thorn in the side of the South Africans – said that the sale to South Africa '. . . would undoubtedly have a profound effect on the tense political and military situation in that region. It could also have an adverse effect on world security. Such a sale should not be approved without a thorough congressional review of its possible consequences.' He singled out for criticism the financial support to be offered through the US Export-Import Bank.[104]

Another legal initiative filed on the same day as the NRDC's, and which is still pending at the time of writing, is the petition against sales of fuel for the Koeberg reactors filed before the NRC by the Lawyers' Committee on Civil Rights under Law on behalf of a number of individuals and organis-ations, including ten members of the Congressional Black Caucus amongst whom are Congressman Charles Diggs and Congressman Andrew Young, now US Ambassador to the United Nations.[105] This petition has probably succeeded in delaying considerably the issue of an export licence for the fuel, and with the change in Administration has clearly become a serious consideration for the Executive branch. As already noted, officials of the State Department and other bodies are far from sure that the licences will be approved in the face of this challenge.

It is of interest that Andrew Young reversed his position on nuclear assistance to South Africa after joining the Carter Administration. In less than eighteen months he switched from opposing the sale of enriched uranium to South Africa to claiming that such a step 'would only encourage separate development of South Africa's own nuclear potential.'[106] This contradiction epitomises the split between hawks and doves in the Administration on this issue.

THE HIDDEN LOBBY

The South African Government maintains a considerable propaganda establishment, aimed increasingly at the United

States. This establishment has been responsible for a number of the US Government's decisions in favour of the minority regime, particularly in the military field. The network includes public relations firms, all-expenses-paid tours for US editors, journalists, politicians and military personnel, films for cinemas and television, the placement of articles and advertisements in major newspapers, propaganda leaflets, books and other publications, lobbying of the Executive branch and key members of Congress, and high-level conferences specially designed to promote the idea that South Africa is of vital strategic importance to the West. The propaganda budget for the US alone is at least $1.3 million.[107]

The most striking feature of the operation is its secrecy, a major reason for its extreme effectiveness. The South African Secretary for Information has admitted that they were aiming for a time '... when 50 to 60 per cent of the Department's methods would be "hidden".'[108] The Johannesburg *Star* has commented that the Department '... is now associated with "unconventional diplomacy", with secret contacts, with underground channels of communication, with subterfuges and all the other things that go by the name of "dirty tricks".'[109]

When the State Department refused to issue a visa for Admiral Biermann, the South African military chief, the decision was reversed by a combination of 'totally unprecedented' arm-twisting by Donald DeKieffer, the main public relations firm representing South Africa,[110] and Mr John McGoff, a powerful Midwestern businessman who had been a close friend of President Gerald Ford 'since he was a little old congressman', and who works closely with the South African Department of Information.[111]

Another route for high-level contacts with South Africa is through the South Africa Foundation, an organisation sponsored by the corporate and business interests operating in South Africa. The Foundation has been responsible for contacts with numerous members of Congress and the

Administration, discussing among other things the naval base at Simonstown. One notable contact in 1975 was with Dr Robert Goldwin, a Special Adviser to the President, and representative of the US Army War College.[114]

Many of the detailed lobbying activities are untraceable; however, the high level of the defence contacts, and their very long-standing nature, are apparent from what has come to light so far. Admiral Biermann, the South African Chief of Defence who visited the US in 1974 despite his initial visa refusal by the State Department, met with US Admiral Thomas Moorer, chairman of the Joint Chiefs of Staff, and J. William Middendorf, Secretary of the Navy-designate, and the Pentagon told inquirers that all the visits with military officials were the result of personal acquaintance or friendship among the people concerned.[115] The visits had in fact been facilitated by Congressman Robert Bauman. Another obliging representative, Senator Louis Wyman, secured a high-level Pentagon meeting for the then South African Minister of Information, Connie Mulder.[116]

The network of contacts pervades much of the more conservative part of the Establishment in the US, including a number of 'think-tanks' such as the right-wing Hudson Institute and the Georgetown Institute for Strategic Studies. It has been deeply entrenched for many years, and in fact some of the most outspokenly pro-South African advocates are retired military personnel, who maintain key contacts in the overall defence establishment, often have connections with relevant industries, including weapons manufacturers, and are free from official restrictions on their activities. General Marshall, a former US Chief of Staff, who visited South Africa on a number of occasions, testified on its behalf in the Namibia case before the International Court of Justice, and made it quite clear that in his opinion, 'Seeing the globe as a whole, the Cape is an anchor position.'[117] In a paper for the right-wing American-African Affairs Association entitled *South Africa: The Strategic View*, Marshall quotes Admiral Arthur W. Radford, former Chairman of the Joint Chiefs of

Staff, who on visiting Cape Town in October 1967 said to the South Africans, 'You are now at the crossroads of the world both economically and militarily.'[118]

The multinationals: what price on the grey market?

An important part of the lobby for South Africa, but also important in their own right as dealing directly with the Government there, are the US-based multinational corporations dealing in nuclear power. Much fissionable material is owned and traded privately in the US, unlike many other Western countries, and the trend has been to increase private control, at least under the Nixon and Ford Administrations.[119] Deals in enriched uranium from the US are increasingly in the hands of corporations. Many of these deals go through the World Nuclear Fuel Market, a company based in Atlanta, Georgia, since 1974. This is subject to no governmental controls whatever, and is not even registered with the US Securities and Exchange Commission or the Interstate Commerce Commission. In 1974 alone, one billion dollars' worth of transactions – including 'loans' of plutonium – were made through this organisation. It is owned by a multinational, the Nuclear Assurance Corporation.[120]

Even the critical parts of the fuel cycle such as enrichment and reprocessing have been put on offer to the corporations – although various proposals to turn these processes over to private interests were ultimately turned down by the companies on the grounds that they did not offer any prospect of reasonable profits. In fact considerable losses have been made in the US in unsuccessful attempts to build a commercial reprocessing plant.[121] However, companies do possess the technology involved, and would be in a position to sell it to the highest bidder if this were found to be in their interests. Administration representatives have confirmed that private companies can provide reprocessing capability. About 20 US companies were also granted access to classified enrichment technology by ERDA in the hope of bringing them into the commercial enrichment business.[122]

It is important to recognise that the corporations are not doing well financially out of nuclear power. In fact they have very serious losses to be recouped at some stage. General Atomic Co., a joint subsidiary of Gulf Oil and Royal Dutch Shell, having lost about $500 million on reactor orders with prospects of further losses if outstanding contracts at the fixed price were enforced, has left the commercial reactor business altogether.[123] Westinghouse, the world's largest reactor vendor, is now facing the possibility of incurring $500 million to $2 billion in fuel contract losses arising out of rapidly increasing uranium prices. GE is estimated to have lost between $500 and $600 million on 'turn-key' reactor sales. Babcock and Wilcox and Combustion Engineering are estimated to have suffered combined losses of over $150 million. A EG Telefunken, a major partner in KWU of West Germany, has withdrawn from KWU, citing 1974 reactor losses on reactor sales of between $274 and $287 million, with every indication of the trend being permanent.[124]

The European-based companies are in fact even more anxious for some profitable business because of their more limited domestic markets – particularly in view of massive public opposition to nuclear power. The desperate competition for orders is a basic reason for the short-cuts, with regard to nuclear non-proliferation safeguards, which are causing such concern, particularly in Washington. Examples include the West German sale to Brazil of the complete fuel cycle, and the French sale of a reprocessing plant to Pakistan. The core of the market for nuclear power and related processes at the moment, and for the foreseeable future, are countries with an interest in obtaining a nuclear weapons capability.

It is important to note that while fierce competition rages between the various reactor vendors, it is the US-based companies which are largely responsible for the regulation of the international market. In the first place, four US multinationals control over 70% of total world orders for nuclear reactors: they are General Electric, Westinghouse, Babcock and Wilcox, and Combustion Engineering. G E and Westinghouse alone are as large as the rest of the 'competition' put together.[125] National jurisdictions mean relatively little, except insofar as one country may have fewer controls on nuclear exports than another. The entire range of reactor vendors, except for the Swedish ones, have direct relationships with the four major US firms through licensing, subsidiary or shareholding arrangements.[126]

The companies make little or no attempt to take proliferation dangers into account in their sales of nuclear equipment and materials. Arms control officials in Washington are aware that the corporations are trying hard to sell the very items that the US Government is trying to limit, and that there are some companies that will do anything – regardless of the consequences – unless it is specifically barred by law. They also recognise the fact that some companies make an effort to circumvent the controls and take advantage of any loopholes in the letter of the law.

The tendency is well illustrated in the case of Bechtel's offer to Brazil to build an enrichment plant there, even though official US Government policy was to prevent Brazil from obtaining such a facility and to dissuade the West Germans from making any such deal. Myron Kratzer, formerly of the State Department's Bureau of Oceans and International Environmental and Scientific Affairs, charitably described the conflict as an 'honest misunderstanding on the part of Bechtel as to US policy'.[127] However, it was apparent that his extremely pro-nuclear predecessor, Dixy Lee Ray, also a former head of the AEC, had quietly encouraged the deal with her position that US companies ought to be able to build an enrichment plant in Brazil.[128]

Arms control officials had not learned of the Bechtel offer until after the return of a State Department delegation to

Bonn, which conveyed official concern about Bechtel's proposed deal with Brazil; moreover, the German Government had known about it for some time, and this greatly strengthened its determination to resist official US Government pressure which it saw as merely protecting US commercial interests rather than being an expression of genuine concern over nuclear proliferation. Some American officials were deeply angered by Bechtel's role, one of them describing it as 'totally unauthorised' and 'way out of line'. Asked whether it contributed to German intransigence in the matter, he replied brusquely, 'Draw your own conclusions.'[129]

Bechtel has also been involved in South Africa, providing assistance in the design of a full-scale plant to extract uranium from heavy metal concentrates.[130] Another American company involved is Union Carbide, which participated in the construction of SAFARI-1 and is now exploring for uranium with Anglo-American in Namibia. Several other American companies are also prospecting for uranium in South Africa.[131] It is likely that photographs produced by the American Government's Earth Resources Technology Satellite, shown to the corporations, have led to this deep involvement in South African uranium exploration.[132]

The Brazilian case illustrates the antagonism and non-communication between different agencies of the Executive in Washington, as well as the reliance of the more aggressive nuclear sales representatives on the support of their governmental allies. As Dixy Lee Ray herself put it, 'They [the arms control officials] knew my position and they made every effort to keep me and my bureau out of it.' She contended that ACDA simply does not belong in the field of nuclear export policy. 'They're meddling in areas where they don't belong, and they've made a real mess of things... They're trying to bottle up nuclear technology.' Asked to comment on the dispute, a Bechtel Corporation official at the California headquarters said, 'I'm sure the State Department knew what we were doing every step of the way... My impression is that Dixy Lee Ray was kept apprised, totally.'[133]

A senior employee of General Electric, Dale Bridenbaugh, who had been 22 years in the company and resigned while manager of performance evaluation and improvement, complained about GE's attitude toward sales of reactors to Egypt and Israel: 'I came back and asked my boss how we rationalize these sales to countries at war with each other. He said that wasn't our responsibility.' Richard Hubbard, another GE employee who resigned at the same time as Bridenbaugh after 16 years as manager of the quality assurance section in the nuclear energy control and instrumentation division, added that, when he raised questions: 'My boss characterized my attitude as paternal. Who are we to question who we sell our products to? If they want to buy them, that's their business.'[134]

Not content with this cavalier attitude toward major concerns over nuclear proliferation, the corporation actually put considerable sums into lobbying Congress and the Executive for more relaxation of controls on their nuclear exports. The Atomic Industrial Forum (AIF), which is the major lobbying group for the nuclear industry, responded to growing concerns in Congress about nuclear exports to India and elsewhere by quickly establishing a nuclear export committee, with high-level representation from the corporation involved in the nuclear industry, to lobby on the nuclear export question. At the American Nuclear Energy Council, the lobbying group on the issue of nuclear export policy – especially its effect on the US domestic scene – is becoming increasingly important.[135]

In November 1976 a nuclear conference was held in Washington DC by the AIF, together with the American Nuclear Society and the European Nuclear Society. They offered participation on the International Advisory Committee for the conference to Dr Roux, as representative of the Republic of South Africa.[136]

It is clearly in the interests of the corporations that proliferation be accepted as an unavoidable fact of life, since this would allow them to deal openly in the full range of equipment and nuclear materials with all near-nuclear states. It is probably no accident that within the US Energy Research and Development Agency (ERDA), an agency very closely linked with corporate interests, there is a strong feeling that nuclear proliferation is inevitable, and it is therefore useless to do anything except try to smother the market with US goods – the argument being that US suppliers are more responsible than others – in order to retain some degree of control over the fuel cycle.

In their dealings with foreign governments, the corporations seem inclined to use highly questionable sales tactics. Westinghouse reported to the US Securities and Exchange Commission in 1976 that its subsidiaries had been making 'questionable payments' to foreign officials. In addition to five such cases reported, Westinghouse also revealed suspect 'commission' payments to four sales representatives whom it uses for overseas business, including an 'improper allowance' of $150,000 apparently destined for the bank account of an unnamed foreign official.[137]

Considerable concern is now being voiced about the emergence of a black or 'grey' market in nuclear materials and equipment made available to near-nuclear countries or even to non-governmental private groups by corporations anxious to recoup their investments in nuclear hardware and technology.[138] As one critical observer summed up the situation:

> 'With the sharp decline in demand for nuclear power reactor orders coupled with an obvious increase in demand for nuclear weapons amongst third world nations, the possibility of a 'grey market' for nuclear materials evolving has increased significantly. This situation is further enhanced by the presence of un-regulated private transfer mechanism dealing in surplus nuclear material obtained primarily from US government contracts. . .
>
> The result has been the evolution of an international market where state-backed firms vie for export contracts primarily with weapons-seeking states.'[139]

The actual or potential existence of a 'grey market' in nuclear fuel and technology is obviously a major factor in considering the help which South Africa might get from US sources – corporate as well as governmental. In this respect South Africa has a number of cards to play in dealing with the multinationals. One of them is the existing investments of both General Electric and Westinghouse in the South African economy, where they are subject to stringent control – as are all investors in that country. General Electric in particular has an estimated $55 million invested in South Africa and supplies a wide range of electronics, locomotives and other hardware to the South African Government and its agencies.[140] Another major investor in South Africa is Union Carbide, which is a major contractor in the US for the Atomic Energy Commission and its successors, and also runs the Oak Ridge nuclear research centre in Tennessee.

One important card for South Africa to play in dealing with multinationals is its substantial uranium reserves; these could become even more significant if the commercial enrichment plant is built. It is worth noting that the French Government used its own uranium reserves to telling effect in dealing with Westinghouse, when that company was caught in a squeeze between fixed contracts to supply fuel for its reactors, and rapidly rising uranium prices together with a cartel that was withholding uranium from the reactor vendors. In exchange for 500,000 pounds of uranium annually, Westinghouse was persuaded to divest itself of a large proportion of the stake in the French nuclear industry which it had painstakingly built up over the years.[141]

A number of US corporations are already involved in aspects of the South African nuclear programme. In addition to the Foxboro computer sale already outlined, examples include Arthur G. McKee, whose Western Knapp Engineering Division is the main contractor building the Rössing mine, and Du Pont Chemical Co., which has sold technical information to Fluoro Carbon Products Ltd, a new company formed to make South Africa independent in supplies of

Teflon – which has important applications in nuclear technology. There could well be a direct involvement in the West German provision of a licence and components for the South African enrichment plant, since Allis-Chalmers, the company which supplied SAFARI-1, is closely involved in the production of compressors marketed by its associate company MAN of West Germany. MTU (Motoren-Turbinen-Union, Munich), a central participant in the enrichment deal, is also closely linked with G E.[142] American companies alleged by the West German Anti-Apartheid Movement to be involved in the commercial enrichment plant include Leeds and Northrup, supplying electronics; Federal Products, providing precision measuring devices for the treatment of aluminium; and SWF (a subsidiary of ITT).[143] *Nuclear Fuel* reported in August 1977 that the United States Government was in favour of American participation in this venture.[144]

The prospects of the US Government controlling the proliferation of links of all kinds promoted by its multi-nationals can be evaluated by considering the record on the US arms embargo on South Africa. One State Department official who drew this parallel is quoted as saying: 'You get the same arguments, but there the industry has largely prevailed. And there are essentially no controls.'[145] This is borne out in a study by Michael Klare, who found a number of illegal and indirect sales, including the licensing of technology to the South Africans, often through third parties such as Italy.[146] US Government guidelines on the arms embargo, as on nuclear exports, are obscure and often conflicting: one senior State Department official described the 'grey areas' guidelines as a 'nightmare'.[147]

As in the case of nuclear exports, licence applications for arms exports are dealt with out of context: thus dual-use items like light aircraft are licensed by the US Government, despite the fact that all such aircraft in South Africa form part of the Commando reserve force which can be called up at any time and are of particular use in any operation to crush an African uprising. As regards nuclear exports, the Nuclear

Regulatory Commission has information on the specific safeguards applicable to the export item in question only; its decision does not involve the safeguards situation relating to the importing country's entire fuel cycle. Issues such as procedures to be followed if US-supplied fuel is reprocessed are left vague. Even details of the precise IAEA safeguards procedures relating to the item in question are not disclosed to the US since the IAEA treats this as 'Safeguards Confidential Information'.[148] The decision to license the exports, therefore, is made on inadequate information.

In relation to the arms embargo applied somewhat erratically by numerous countries, the South African Defence Minister openly welcomed the trend towards licensing and financing arms production in South Africa itself, in joint ventures with local firms or the State-owned Armaments Development and Production Corporation of South Africa (ARMSCOR), since it would save the country much of the cost of armaments research. 'It is our duty to be ready for anything in the light of continual meddling in South Africa's domestic affairs, threats of sanctions and boycotts, and open animosity in certain circles.'[149] The establishment of UCOR, based on West German technology, is on the same pattern. It may be no coincidence, either, that the major corporations dealing with nuclear technology and supplies – including General Electric, Westinghouse and Bechtel – are major arms manufacturers in their own right, with ample experience in high-pressure sales tactics as well as in evading inconvenient embargoes on the export of their hardware or of the technology to manufacture it.

9

The Urenco 'Troika'

The background

The British and Dutch Governments became intimately
involved with the West German nuclear network through the
Urenco-Centec consortium for uranium enrichment based
on West Germany's war-time research on the gas centrifuge
process. This important element in the diversification of their
nuclear interests and partnerships, using South African
uranium as raw material, is helping the Germans to move
further toward their goal of closing the nuclear fuel cycle
while retaining international respectability.

The German technology, located just across the border in
Holland, is also linked with the British centrifuge plant at
Capenhurst in a complex arrangement based on the exchange
of expertise and production capability in advanced weaponry.
However, the British operation is somewhat separate from
the German/Dutch one.

Germany's war-time involvement in the development of
nuclear fission has been outlined in Part I. Part of the

299

Organisation of the Troika, 1975

Source: Peter Boskma, Wim Smit, Gerard de Vries, *Uranium Enrichment: History, Technology, Market*

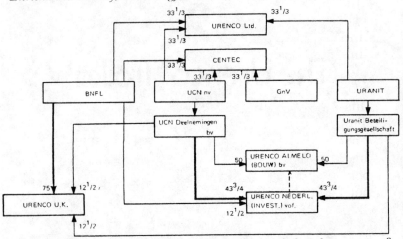

programme was composed of research and development of the gas centrifuge for uranium enrichment, a process which is the direct forerunner of that being used at Almelo.

A number of individual West Germans involved in the Troika had personal roots in the military-industrial complex of the Third Reich. One is Prof. Karl Winnacker, who during the war was a Director of IG Farben. Winnacker now chairs the Supervisory Board of Hoechst AG, is a former member of the Supervisory Board of Gelsenberg, another key company, and also sits on the Degussa Board. He has great influence with Nukem and with Uranit, the West German partner in Urenco which is owned by Hoechst, Nukem and Gelsenberg. Another important individual is Dr Felix-Alexander Prentzel, who was also a Director of IG; he became Chairman of Degussa's Managing Board from 1959-73, and is a member of the Metallgesellschaft Supervisory Board, Deputy Chairman of the Hoechst Supervisory Board and former member of the Gelsenberg Supervisory Board. Prentzel therefore has influence in all three Uranit constituent firms. He also has important financial connections, particularly with the Dresdner Bank.

Another key appointment, which was instrumental in giving Degussa working control over the centrifuge process, was that of Dr Heinz Schimmelbusch to the Chairmanship of the Urenco Board of Directors in August 1971. A mining engineer who was with Degussa all through the war, Schimmelbusch took the initiative in the early fabrication of reactor fuel elements,[1] and was instrumental in the establishment of Nukem, of which he is now Chairman. Last but by no means least, there is the vast German-based financial empire backing the enormously expensive nuclear ventures; perhaps the most noteworthy of the bankers is Hermann J. Abs, former Chairman of the Deutsche Bank, who is on the Supervisory Board of Metallgesellschaft, Chairman of the Supervisory Board of RWE, and holds prominent positions in BASF, Gelsenberg and Siemens (which now owns the whole of KWU, the reactor manufacturer).

During the period of West Germany's Fourth Atomic Programme of 1973-76 alone, a sum of DM 529 million was provided by the Federal Government for gas centrifuge pilot-plant work.[2] The German firms participating in the construction of the enrichment facility at Almelo were promised DM 700 million in governmental subsidies.[3] As explained officially by the Federal Government:

> 'Because of the considerable development risks and the long time period required for economical operation, government aid in the development of enrichment processes in the Federal Republic, and in the construction of test and prototype facilities has been indispensable. The state assures an appropriate share for German industry in the construction of new enrichment capacity.'[4]

Enrichment capability at Almelo is also a vital ingredient in the sales drive behind West German reactors; marketed by KWU to Brazil and elsewhere.[5] KWU's contract with the Government of Iran, for example, includes the delivery of enriched fuel for the reactors.[6]

301

German domination of the Dutch enrichment plant

The 'Troika' was founded by the Almelo Agreement of 4 March 1970,[7] and ratified promptly by the three governments, with a minimum of parliamentary debate within sixteen months. The tripartite collaboration came into effect with the formation of the two 'joint industrial enterprises', as called for by Article 1.2. The first is the plant-builder Centec (Gesellschaft für Centrifugentechnik mbH), established under West German law at Bensburg, near Cologne, on 29 July 1971. The second is Urenco Ltd, founded in August 1971 with head offices at Harlow, Buckinghamshire. Centec's principal functions include integrating all centrifuge research and development in the three countries, and to 'coordinate and control' the future integrated research and development programme.[8] It was decided in late 1973 that the managements of the two companies would be combined under a single chief executive, with the same people sitting on both boards.[9]

The first phase of the pilot facility of UCN (Ultra-Centrifuge Nederland NV) went into operation at Almelo in 1972. The first enriched uranium produced there was loaded into the Dodewaard nuclear power station in April 1974. This probably represents the first non-laboratory production of enriched uranium outside the existing nuclear-weapons states. By 1974, three pilot plants were finished: British Nuclear Fuels' plant at Capenhurst, and the West German Uranit plant adjacent to the UCN one at Almelo. These two each have a capacity of 25 tons of uranium a year, and BNFL's has a 14-ton capacity.[10] Construction costs were met 'primarily by the three governments', according to a West German paper.[11] A 200-ton demonstration plant is now in operation at Almelo.

Following a decision to 'further commercialise'[12] and bifurcate the Troika, the organisational structures were changed to give BNFL greater participation in the Capenhurst plant through the new Urenco UK, with minority shares held by UCN and Uranit (representing West German interests in the project); UCN and Uranit likewise took over greater proportions of the Almelo venture with the formation of Urenco Nederland. Urenco Ltd ceased to be the managing partner at both sites, as it had originally been. The extremely complicated network of new companies (as shown on p.300) and the cross-shareholdings established, combined with a great reluctance on the part of all three governments to disclose information about the basic workings of the projects, make the operation of the Troika increasingly obscure and outside the scope of parliamentary supervision; however, it is evident that West German influence over the Almelo operations has become greater than originally envisaged.

The United States was less than enthusiastic about the formation of the Troika, partly because of the access to enrichment which it gave West Germany. It seemed to accept the proposed arrangement after West Germany had signed the NPT on 28 November 1969. A key factor in international acceptance of the prospect of West Germany acquiring an enrichment capability was the decision to build the plant at Almelo, on the Dutch side of the border. The 'balancing' participation of Britain was also important in making the arrangement acceptable to Dutch public opinion as well as to the Americans – although, despite considerable progress within the Netherlands on centrifuge research, doubts could not be suppressed as to whether, given its small size and weight relative to West Germany, and the dominance of the Germans in the technical field, this was really a union of equal partners. In fact some Dutch researchers came to the conclusion that with the division of the Troika into a pre-dominantly British and a predominantly West German-Dutch branch, Dutch interests would be overwhelmed:

'Formally, each of the three countries has veto rights in the Joint Committee, but the question arises if, with a

303

continuing dividing up of the project, this veto will continue to exist de facto. At any rate, it is certainly possible for Britain to pursue an independent policy. In this case, West Germany would undoubtedly gain the upper hand in the other branch.'[14]

The role of the Dutch Government has been crucial to West Germany, however, in setting up the latter's access to enrichment capability. As two British journalists commented in 1969:

'Germany's principal problem, of course, was that world opinion would be greatly offended by any plans to build an enrichment plant on German soil. The solution it sought from collaboration with Holland was joint ownership of a Dutch plant.'[15]

The pressure which the Germans brought to bear on Holland to set up the Troika was persistent, as described in 1967 by a Dutch advocate of the internationalisation of centrifuge, Deputy A. De Goede:

'The Germans have repeatedly demonstrated their wish to work with the Netherlands... In 1965, the question arose of Dutch-German consultations on cooperation, and the political and any military aspects involved were very carefully weighed up... To my knowledge, Germany pressed the point of cooperation with the Netherlands a few months ago... Politically, it was thought extemely unlikely that Germany would be able to begin enrichment by itself.'[16]

The key intermediary role in setting up the Troika with Britain, and in gaining US acceptance, was said to have been played by former Dutch Foreign Minister and current NATO Secretary-General Joseph Luns – who is well-known to those interested in southern Africa as a major force within NATO advocating closer ties with the former Portuguese colonial regimes and the minority-ruled countries of southern Africa.

The participating Dutch companies – vast multinational conglomerations which happen to be based partly in Holland – also form an important part of the Almelo project, as well as having their own very strong links with West German and British commercial interests. The companies include Royal Dutch/Shell, the largest corporate grouping outside the United States, whose sales exceed the total Dutch Government budget. It has recently diversified into mining and metallurgy, with a $100-million expansion plan in that area for South Africa alone.[17] There is also the highly integrated, electronics-based Philips Group of Eindhoven, whose wide range of interests includes telecommunication, data- and defence-systems. Both Shell and Philips have seconded leading personnel to Urenco/Centec.

Another important Dutch-based participant is the manufacturing group VMF (Verenigde Machinefabrieken-Stork) which is considered highly vulnerable to 'alien' penetration because of its relatively narrow Dutch capital base. During the Almelo negotiations VMF was subject to attempts by foreign groups to obtain a majority shareholding. These bids were unsuccessful because of aid from two Dutch banks, but the episode showed the vulnerability of the Dutch industrial presence within the Troika.[18] The outside forces may have included American companies, but could equally well have involved the powerful West German atomic industry. This has in any case developed some extremely strong links with the major Dutch-based companies. Among the cross-directorships is that of Hans L. Merkle, a Director of Shell, who also holds Supervisory Board positions with the Deutsche Bank and RWE, as well as BASF and other West German companies. Bernhard Timm, Deputy Chairman of the Supervisory Board of Deutsche Shell, is a former wartime collaborator of Wurster at IG Ludwigshafen and BASF, and is also a leading representative of the Deutsche Bank's chemical wing.

There is reason to foresee a diminishing independent Dutch role at Almelo, even though it is on Dutch territory. At

305

the end of 1976 it was reported that the Dutch industrial partners, including even Shell, wanted to pull out of Urenco altogether because of the uncertain political and commercial prospects.[19] On a technological plane, too, the Dutch are being left behind. The Amsterdam business newspaper *NRC/Handelsblad* summed up the situation in March 1973:

> 'Only the German centrifuge model is being considered for large-scale operations. Dutch scientists are involved in the development of this apparatus: a fact which could be of interest to Dutch industry. But nowhere is this interest exactly defined. For the time being there is a Dutch contribution and we will have to wait and see if it is really going to amount to anything. Experience shows that there is a real danger of the Netherlands being squeezed out by German and British interests.'[20]

In July of the same year, Philips and VMF discontinued their participation in centrifuge construction at Almelo, to the accompaniment of newspaper headlines such as 'Super Centrifuge Outmodes Dutch Centrifuge.'[21] It was reported that the centrifuge design choice, which gave prominence to German developments, was made by Urenco in the UK and Centec in West Germany.[22]

Holland, in fact, has little room for manoeuvre at this point, and the general public there is realizing for the first time what could be involved in the partnership with West Germany. A major political crisis blew up at the end of 1976 over the use of Almelo to supply 2,000 tons of separative work units for enriched-uranium between 1984 and 1994 as the selling-point for Germany's deal with Brazil, a deal which many Dutch people felt was a serious threat to non-proliferation efforts. The *Volkskrant* reported: 'At Urenco the direct connection is pointed out between the delivery of German nuclear power stations to Brazil and the order for the required nuclear fuel from the Urenco consortium.'[23]

A demonstration at Almelo by an estimated 30,000 people (a fifth of them German) in March 1978 reflected the fact that opposition to nuclear power has become the most important unifying issue for the left-wing in Holland and West Germany. The Dutch Parliament has acknowledged the strength of feeling on the issue by voting to restrict the supply of uranium to Brazil. This is not binding on the Dutch Government, but it is a strong influence.[24] The issue remains a difficult one for the Dutch, subject as they are to strong German pressures and a distinct indifference to their case on the part of the British.

Deals with South Africa, such as the procurement of uranium oxide for enrichment, could not be very effectively opposed by the Dutch Government, particularly if the British Government agreed with the Germans over this issue. The Dutch, in fact, have registered the strongest opposition to dealing with South Africa in the nuclear field with the Government decision, taken at the height of a governmental crisis at the end of 1975, to withdraw support for Dutch participation in the American-led consortium to build the Koeberg reactors. Opponents stressed the dangers of South Africa becoming a nuclear power, and some added that the Dutch government and companies would, over a period of many years, '. . . have a vested interest in the South African Government maintaining a sturdy position and order in the country at all cost [sic].'[25]

It is ironical, in the light of this decision, that one of the Troika's main links to South Africa is through a Dutchman, Professor Jacob Kistemaker. Kistemaker paid an official visit to South Africa in 1975, in connection with nuclear energy.[26]

It is evident that a substantial proportion of the uranium to be enriched at Almelo will be of South African or Namibian origin. West Germany already has large contracts for South African uranium with 40% of its total uranium imports coming from there in 1975.[27] Through Urangesellschaft – which is jointly owned by Metallgesellschaft, Gelsenberg and Steag – the German atomic industry is not only buying Rössing uranium but has a financial stake in the mine, as outlined in Chapter 4. British participation also contributes to the emphasis on South African and, more particularly, Namibian uranium oxide, because of its massive contract for 7,500 tons of Rössing uranium and RTZ's leadership of the whole venture. The enriched uranium fuel for the advanced gas-cooled reactors (AGRs), which are starting to come on stream in Britain now, will come through Almelo, according to sources in the British nuclear industry.[28] With Rio Tinto Zinc of Britain as heavily committed as it is to Rössing – and to Namibia and South Africa in general – it is worth noting that RTZ has important links with the successors of IG Farben, including an 18% founding share in the key company Nukem.

One reason for Britain's enriched uranium for commercial purposes being processed at Almelo instead of at Capenhurst, which has a gaseous diffusion plant as well as the pilot centrifuge project, is that the British partner in the Troika is free to use its participation for the manufacture of nuclear weapons. Many of the British centrifugists have weapons backgrounds,[29] and British participation in the Troika was based to a large extent on increased co-operation with the West Germans in military research and development.[30]

Britain is free to combine gaseous diffusion with the centrifuge in order to produce highly-enriched or weapons-grade uranium, as specified in Article VI, paragraph 2 of the Almelo Agreement. The British Government has apparently told its partners that this option would be carried out from a Troika installation situated within the UK.[31] Given the military use to which British participation in the Troika can be put, and with the economic and political dependence of the British Government on German good-will in the EEC and in general, it may be that their concern over possible German intentions with regard to uranium enriched at Almelo is deliberately muted.

It would be simple enough for the Germans to follow the British model and combine the centrifuge and jet-nozzle enrichment processes to produce weapon-grade uranium, either on German soil or in a country such as South Africa, or, prospectively, Brazil. A number of West German companies involved in centrifuge have also participated in building South Africa's enrichment plant. MAN, for example, which supplied the compressors for South Africa, is also a major reactor manufacturer in the Federal Republic. An alternative would be to use enriched uranium, from Almelo or the new Troika plant which West Germany is now proposing to build, to feed into German reactors for the production of plutonium. West Germany already has a pilot reprocessing plant at Karlsruhe, and is planning to build a full-scale plant there also. Although built at Government expense, the pilot plant has been taken over by KEWA, a company in which the Urenco participants Hoechst, Gelsenberg and Nukem have shares. The West Germans will also be providing a reprocessing plant for Brazil. South Africa, as indicated in Part II, already has a small reprocessing plant.

There seems to be little prospect of preventing the West German atomic industry, backed by the Federal Government, from building up a major enrichment capacity of its own, on the basis of the Almelo Agreement. Article III.1 of the Agreement states that:

> 'The joint industrial enterprises shall use their best endeavours to meet all orders for uranium enrichment services placed with them by customers in the territory of any of the Contracting Parties, whether or not the fulfilment of such orders would involve the installation of new enrichment capacity. The joint industrial enterprises shall, however, be bound to meet such orders of the Contracting Party concerned if entities within its territory agree to provide such portions of the extra finance involved as are not forthcoming from the joint industrial enterprises and the other Contracting Parties.'

What this means, in effect, is that any of the three participating national groups, if they are dissatisfied with the level of activity provided by the Troika, can force an expansion of enrichment capacity simply by providing the necessary finance and placing orders for the enrichment work.

This is precisely what the West Germans are doing. In the wake of the controversy over enrichment of uranium for the German deal with Brazil, the Dutch have been debating whether or not to agree to German demands for the expansion of enrichment capacity at Almelo to 1,000 tons by 1982. Uranit, the West German partner, has announced plans to build a 1,000-ton centrifuge plant in Jülich or elsewhere in the Federal Republic. If the Dutch Government decides not to allow the expansion of Almelo, then the West Germans '. . . will be obliged to proceed with a site in their own country.'[32]

Indeed, with 1980 approaching the Dutch may lose all their influence through the dissolution of the Troika. Article XV of the Agreement states: 'After this Agreement has been in force for a period of ten years, any Contracting Party may give one year's notice in writing of withdrawal from this Agreement.' With centrifuge technology well-established as a result of the co-operation and joint financing of the Dutch and British, and with German dominance at Almelo a *fait accompli* regardless of the existence of the Agreement, the Germans may well decide in the next few years that they no longer need the respectable cover provided by Dutch and British participation with them in a Troika. With the West German nuclear fuel cycle closed at last, the ambition of German scientists and industrialists ever since the war is within reach. With new partners, such as South Africa and others, the political and diplomatic constraints on the right wing in Germany in its attempts to get nuclear weapons for the homeland may yet prove to have been made inadequate because of the technical and industrial advances resulting from the Almelo Agreement.

10

How Israel joined the nuclear club

'I went to Israel recently, and enjoyed every moment there. I told the Prime Minister when I got back that as long as Israel exists we have a hope. If Israel should, God forbid, be destroyed, then South Africa would be in danger of extinction.'

General Hendrik Van den Bergh, head of South Africa's Bureau of State Security (BOSS).[1]

In June 1976, the West German semi-official military monthly *Wehrtechnik* wrote:

'What until now had often been regarded as a pure speculation turned out to be a hard fact: Israel possesses an atomic bomb, more precisely, thirteen bombs, each of them having an explosive capacity of 20 KT, which is equivalent to one of the bombs dropped on Hiroshima or Nagasaki. These bombs can be delivered to the target by the Israeli Kfir and Phantom fighters which had been specially equipped for this purpose.'

The origins of the Israeli atomic bomb go back to the time when Israel, assisted by the Western nuclear powers (in this case primarily by France), entered the nuclear age.

311

The origins of Israel's nuclear programme

The decision to build the 24 megawatt reactor at Dimona was made in 1957 after the Suez war in October 1956. Despite its military success in overrunning the Sinai and the Gaza Strip, Israel was forced by UN, US and Soviet pressure to withdraw from all the territories conquered during the war, without any political settlement or guarantees, except an assurance by the US State Department and the stationing of a UN emergency force at Sharm-el-Sheikh and in the Gaza Strip to guarantee the passage of Israeli shipping and prevent fedayeen border crossings. Israel saw herself in a situation of political isolation, faced with a US embargo on arms supplies at a time when her Arab neighbours were rearming rapidly with the help of the Soviet Union.

In 1957 the Israeli Cabinet approved a proposal by David Ben Gurion, then Prime Minister and Minister of Defence, to build the 24-megawatt nuclear reactor at Dimona.[2] The public was kept in total ignorance of the decision until Ben Gurion disclosed it in the Knesset on 21 December 1960, in response to a request by US Secretary of State Christian Herter to clarify the rumours that Israel was building a plutonium-producing reactor that would enable it to manufacture atomic weapons. Ben Gurion categorically denied the rumours that Israel was producing an atomic bomb; he stated that the Dimona reactor would serve the needs of industry, agriculture and health, and would train scientists and technicians for the future construction of nuclear power stations.

This statement did not dispel the rumours in the West, nor calm the anxieties in Arab countries.[3] Neither did they dispel certain doubts and anxieties in Israel, which intensified when the news leaked out that six of the seven members of the Israeli Atomic Energy Commission, nominated by the Minister of Defence in 1952, had resigned because of their disagreement with the work of the commission. This left only the chairman of the IAEC, Professor E. D. Bergmann, in charge of the nuclear programme.[4]

When Israel's nuclear programme was launched its main source of supply of military material and major weaponry was France, which at that time had close ties with Israel. France was then facing a revolt in Algeria which was supported by Egypt (then called the United Arab Republic). For both France and Israel, Egypt was the common enemy, and President Nasser was, in Israeli eyes, the arch-enemy trying to mobilize the Arab world for a total war against the Jewish state. The Franco-Israeli collaboration included scientific co-operation. The high level reached by Israeli scientists, in physics and nuclear energy, was an asset for the French in developing their *force de frappe*. The Israelis too took full advantage of the cooperation: the Dimona reactor was conceived, planned and executed with French help. The exact contracts, commitments and objectives of this deal have never been made public.[5]

West Germany also gave Israel a hand. Scientific co-operation between the two countries started in the late 1950s, when West German atomic scientists made contact with scientists of the Weizmann Institute in Israel. West Germany sponsored research projects and financed installations like the 6-MV-Tandem-van-de-Graaf accelerator, worth DM6 million, which enabled Israel to set up a department of experimental nuclear physics at the Weizmann Institute in Rehovot. In 1963 the informal scientific relationship was consolidated by a joint agreement, and the Minerva Society, a subsidiary of the West German Max Planck Gesellschaft, was founded to sponsor scientific co-operation with Israel. West German foundations like the Stiftung Volkswagenwerk or the Deutsche Forschungsgemeinschaft also have special programmes for sponsoring the exchange of scientists and for joint West German-Israeli research projects.

Israel's nuclear programme gave rise to considerable anxiety which was expressed repeatedly in the Knesset. As a result, at the end of 1961 the Committee for the Denuclearization of the Israeli-Arab Conflict was established. It consisted of prominent scholars and scientists concerned with Israel's political and security problems, and known to have close contacts with Cabinet Ministers and leaders of the major political parties. The Committee wanted Israel to announce a plan for the denuclearization of the Middle East.

The Committee's efforts met with considerable sympathy. Some of the leading members of the major political parties, both in the Government coalition and the Opposition, expressed their support for the idea; one coalition party, Mapam, adopted as its official policy the denuclearization of the Middle East with international guarantees.

In April 1962 the Committee published the following demands:

1. That Middle East countries refrain from military nuclear production, if possible by mutual agreement.
2. That the UN be requested to supervise the region in order to prevent military nuclear production.
3. That the countries of the Middle East avoid obtaining nuclear arms from other countries.

Zalman S. Abramov, leader of the Gahal Party, and members of the Knesset took issue with the Israeli Government's rejection of a proposal by Swedish Prime Minister Tage Erlander that the non-nuclear countries abstain from producing nuclear weapons.[6] And in a Knesset debate, the leader of the Mapam, Hazan, called for an Israeli initiative to prevent the introduction of nuclear weapons into the Middle East. The Government rejected all plans for a nuclear-free zone as irrelevant to the real threat originating from the conventional arms race. Its stand was that there were no nuclear weapons in the Middle East and Israel would not be the first to introduce them; however, Israel could be destroyed by conventional weapons, so the emphasis should be laid on conventional disarmament. Ben Gurion kept his nuclear-option policy, and this increased the strain in US-Israeli relations.

314

US-Israeli nuclear co-operation

From the beginning the US took a serious view of Israel's nuclear initiative, expressed grave concern about the introduction of nuclear weapons into the Middle East, and exerted heavy pressure on Israel to renounce any such intent. The repeated statements by Israeli spokesmen that the reactor would serve only peaceful purposes did not alleviate official US suspicions. The demand for verification through inspection by US experts grew stronger, in particular after John F. Kennedy became President in January 1961.

In a meeting with President Kennedy in May 1961 in New York, during Ben Gurion's private visit to the US, the Israeli Prime Minister had to accede to the demand that American scientists be allowed to visit the reactor. Two such visits took place before Ben Gurion resigned in 1963. President Johnson, who took office the same year, continued to exert pressure. The first step towards an agreement was made in a meeting between Johnson and the new Prime Minister, Levi Eshkol, who was officially received in Washington in May 1964.

Johnson appeared reassured that Israel did not plan the separation of plutonium and the manufacture of weapons. A system of regular visits by US experts was agreed upon. Another outcome of this meeting was the sale of Hawk missiles to Israel – the first official and direct sale of American arms to Israel.[7]

In June 1965, apparently alerted by the report of Roswell I. Gilpatrick, Chairman of a special committee to study the spread of nuclear weapons in the Middle East and Asia,[8] Senator Robert Kennedy rose in the Senate to declare: 'The need to halt the spread of nuclear weapons must be the central priority of American policy.' He cited Israel and India as having supplies of 'weapon-grade fissionable material and [as being able to] fabricate an atomic device within a few months.' He proposed an immediate joint initiative with the Soviet Union to obtain a pledge from Israel and other nations

315

that they would neither acquire nor develop nuclear weapons, in return for which the US and the USSR would provide guarantees against nuclear aggression and blackmail.[9]

Israeli reaction was summarized in an editorial in the London *Jewish Observer*:

> '... the limited attempt to restrict proliferation in the Middle East may increase rather than lessen the danger of war ... The threat comes from the acknowledged aim of Nasser...to destroy Israel... The halting of such nuclear development as exists would not change this situation for the better. It might, on the contrary, encourage aggession by conventional weapons.'[10]

Continued American pressure and Israeli resistance resulted in hard bargaining. Israel claimed that it would not introduce nuclear weapons but needed nuclear power for urgent development projects such as the desalination of sea water.[11]

Israel was asking for real assurances of its security as the price of non-nuclear status. On 7 January 1966 the *New York Times* reported that Israel had entered into a secret contract to purchase a large number of medium-range ballistic missiles developed in France. The purchase of these missiles, able to reach Egyptian targets from Israeli launching-pads, was seen as indicative of Israeli intentions to develop atomic warheads.[12] This might explain Nasser's effort to obtain nuclear facilities from the USSR. The Soviets refused, but agreed to a guarantee of Soviet nuclear protection if Israel developed or obtained access to nuclear arms.[13] In February 1966 US government circles made it known that they were considering a plan to build atomic reactors for desalination as a means of heading off a nuclear arms-race in the Middle East. Aside from the economic benefits, the major strategic purpose would be to cut through the atmosphere of suspicion and distrust concerning the nuclear issue and to establish an international inspection system for nuclear facilities in the Middle East.[14]

The US desalination project caused bitter controversy in

Israel. Economic experts doubted the competitiveness of nuclear-power desalination and claimed that other methods were cheaper. The supporters of a nuclear deterrent rallied to oppose the plan on the grounds that the submission of the Israeli reactor to US supervision (which was understood to be a condition for US aid) was an infringement of Israeli sovereignty.

On 7 March 1966, the US Department of the Interior published a report confirming the feasibility of a 1,250 megawatt reactor in Israel which would generate 200 megawatts of saleable electricity and power for a desalination plant producing 100,000,000 gallons of water a day. Such a plant could be in operation by 1972 at a cost of $200 million. Shimon Peres, then Deputy Defence Minister, declared that he was opposed to seeking financial aid from the US, and that Israel should attempt to raise the money in France. It seems that the issue was resolved with the announcement of Prime Minister Eshkol that he was about to start negotiations with the US President on ways and means to implement the project.[15] There can be no doubt that the plan for a joint Israeli-US nuclear undertaking was based on a tacit agreement about Israeli nuclear policy in the near future, because two weeks later came news of the resignation of Professor E. D. Bergmann, Chairman of the IAEC, and the formation of a new Committee headed by the Prime Minister himself.

Replying to a debate in the Knesset in May 1966, Premier Eshkol rejected, as before, the idea of a nuclear-free zone in the Middle East. But he nevertheless hinted that while regional disarmament seemed out of the question now some 'understanding' would have to be reached lest nuclear weapons be introduced in the area.[16]

Unknown to the public (but not unknown to Eshkol), Israeli scientists were by that time making rapid progress in their efforts to construct Israel's atomic bomb. According to *Wehrtechnik* (June 1976, quoted above) Premier Eshkol, soon after he succeeded Ben Gurion, discovered that Moshe Dayan, then Minister of Defence, had defied the decision of

317

Israel's National Security Council which vetoed the military nuclear programme, and ordered the manufacture of the bomb under conditions of the strictest secrecy.

According to *Wehrtechnik* Eshkol had to endorse Dayan's view that 'Israel had no other choice.' Dayan did not believe that Israel could go on manning tanks and planes indefinitely. The nuclear deterrent was to provide the answer. The *Wehrtechnik* article also refers to the rumours circulating in 1963 that Israel had conducted a small underground test in the Negev, after which large-scale preparations started for a military atomic energy programme.

It gives the following account of these further developments:

> 'Western experts believe that by 1963 Israel had staged a subterranean test in the Negev and that soon afterwards preparations of A bomb materials started. In 1969 everything was settled but production of the bomb did not start right away. Israel's scientists concentrated on the development of new methods to cut production time of the bomb. Dimona in the Negev is not only guarded by troops, but also has a highly developed electronic system and radar screens, working around the clock. It is strictly forbidden to all planes, including Israeli war-planes, to fly over the area. During the Six-Day War an Israeli Mirage III went astray in the area. The plane was ruthlessly shot down by an anti-aircraft missile fired by their own people. When in 1973 a Libyan civilian aircraft inadvertently approached the area Israeli fighters tried to force the plane to change course. When this proved to be of no effect the plane was shot down. 108 of the 113 passengers were killed.'

Israel's bomb

Israel's nuclear effort has been the subject of numerous reports and comments in the international Press, of serious studies by atomic experts and research institutes, and of much conjecture and speculation. There are, however, no official statements or reports by the Israeli Government or the Israeli Atomic Energy Commission on the nuclear programme, its budget, or its objectives. The Government has also consistently refrained from commenting on the evaluation of Israeli nuclear capability by foreign experts who, on the whole, have a high opinion of Israeli achievements.

According to George H. Quester, Assistant Professor of Government at Harvard University:

'After some years of French assistance...the Israeli program includes a reactor at Dimona which can produce five to seven megawatts of electrical power, or five to seven kilograms of plutonium (about enough for one bomb) a year...Israel is now approaching self-sufficiency in nuclear technology, perhaps capable of undertaking the construction of larger power reactors and of separation plants for reprocessing plutonium as well as more advanced projects. The assembly of fissionable materials into weapons is not beyond Israeli competence either, allowing a relatively small period to complete the basic research. It is only in the supply of uranium that Israel might remain for some time dependent on outside sources, although enough such material might yet be extracted from phosphates to support the Dimona reactor.'[17]

Most foreign experts have made a similar assessment of Israel's nuclear development. Leonard Beaton considered the possibility of some separation capability at the laboratory level,[18] while S.B. Bell suggested the possibility of some clandestine separation facility.[19]

Professor Shimon Yiftach, a member of the Israel Atomic Energy Commission and teacher at Haifa Technion, stated in an article in *Maariv* on 15 March 1973, that

'...Israel is on the threshold of a revolution in the production of electrical power. Natural uranium will increasingly be replacing oil. As a result of this revolution, Israel is about to start a nuclear industry. The

319

Israeli government is closer today than it has been in the
last 10 years to a decision to order a nuclear reactor of
400-600 megawatts to generate electricity.

Since the planning and construction of a nuclear
power reactor requires seven to eight years, nuclear
power will make its entry into Israeli industry only at the
end of this decade.'[20]

According to Professor Yiftach, Israel will have ten nuclear
reactors by the year 2000, of which four or five will be built by
1990. Future reactors will use natural uranium extracted
from phosphates. In his opinion, the reserves of phosphates
in Israel amount to 220 million tons but they are poor in
uranium. The maximum amount of uranium in them is
estimated to be 25,000 tons.

Professor Yiftach estimates the amount of extractable
uranium as being 50-60 tons a year. 'This quantity can serve
to fuel three nuclear reactors, if the uranium is stored from
now on.' He also made it known that the nuclear reactor in
Nahal Sorek, which uses enriched uranium and natural
water, will be increased from five to ten megawatts.

According to the data available to SIPRI, Israel's nuclear
programme is still at the 'research' stage; the reactor at
Dimona is registered as 'research plant', while the reprocessing
plant is still assumed to operate on a small scale only.[21] The
SIPRI data is that released and carefully screened by the
Israeli Atomic Energy Commission.

The discovery that Israel had a nuclear bomb was,
according to the article in *Wehrtechnik*, made by the US
reconnaissance plane SR-71 Blackbird in 1973. Two Israeli
Phantoms were instructed to shoot down the Blackbird, but
they could not reach its 85,000 ft altitude and it made an easy
escape.

One explanation of how Israel acquired nuclear weapons
has been put forward by Howard Kohn and Barbara Newman
in a sensational story which appeared in the American
magazine *Rolling Stone* in December 1977. The authors claim
that Israel has up to 15 large nuclear bombs, and they

describe how some 8,000 pounds of uranium and plutonium have 'gone missing' in the United States and found their way to Israel. Almost 1,000 pounds of the missing material was highly-enriched (weapons-grade) uranium. Only 22 pounds are needed to manufacture a functional bomb.

The company that 'lost' the largest amount of enriched uranium was the Nuclear Materials and Equipment Corporation (Numec) of Pennsylvania, whose then President, Zalman Shapiro, has close ties with Israel. Numec began operations in 1957, and when officials of the Atomic Energy Commission inspected Numec's inventory in 1965 they found 194 pounds of enriched uranium missing. Shapiro claimed the material had been accidently buried in waste pits at the plant. The AEC demanded that the pits be dug up, but only 16 pounds of uranium were recovered there.

Taking a closer look at Numec's records, the inspectors found that 382 pounds of uranium were in fact unaccounted for. The AEC investigator in charge of the inquiry was preparing a 'blistering' report on Numec when he suddenly decided to resign and accept a job offer from Shapiro.

Other AEC officials tried to get the Justice Department to investigate whether the material had been stolen, but then, according to *Rolling Stone's* informants, President Johnson ordered an official cover-up of the issue. This was because the CIA, which was also watching Numec, was afraid that its own role in supplying Israel with nuclear material and technology during the 1950s (in direct opposition to the official position of the Government) might be exposed.

The official inquiry dried up, but in 1967 the AEC reinventoried the plant and this time found another 190 pounds of enriched uranium missing, making a total of 572 pounds. *Rolling Stone's* sources said this was because Shapiro was continuing to supply Israel with uranium. The CIA tailed Shapiro during his visits to Israel, and was certain that he was part of a nuclear-material smuggling ring.

Shapiro had to pay the AEC $1.1 million in penalties for the missing uranium. He then sold the plant, and Israel lost its major source of weapons-grade uranium.

Kohn and Newman did not name their major informants but wrote that '...both are experts in Mideast affairs; one is a highly–valued Pentagon consultant, the other a former National Security Agency official. Their information was corroborated, in part, by a former top CIA official, a former White House aide and other government officials.'

The sources told *Rolling Stone* that after Israel lost its Numec contact it resorted to desperate tactics. A special commando unit was trained by Israel's intelligence service to raid the nuclear powers. The unit's first foray was highly successful. 'In France the Israeli hijackers attacked a 25-ton truck that was ferrying a load of Government uranium. They fired tear-gas into the truck's cab to disable the driver, then escaped with the truck and smuggled the radioactive booty to clandestine military bases in the Negev desert.' Another raid was successfully carried out in England, but Israeli scientists were disappointed to discover that what had been stolen was a load of low-grade 'yellowcake' uranium.

Rolling Stone was told that all these incidents, involving the USA, France and England, were hushed up by the respective governments for nearly a decade because they feared the political consequences. Israel's intelligence service, Mossad, had counted on this when it planned the hijackings.

Israel apparently also obtained nuclear material from West Germany and France, but because these governments did not want to be seen collaborating with Israel, fake hijackings were arranged. By 1968 Israeli scientists were perfecting their technique of enriching low-grade material, and a deal was made with West Germany to trade certain expertise for 200 tons of yellowcake. While idling in the calm seas of the Mediterranean in November 1968, the West German freighter *Scheersberg* was 'raided' by commandos, and 200 tons of yellowcake were removed.

The *Rolling Stone* story is supported by the *Jerusalem Post*.[22] It quotes the Israeli Government explanation that 'Israel obtains fuel for the Dimona reactor by extracting uranium as a by-product of phosphate plants along the Dead Sea'; but it clearly attaches more credibility to the statement made by Paul Leventhal, former Counsel to the Senate Government Operations (now Governmental Affairs) Committee which handles legislation dealing with the spread of nuclear weapons. Paul Leventhal told the *Jerusalem Post* the following story:

> 'The German-registry ship and its cargo disappeared in the Mediterranean while *en route* from Belgium to Italy. A few weeks later it reappeared with a new name, new registry, a new crew but no uranium; it is assumed that it was unloaded in Israel.'

Leventhal said that he had made the disclosure to nuclear-power opponents in an effort to show the need for greater safeguards in the handling of nuclear materials:

> 'The important thing is not the country that got it, but that nuclear material could be stolen. The 200-ton shipment was enough to cover Israel's nuclear fuel requirements for 20 years. It is assumed that Israel has the materials to make nuclear weapons. That is no surprise.'

Asked whether he was absolutely certain that the 1968 uranium shipment was redirected to Israel, Leventhal said: 'No, I only related what I heard from several different sources, all of them authoritative. My primary information came from a non-US source, but in each of the versions I heard Israel was the country that got it.'

Leventhal understood that the uranium was raw ore, milled into natural uranium concentrate or yellowcake in Belgium. Uranium can be used in this form in the Dimona reactor.

Money and scientific information were also traded for uranium from France in 1969, according to *Rolling Stone*. The uranium was once again 'stolen' by the commandos.

America's Atomic Energy Commission investigators learned

of President Johnson's cover-up of Numec/Israeli dealings only late in 1977 when they were briefed by senior CIA official Carl Duckett. He told the AEC men that Johnson had ordered CIA director Richard Helms to let Israel keep the contraband uranium, to drop its investigation, and not to inform other Federal agencies. There was no more official action, said Duckett, until the House Commerce Committee, researching the general problem of nuclear security during 1976, learned of the missing Numec uranium.

Duckett was then called in to brief the Nuclear Regulatory Commission (NRC), which had replaced the AEC in 1975 as the official nuclear watchdog. He told the Commission that the CIA was virtually certain that the Numec uranium had gone to Israel. The Ford administration then ordered an FBI investigation, which also concluded that Numec had been channelling enriched uranium to Israel. No action was taken because the then Attorney General felt the evidence would not be admissible in court. The file was passed on to the Carter administration early in 1977 and the case is still surrounded by official secrecy.

Howard Kohn, in a separate article in the same issue of *Rolling Stone* which revealed the clandestine supplies of uranium to Israel, says that several other nuclear plants in the USA are unable to explain significant discrepancies in their inventories. Some investigations have been conducted to ascertain whether uranium or plutonium was sold to Israel, but no conclusive evidence has been revealed.

Because of a variety of incidents at nuclear plants involving bomb threats, extortion attempts and security breaches, the NRC, he says, may resort more and more to police-state measures:

> 'A secret report commissioned by the NRC in 1975 raises the likelihood of a special nuclear police force empowered to conduct domestic surveillance without a court order, to detain nuclear critics and dissident scientists without filing formal charges and, under certain circumstances, to torture suspected nuclear terrorists.'

Israel and South Africa

After the US and Soviet spy satellites discovered in August 1977 that South Africa was ready to test its first atomic bomb in the Kalahari Desert, *Newsweek* wrote that '...some US intelligence analysts concluded that the bomb the South Africans had planned to set off actually had been made in Israel.'[23] A high-ranking Washington official was reported as saying, 'I know some intelligence people who are convinced with near certainty that it was an Israeli nuclear device.'

There are a number of reasons for which Israel is justifiably suspected to have a hand in the South Africa nuclear programme. The bonds between the two countries were aptly summed up in 1968 by *Die Burger*, the Cape Province paper of the National Party:[24]

'Israel and South Africa have much in common. Both are engaged in a struggle for existence, and both are in constant clash with the decisive majorities in the United Nations. Both are reliable foci of strength within the region, which would, without them, fall into anti-Western anarchy. It is in South Africa's interest that Israel is successful in containing her enemies, who are among our own most vicious enemies; and Israel would have all the world against it if the navigation route around the Cape of Good Hope should be out of operation because South Africa's control is undermined. The anti-Western Powers have driven Israel and South Africa into a community of interests which should be utilised rather than denied.'

Die Burger's advice was not lost on either Tel Aviv or Pretoria, which were burying their past differences[25] and beginning to improve their relations through the growth of economic and military co-operation. When Black Africa broke off diplomatic relations with Israel in 1973,[26] South Africa compensated by opening its domestic market to Israeli goods. In a Press interview in 1974, Yitzhak Unna, Israel's first

325

Ambassador to South Africa, disclosed that South Africa had raised the ceiling for direct investment in Israel from R7 million to R20 million. He said that in 1973 Israel had imported goods worth $32 million from South Africa, mainly sugar and cement, and that exports had totalled $12 million. Between 1971 and 1975 total trade between the two countries had almost tripled, from R13 million to R37 million.[27]

Israeli-South African military co-operation (see Appendix 8) has been even more extensive.[28] Relations between the countries reached a peak in April 1976, when South African Prime Minister John Vorster visited Israel. The *Jerusalem Post Weekly*[29] reported on the dinner given in Vorster's honour by Israeli Prime Minister Rabin. After commending 'South Africa's long support of Israel as a free and independent Jewish state' Premier Rabin said:

'I believe both our countries share the problem of how to build regional dialogue, coexistence and stability in the face of foreign-inspired instability and recklessness ... This is why we here follow with sympathy your own historic efforts to achieve détente on your continent, to build bridges for a secure and better future, to create coexistence that will guarantee a prosperous atmosphere of cooperation for all the African peoples, without outside interference and threat.'

Officials in Jerusalem firmly denied that Vorster was in Israel to buy arms or to set up an 'anti-Communist alliance', as the South African press was speculating. Vorster himself told newsmen that talk of an impending arms deal was 'utter nonsense'. Israel's ambassador to Pretoria, Yitzhak Unna, told a radio interviewer that Israel's and South Africa's defence needs were quite different. Unna said that the visit had no specific diplomatic goal: it signified the improving relations between the two countries. South Africa had been one of the few countries that had not turned its back on Israel after the Yom Kippur War, the envoy noted, and relations had become increasingly normalized since the two countries had raised the level of their representation to full embassies following the war. South Africans regarded Israel as a bulwark against Soviet expansionism. Ambassador Unna added that the invitation to Mr Vorster by no means implied approval for South Africa's *apartheid* system. Israel maintained normal relations with many states whose internal systems it did not approve of — and vice versa.[30] Vorster was quoted by Israeli Radio as commenting during a tour to Sharm-el-Sheikh that '...relations with Israel have never been so good.'[31] However, Vorster's announcement that relations between the two countries would be expanded in a number of areas, including scientific and technical co-operation,[32] triggered off speculation that this involved a deal in the field of nuclear energy.

Very little is known about the nature of this co-operation; both South Africa and Israel operate in conditions of the strictest secrecy. However, now that Israel can no longer rely on 'hijacking' uranium shipments, there can be little doubt that the Israelis regard South Africa as a vitally important source of uranium, both natural and enriched, free of any international safeguards or inspection.

There is also a similarity between Israeli and South African defence strategies. The Israelis have always held the view that Arab superiority in numbers was outweighed by Israeli superiority in arms, military technology and 'heroism'. This Israeli school of thought was explained by the former Foreign Minister and Prime Minister Moshe Sharett as follows:

'The only language the Arabs understand is force. The State of Israel is so tiny and so isolated ...that if it does not increase its actual strength by a very high coefficient of demonstrated action, it will run into trouble. From time to time, the State of Israel must give unmistakable proof of its strength and show that it is able and ready to use force in a crushing and highly effective manner. If it does not give such proof, it will be engulfed and may even disappear from the face of the earth.'[33]

327

In similar vein Vorster warned President Kaunda in 1970 that Zambia would be 'hit so hard that she would never forget it'.[34]

But the myth of Israeli invincibility was shattered by the 1973 war, and South Africa's military fiasco in Angola in 1976 did serious damage to South Africa's image as a great military power in Africa.

The military and psychological setbacks to both countries contributed to the importance of a strategy based on a nuclear 'deterrent', which the military planners of both Israel and South Africa see as the only solution to offset an inferiority in manpower.

11
Iran and Brazil

The Shah's finger on the nuclear trigger?

Iran is only at the threshold of the nuclear age, but by 1990 it plans to have a network of twenty nuclear power stations; these would treble present energy capacity and put Iran's reactor construction programme in the category of countries like France, West Germany and Japan. A total of some 100,000 tons of uranium would be required for the expected lifetime of the reactors; and this is where South Africa comes in.

Washington Post correspondent Thomas O'Toole reported on 12 October 1975 that Iran was about to sign an agreement to buy $700 million-worth of uranium from South Africa and put up part of the money for a huge uranium-enrichment plant to be built in South Africa. The Iranian bid may explain why the South African Uranium Enrichment Corporation (UCOR) pulled out of the deal with the West German Company Steag, described in Chapter 3 (pp. 83-84). The Iranian offer had an additional attraction which the Germans could not match: oil. Iran is the largest and most important exporter of oil to South Africa and the only one which, until now, has ignored the OAU's appeal for an oil boycott of South Africa.[1]

The Iran-South African deal was reported to involve the purchase by Iran of 14,000 tons of uranium oxide from South Africa; this was described by the *Washington Post* as '... the largest single sale of its kind', and it sets the seal on economic relations between the two countries. For Iran the agreement holds the prospect of uncontrolled use of imported uranium, since South Africa is not a signatory to the NPT nor is its uranium mining subject to any international safeguards or inspection.

THE IRANIAN ROAD TO NUCLEAR CAPACITY

The Iranian nuclear programme began very modestly and in strict compliance with the safeguards obligations imposed on the signatories of the NPT, which Iran signed in 1968 and ratified two years later. There were two reasons for this careful approach:

1. Unlike Israel and South Africa, Iran lacked both the technological basis and the qualified personnel even for a nuclear research programme. There was absolute dependence on 'know-how' from the nuclear powers assisting Iran to overcome this drawback: the USA, France and West Germany; this excluded any possibility of the secrecy in which the Israelis and South Africans would be able to operate.

2. Prior to the 1973 oil crisis which catapulted Iran into the world of the super-rich, the Shah, a statesman with a shrewd grasp of *realpolitik*, was constantly aware of his 'big brother', the neighbouring Soviet Union, and was worried about Soviet military support for some of the Arab countries in the area, notably Iraq. He knew that any wrong moves concerning nuclear developments could provoke Soviet counter-measures, such as supplying nuclear warheads to the military rulers in Baghdad.

Within barely five years the Shah's position changed dramatically: Iran had acquired the most modern and sophisticated weaponry, finding the Western powers, notably the United States, to be willing suppliers. In order to minimise any suspicions the Soviets might have had about his shopping for arms in the West, he also placed a $414-million order with the Kremlin for the delivery of an unspecified number of Soviet missiles including SAM-7, SAM-9, VCIBMP-1, and ASU-85 and 2SU-23-4 guns. The agreement was signed by the Shah's War Minister, Toufanian, in Moscow in November 1976.[2]

The purchase of two 900MW French nuclear power stations for 10 billion francs, which was announced on 15 June 1977, meant the elevation of Iran to one of the Third World's nuclear powers. This Iranian-French deal is believed to have been initiated by the Shah in June 1974 when he came on a State visit to France to discuss the expansion of nuclear co-operation between the two countries. This was a follow-up of the 1973 agreements, according to which France was to supply several small nuclear reactors with a capacity of 5,000 MW at a cost of $1.2 billion.[3] His programme included a visit to the French Nuclear Research Centre at Sacloy – the same institution which, in 1955, helped Israel start its top-secret nuclear programme at Dimona in the Negev desert.

Iran became interested in nuclear research in 1958. It started on a small scale under the auspices of the National Iranian Atomic Energy Commission in the Ministry of Economics, and with the assistance of the US Atomic Energy Commission. Iran signed bilateral agreements for the peaceful utilization of nuclear energy with the USA (27 April 1957), France (26 June 1974), and West Germany (4 July 1976).

Since 1958 the nuclear centre at Teheran University has been conducting research in nuclear physics, electronics, nuclear chemistry, radio-biology, and medical physics. It has also been training Iranian experts in these fields under the direction of Dr H. Rouhaninejad, an Iranian nuclear physicist. Teheran University's 5MW pool-type research reactor supplied by the United States went into operation in November 1967. Since 1972, it has also utilized a 3MV van de Graaf-type accelerator.

On 21 May 1973 *Newsweek* reported the Shah as saying that Iran was going to save its oil resources for its own future petroleum and chemical industries, and would therefore introduce nuclear power as an alternative source for its own energy needs. This might seem a credible explanation of Iranian nuclear intentions were it not for the fact that at the same time Iran has also embarked on perhaps the most spectacular armaments programme ever undertaken by a Third World country.

In 1974 Iran's military expenditure jumped from the already impressive figure of $1.8 billion in 1973 to a staggering $4.1 billion. This figure rose further in 1975 to reach $7.3 billion, which is over $1 billion more than the entire expenditure of the whole African continent (excluding Egypt but including South Africa).[4] On 6 February 1978 the Prime Minister presented a new budget which allocated $10 billion for arms purchases: this was 17 per cent of the total budget, and 25 per cent more than in 1977. The following are a few examples of Iran's military acquisitions in 1974-76: 30 helicopters (Aéro Spatiale AS.11/12) and 12 missile boats (Kaman class) from France; almost 300 military aircraft from the USA, including 80 Grumman F-14A Tomcat fighter-interceptors and seven Boeing 707-396C super air-tankers used by the US Air Force for refuelling in the air; and from Britain, 1,500 Chieftain tanks called 'Shir Iran', 360 Alvis Scorpion light tanks and 175 Vickers armoured recovery vehicles.[5]

The nuclear power-stations followed the 'stratotankers' and other weapons. On 29 July 1975 the *Suddeutsche Zeitung* confirmed the sale of two West German nuclear reactors of the Biblis type with a capacity of 1,200 megawatts. The agreement was signed during the visit of Minister of Economics Friederichs to Teheran. Within a year the French newspaper *Le Figaro* reported that 1,500 West Germans with their families were living in a secluded German village, near the site of the two reactors at Bushez, where 4,000 Iranian workers were also employed.[6]

On 3 November 1975 West German Chancellor Schmidt concluded his visit to Iran by signing an agreement for the establishment of a university staffed by Germans at Rash, in the Caspian Sea region, to teach some 10,000 Iranian students medicine and technology. The London *Times* quoted the Iranian Prime Minister Amir Abbas Hoveida, as saying that '. . . although financed by Iran, West Germany will take full responsibility for the new university's curriculum, applied research and management.'[7] What is not mentioned in the press reports is that nuclear technology is one of the main parts of the curriculum of the new university. More specific was the little notice in *Atomic Energy Week* reporting an agreement between the Atomic Energy Organisation of Iran (AEOI) and the Karlsruhe Society for Nuclear Research to train 25 Iranian nuclear physicists.[8] The *Financial Times* reported that Britain had joined the Iranian atomic venture by undertaking responsibility for a control system for the German and French nuclear reactor deliveries and installations.[9] But perhaps the most significant announcement was the proposed Iranian co-operation with France in the production of enriched uranium, announced in Paris on 2 January 1975 by the French Industry Minister Michel D'Ornano. He said that Iran had agreed to set up a joint company which would take a 25% share of the French holding in the Eurodif uranium-enrichment project.

Eurodif's plant at Tricastin in southern France, a joint commercial uranium-enrichment enterprise of France, Belgium, Italy and Spain, is scheduled to start production by December 1978; its full capacity will be eleven million separative work units (SWU) per year from 1982 onwards. Although details of the Iranian-French financial arrangement have not been disclosed, it is assumed that Iran had sought and received guarantees of supplies from the enriched uranium produced by Eurodif.

Against this background it is difficult to accept the repeated Iranian denials that a nuclear bomb is the final objective of its nuclear programme. In June 1974 the Paris economics magazine *Les Informations* published an interview with the Shah in which he was asked: 'Do you think that one day you will possess the nuclear weapon?' 'Undoubtedly, and sooner than it is believed,' he replied, adding: 'Unlike India, we have thought first of our people and after that of technology – look at the result today.'[10] Asked about arms purchases in general the Shah said that '... the Persian Gulf is a strategic zone of prime importance. The slightest spark and everything goes up. The very sophisticated army that I maintain is my deterrent and striking force.'

The following day the Iranian Embassy in Paris denied that the Shah had suggested in the interview that his country intended to develop nuclear weapons. The Embassy stated that it was '... in a position to say that His Majesty the Shah never made any statement that could be interpreted in this way.' An embassy spokesman said the statement was a 'pure invention'. He also denied the statement about India. What the Shah had said, he claimed, was that Iran was not thinking '... of acquiring nuclear arms but that if the small states equipped themsleves with such armaments then Iran would revise its policy in this domain.'[11] This kind of blanket denial of official boasting about a nuclear capability is strikingly similar to the South African case (see Chapter 5).

South Africa, Iran and France

On 12 November 1970 the South African Department of Foreign Affairs announced that Iran had agreed to an exchange of Consuls-General between the two countries. This was described as '... the first fru... of the specifically diplomatic offensive aimed at countering the international isolation of South Africa. Iran's wealth in oil and increasing trade with South Africa would almost inevitably mean increased economic contacts'.[12]

This prediction proved correct. Today Iran is South Africa's main supplier of oil and is a partner in the Sasol (oil from coal) project with the National Iranian Oil Company owning 18 per cent of the Sasol shares.

South Africa's nuclear deal with Iran was reported in South Africa to be part of a tripartite arrangement including France. During the Shah's visit to Paris in 1974 to negotiate the purchase of nuclear reactors, three South African Cabinet Ministers were also in the city.[13] A spokesman for the French Foreign Ministry, asked about safeguards for the reactors sold to Iran, said that the agreement with Iran had 'implied safeguards' – an admission that the necessary specifics had been ignored.[14]

The tripartite deal clearly has advantages for all three countries concerned. Iran supplies both money and the oil which South Africa and France, having no domestic sources of oil, need desperately. France supplies the sophisticated technology for Iran and South Africa, without the level of safeguards imposed by her American rivals. South Africa provides uranium, the basis for the Iranian and French nuclear ambitions, also apparently free of the safeguards imposed with increasing stringency by the Americans, Canadians and other major suppliers.

South Africa, Israel and Iran

Iran is Israel's chief source of oil. Iranian oil is also transported through the Eilat-Ashkelon pipeline *en route* to Western Europe. Although there are no diplomatic relations between the two, unofficial co-operation is considerable. Apart from the co-operation of Israeli construction and industrial corporations providing the Iranian economy with technical know-how, there is also a significant level of military co-operation. Details of this have been reported by the US quarterly *Orbis* (published by the American Institute for Foreign Policy) in its 1975 Winter issue. According to the article, hundreds of young Iranian officers have undergone training in Israel, as have Security Service personnel and anti-insurgency units. A pact signed in May 1969 provided for the supplying of the Iranian army with Israeli electronic equipment and the training of Iranian operators. In 1972 it was estimated that Iran purchased military equipment from Israel, including Gabriel missiles, worth $100 million.

The reasons why South Africa, Israel and Iran are striving for the acquisition of their own nuclear capability are strikingly similar and can be summed up as follows:

1. Each of the three countries is situated in an area of political tension and feels seriously threatened by its neighbours. Iran borders on the Soviet Union and Iraq: although the hostility of Iran's neighbours is not comparable with that of Israel's or South Africa's, both of which are permanently on the brink of war with the countries surrounding them, all three states believe themselves to be in a state of siege externally. South Africa is also in a state of continual ferment internally, and all three face serious domestic opposition.

2. Each of the three countries has a special relationship with the Western Powers. Iran plays upon its importance as a key oil-supplier to the Western World, and one that can keep the Gulf open for the movement of oil to the West from the other oil producers in the region. Israel has turned the issue of its own survival into a corner-stone of US policy in the Middle East. South Africa poses as a protector of shipping around the Cape. All three present themselves as a bulwark against 'communist subversion' in their regions.

3. South Africa, Israel and Iran are keenly aware of the limitations of their influence on the Western Powers, with whom they have informal alliances. They all feel that possession of nuclear weapons can strengthen their

bargaining position with the Western alliance. At the same time they would like to prove to the West that they have the capacity to offer critical support in any military conflict with the communists.

All three countries have been helped in their nuclear ambitions by France, which has not signed the NPT, and West Germany, a reluctant NPT signatory. There are also similarities between the West German-South African deal and that between West Germany and Brazil.

A West German bomb for Brazil?

On 26 June 1975 West Germany signed a $2 billion nuclear deal with Brazil: an agreement to supply that country with a complete nuclear industry ranging all the way from uranium prospecting to the use of nuclear power to generate electricity.

The agreement also set out the guidelines under which individual German companies would arrange contracts with Brazil. Its first section deals with the prospecting for and the exploitation of uranium deposits in Brazil, together with production of natural uranium concentrates and compounds. West Germany's co-operation here is intended to give it access to Brazil's uranium in the future. Another section of the agreement deals with the enrichment of uranium. The Brazilian nuclear company, Nuclebras, will engage in a joint programme with Steag of Essen to develop the latter's jet-nozzle enrichment process, perfected through Steag-UCOR co-operation.

The common uranium-enrichment technology – the jet-nozzle system – is not the only link between Brazil and South Africa. Until Brazil's own uranium deposits, located in the Mato Grosso rain-forest area, yield their first output, Brazil will be importing South African uranium enriched in the United States. This is already the source of fuel for the 626 MW power reactor at Angra dos Reis supplied and installed by the American company Westinghouse. Interatom of Germany will join Steag and Nuclebras in building a demonstration plant in Brazil, with a capacity of 180,000 SWU per year, that is planned to start operation in 1981. This will be followed by a commercial plant later on.

Under the nuclear reactor section the guidelines lay down the supply of heavy power-plant equipment and the establishment in Brazil of nuclear construction companies and companies to supply the nuclear fuel elements. Kraftwerk Union and its 350 German sub-contractors will thus play an important part in the supply of up to eight nuclear power-stations in Brazil.

Further sections deal with the reprocessing of irradiated fuel elements and with financing. For fuel, Brazil proposes to buy an oxide fuel fabrication plant to be built by the Kraftwerk Union/Nukem/General Electric subsidiary, Reaktor Brennelement Union. Spent fuel reprocessing facilities would be provided by Kewa-Kernbrennstoff-Wiederaufarbeitungs, a joint subsidiary of Hoechst, Bayer, Gelsenberg and Nukem.

The agreement between the Foreign Ministries of West Germany and Brazil provides the framework of safeguards for the peaceful use of the nuclear technology that is being made available to Brazil. However, these assurances, which West Germany as a NPT signatory is obliged to apply to all its nuclear transactions, do not actually prevent Brazil from producing enriched uranium itself. After Brazil reaches this stage, it is a small step to converting enriched uranium into nuclear explosives.

After signing the series of agreements in Bonn, the Brazilian Foreign Minister, Antonio Azevedo da Silveira, declared: 'Brazil has gained new technological and political status on the world scene.'[15]

The West German-Brazilian agreement has come under heavy fire from the US Administration and Congress. Although some of the criticism was undoubtedly motivated by disappointment that the US lost the Brazilian deal to West

Germany, Fred Iklé, Director of the US Arms Control and Disarmament Agency, rightly commented that:

> 'Unhappily, shortsighted commercial interests some-
> times militate against application of effective controls...
> You would think that all nations willing to export nuclear
> materials ,or equipment would be anxious to prevent
> proliferation... Even the largest nations would suffer
> grievously if nuclear explosives became widely available.'[16]

The possibility of conversion of the Brazilian nuclear system, the biggest and the most sophisticated in Latin America, to the manufacture of nuclear weapons has disturbed not only the US but also Argentina, which has long competed with Brazil for Latin American leadership.

James Nelson Goodsell, the Latin American correspondent of the *Christian Science Monitor*, has reported that the signing of the West German-Brazilian agreement gave rise to calls in Buenos Aires for counter-action.[17] Considering the fact that Argentina, which has been operating a 319 MW nuclear-power plant since 1974, has also abstained from signing the NPT, one can easily imagine what kind of 'counter-action' the Argentinians have in mind.

American opposition to West Germany's agreement with Brazil continues under the Carter Administration as outlined in Chapter 8. However, West Germany has refused to bow to American pressure to cancel the agreement. *Der Spiegel* has reminded its readers that this was not the first time the US has tried to inhibit German business with Brazil:

> 'The order from Brazil looked perfect: the Brazilians had
> placed an $80 million order with the Federal Republic of
> Germany for ultra-centrifuges with which they could
> enrich uranium themselves. Nothing came of it; Washington
> banned the deal... This was in 1953... At the time of
> course the USA was an occupying power in Germany.'[18]

But this time the West Germans made it very clear that Washington could no longer tell Bonn what to do.

Critics in the Federal Republic have also been ignored.[19] They are powerless against the official West German argument, the logic of which has long guided the arms dealers of the world: 'If we do not supply it, somebody else will.'[20]

Postscript
West Germany: the generals and the bomb

Atoms for peace and atoms for war are two sides of the same coin. The story of the rise of West Germany as a nuclear power is therefore a story of rearmament.

In the course of its thirty-year history, the Federal Republic of Germany has made remarkable progress in armaments. The first 102 soldiers of the West German Army, the *Bundeswehr*, were heckled by angry crowds of their own people when they first appeared in uniform in January 1956. In twenty years they have grown into a 465,000-strong armed force with 800,000 in reserve, supported by a defence budget of about DM40 billion a year (almost £10 billion). They constitute the largest contingent of NATO ground and air forces, and are also the best equipped. The *Bundeswehr*'s most formidable weapon, the *Leopard* tank, has made it as effective in mobile warfare as the *Tiger* tank (its predecessor) made Hitler's *Wehrmacht*.

Equally impressive have been West German nuclear achievements, due largely to transfers of sensitive technology from the United States. The West German nuclear industry now employs about 100,000 highly-skilled workers. The Frankfurt-based Kraftwerk Union (KWU), a subsidiary of

341

Siemens and AEG-Telefunken, can deliver six ready-to-use nuclear plants a year. It has an order-book worth DM28 billion – bigger than any other West German company. Its pressurised water reactor, Biblis, heads the list of West German exports. Among the thirty-three countries with which the Federal Republic conducts nuclear business are the 'near-nuclear' powers of Brazil, Argentina, Iran, Israel and South Africa – all determined to obtain a nuclear capability, and none of them except Iran a party to the Nuclear Non-Proliferation Treaty (NPT).

Despite this military and industrial progress, the West German generals have not yet realized their ambition to have their fingers on the nuclear trigger. But they are coming very close. The nuclear warheads for the estimated 7,000 rockets and missiles on NATO bases in West Germany, aimed at targets in the Soviet Union and Eastern Europe, are still locked away in bunkers guarded by a special American task-force. But the barrier inhibiting the West German commanders' direct access to these weapons is more political than physical. Surrounded by German troops, the American door-keepers of the nuclear devices are as much hostages as they are guards.

A clash between the American and German units over access to the nuclear store is almost inconceivable. It is also unnecessary. There are such large stocks of plutonium and enriched uranium on West German territory that, if it were converted to military use, the country could have a consider-able number of nuclear bombs of its own. The Stockholm International Peace Research Institute has estimated the theoretical capacity for West German atomic bomb production in 1976 at about 100 bombs, increasing rapidly. The conversion would be an easy technical task for the West German nuclear industry.

Because of the Second World War experience and the present anti-nuclear mood of the West German public, no government there is likely to defy international safeguards and the country's treaty commitments by ordering the production of nuclear weapons inside its territory. This would upset both the Soviet Union and West Germany's Western allies, both sides being anxious to maintain the precarious balance of power. But such political considerations are disregarded by West German military circles, which have for a long time been impatient at the subsidiary role assigned to them by the United States in the event of a Third World War – the role of a dispensable 'tripwire' allowing the rest of NATO time to mobilise.

One document illustrating the nuclear plans of the West German generals is the so-called 'Generals' Memorandum', officially known as *Voraussetzungen einer wirksamen Verteidigung*, which was published in the Government Bulletin No. 155/S.1527 of 20 August 1960.

The Memorandum sums up the deliberations of the commanding officers of the *Bundeswehr* at the *Kommandantsbesprechung*, held from 11 to 13 July 1960 in Kiel. It examines the international political situation and the measures to be taken in order to strengthen the *Bundeswehr* and increase its credibility with West Germany's allies as well as with its enemies.

In the section entitled 'Nuclear Weapons for the Defence Forces'(see Document 1, below), the generals put forward a strongly-worded demand for atomic weapons as an essential prerequisite for the 'effective armament' of the shield forces of which the *Bundeswehr* formed the core. 'The soldiers of the *Bundeswehr* are entitled to weapons of equal effectiveness to those of the enemy', states the Memorandum. It continues, 'As matters stand now, the commanding officers' responsibility for their soldiers under present conditions compels them to demand nuclear weapons. Otherwise a dangerous situation would arise: German soldiers would face problems they cannot possibly handle, and their self-confidence would be shattered.' The Memorandum concludes with the following sentence: 'No effective defence of a country can exclude the atomic armament of the shield forces.'

The Memorandum, which also appeared in the official

military publication *Information für die Truppe* (Information for the Troops), was the work of Inspector-General of the Armed Forces Adolf Heussinger and his Minister, Franz Josef Strauss. Strauss is leader of the Bavarian Christian Social Union (CSU), has a seat on the Board of West Germany's biggest military consortium (Messerschmidt-Bölkow-Blohm), and is spokesman for the right wing in the Federal Parliament.

He is, at the same time, a supporter of the South African Government. He has a close relationship with Dr Roux, President of the South African Atomic Energy Board, and has made frequent visits to the Republic.

Appointed Minister without Portfolio in 1953 by Adenauer, Strauss soon made his way to the key areas of defence and foreign affairs. In October 1955, he became the first Minister in the newly-created Ministry of Atomic Affairs, a position he used for building close ties between the nuclear industries of West Germany and the United States.

When Strauss took over the Ministry of Defence on 15 October 1956, a post he held until 1962, he faithfully followed Adenauer's line that West Germany's survival lay in close alliance with the United States and other Western countries; he nevertheless insisted on full participation by West Germany· in all major Western defence decisions and demanded that West Germany have access to nuclear weapons. He continued to advocate West German access to nuclear hardware after leaving the Defence Ministry.

The controversy reached new heights during the debate in 1968 about the NPT. West Germany accepted the treaty only after a government confrontation with the right-wing opposition in 1969. The *Frankfurter Allgemeine Zeitung* of 3 January 1964 reported that in the previous month, when the NPT was being negotiated between the United States and the Soviet Union, West German Chancellor Ludwig Erhard and Foreign Minister Schröder had told President Johnson that West Germany would not accept any further restrictions on its nuclear ambitions unless it received clear guarantees regarding the sharing of responsibility in nuclear matters.

More recently, there have been revelations about American plans which suggest that, in the event of war, the US would abandon up to one-third of West Germany to Warsaw Pact forces. Another shock to West Germans was the revelation that French missiles carrying nuclear warheads were actually aimed at targets in the Federal Republic. The proposed use of the neutron bomb on West German territory is also of considerable public concern there. Curiously, Strauss has remained silent about these disclosures.

There are many West Germans who believe that the US would hesitate to deploy its nuclear weapons to defend West Germany from an attack, since this would expose the American heartland to retaliation by missiles launched from Soviet submarines around its coastline. Isolationist sentiments in the US are also reflected in growing opposition in Congress to maintaining troop levels in West Germany, and there are doubts about the American commitment to West European defence in the long term. One may well wonder why, in this situation, Strauss has uncharacteristically said nothing at all.

There is little doubt that the South African nuclear arsenal, having been built up with West German help (see Chapter 3), could be made available to the West German military. As Table 7 shows, the nuclear fuel cycle in South Africa is complementary to that of West Germany; combined, they represent a formidable combination for the production of fissile materials.

In the world of international political brinkmanship, with its precarious balance between strike capability and deterrence, the possibility of West German access to South Africa's nuclear weapons would make the *Bundeswehr* more independent of the United States. It would also raise the status of West Germany with both NATO and the Warsaw Pact; in short, the country could emerge as an important military power in its own right – reflecting adequately, for the first time since the last war, West Germany's economic and financial strength.

Table 7

**West Germany and South Africa:
closing the nuclear fuel cycle.**

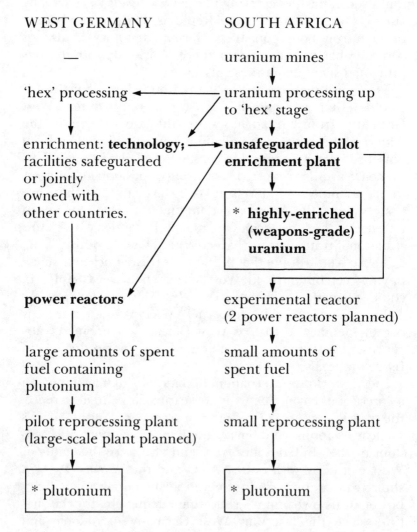

WEST GERMANY   SOUTH AFRICA

—   uranium mines

'hex' processing ◄─────── uranium processing up
                         to 'hex' stage

enrichment: **technology;** ──► **unsafeguarded pilot**
facilities safeguarded   **enrichment plant**
or jointly
owned with
other countries.

                         ┌─────────────────────┐
                         │ * **highly-enriched** │
                         │   **(weapons-grade)** │
                         │   **uranium**   │
                         └─────────────────────┘

power reactors   experimental reactor
                         (2 power reactors planned)

large amounts of spent   small amounts of
fuel containing   spent fuel
plutonium

pilot reprocessing plant   small reprocessing plant
(large-scale plant planned)

┌──────────────┐   ┌──────────────┐
│ * plutonium │   │ * plutonium │
└──────────────┘   └──────────────┘

* fissile materials for nuclear weapons.

Conclusion

Until recently the overriding question about South Africa's nuclear programme was a simple one – could it make a nuclear bomb? It is now evident that it can and that work on testing the bomb has been going on for several years, first under the guise of an interest in 'peaceful nuclear explosions' and finally in the construction of a weapons-testing site in the Kalahari Desert which was identified by American and Soviet reconnaissance satellites.

This development raises the more complicated question of nuclear proliferation and the greatly increased risk that regional conflict could trigger a nuclear war in which the superpowers would become involved.

Will the United States, openly or covertly, extend its nuclear umbrella over South Africa as a guarantee of permanent minority rule there, in opposition to forces of African liberation?

What will the West German Government do with its development of enrichment technology, and the access to special nuclear materials which has arisen from the building of the enrichment plant for South Africa; and how strong are the forces within West Germany that are pressing for rearmament with nuclear weapons?

Will the French and West Germans continue with their plans to give South Africa a 'commercial' enrichment plant and a huge nuclear power station, which together would enable South Africa to produce highly-enriched uranium and plutonium in quantity?

Can the frail nuclear-safeguards system be strengthened to limit the nuclear ambitions of countries such as Israel, Iran, Brazil, Argentina, Taiwan, and others on the fringes of the Western alliance, which are seeking nuclear weapons of their own and would so easily be able to get the necessary materials from South Africa?

Fundamental to all these questions is one unknown which is not directly related to nuclear power or to military alliances as such: it is the future of the South African economy and its ability to attract large amounts of capital from the Western world to finance the enrichment plant, the nuclear reactors and the whole research, development and industrial infrastructure that goes with these. We have already noted the secret decision in Washington to subsidise South Africa and its uranium industry by quietly driving up the gold price. The British Government's contribution is in the form of a massive contract to buy uranium from the Rössing mine in Namibia. The French and West Germans have done the same, buying considerable quantities of uranium from South Africa and investing in Rössing. These and other Western countries encourage trade, investment and direct loans for South Africa, and provide a variety of subsidies for this purpose. South Africa's survival is heavily dependent on such subsidies continuing virtually indefinitely.

The United States, Israel, West Germany, France and Britain: all have helped to provide the hardware, the fissile materials, the technology and the finance which have enabled South Africa to build a nuclear weapon. This is now an established fact, and has helped the South African Government to reject much of the international condemnation and pressure which was aroused by its suppression of the Soweto uprisings in 1976, by the bannings of anti-racist organisations and mass arrests in 1977, and by the torture and killing of detainees (of whom the best-known is Steve Biko). After the virtual collapse of the parliamentary opposition before the 1977 elections, the ruling National Party embarked on still further repression of black resistance.

It is not clear that South Africa can survive as a white minority-ruled régime even with nuclear weapons, if its economy continues to deteriorate and international money-markets remain largely closed to it. The régime may have calculated that once its commercial enrichment plant and its nuclear reactors are in operation, it can underwrite the economy by selling the precious weapons-grade uranium and plutonium, naming its own price to insecure and aggressive régimes around the world.

Documents

Document 1

Extract from West German Government Bulletin no. 155/S.1527 of 20 August 1960 (with partial translation).

Translation

Nuclear Weapons for the Defence Forces (The Shield)

If the Army were to reject nuclear weapons unilaterally, Germany could not be defended. The existence of atomic weapons would make the enemy hesitate to prepare for a sudden and massive attack. If our own troops had to fight without tactical nuclear weapons, the enemy would have a greater chance of breaking through the European defence system. Because of this weakness of our defence forces (the Shield), the Free World would be left with no alternative but to capitulate – the Soviets are counting on this – or to take refuge in 'the Sword' which would mean unlimited nuclear war. To dispense with tactical nuclear weapons means either to use strategic nuclear weapons or to capitulate.

The *Bundeswehr* must have the same weapons as the allied defence forces. NATO's forces must consist of ONE Unit. Otherwise the enemy would choose as the target of his attack the point where the adversary's weapons are inferior.

The soldiers of the *Bundeswehr* are entitled to weapons of equal effectiveness to those of the enemy. As matters stand now, the commanding officers' responsibility for their soldiers compels them to demand nuclear weapons. Otherwise a dangerous situation would arise: German soldiers would face problems they cannot possibly handle, and their self-confidence would be shattered

The enemy's military preparations demand the gradual build-up of a deterrent. This deterrent in turn requires balanced nuclear and conventional arms for the defence forces in addition to strategic air and sea establishments. If the means available for military purposes

Voraussetzungen einer wirksamen Verteidigung

Eine Zusammenfassung militärischer Gesichtspunkte durch den Führungsstab der Bundeswehr

Vom 11. bis 13. Juli 1960 fand eine Kommandeurbesprechung der Bundeswehr in Kiel statt. Hierbei berichteten der Bundesminister für Verteidigung und der Generalinspekteur der Bundeswehr über Lage, Erfahrungen und Aufgaben beim Aufbau der Bundeswehr. Die Kommandeure erörterten den Stand der bisherigen Maßnahmen und die Voraussetzungen einer wirksamen Verteidigung. Der Führungsstab der Bundeswehr hat nun in knapper Form die „Voraussetzungen einer wirksamen Verteidigung" niedergelegt. Die Schrift ist jetzt als Sonderbeilage zu Heft 8/60 „Information für die Truppe" erschienen. Es handelt sich um eine Zusammenfassung militärischer Gesichtspunkte für die Verteidigung der Bundesrepublik. Sie lautet wie folgt:

Die Lage

Das deutsche Volk hat durch seine berufenen Vertreter der Bundeswehr den Auftrag gegeben, sein Recht und seine Freiheit zu schützen. Sie sind ständig bedroht durch den militanten Kommunismus, der sich in den Armeen des Ostblocks ein gefährliches Machtinstrument geschaffen hat. Die kommunistischen Führer haben immer wieder die Weltrevolution als ihr Ziel proklamiert. Gewaltanwendung ist in ihren Augen nicht verwerflich, der Krieg nach ihrer Lehre die „Hebamme der Geschichte". Nur Freiheitsmut, Vaterlandsliebe und eine für alle Fälle gerüstete Verteidigungsmacht können diese tödliche Gefahr bannen. Der Bolschewismus respektiert nur die Macht, sonst nichts.

Eine schlagkräftige Bundeswehr ist ohne Wehrpflicht undenkbar. Das Ja zur Landesverteidigung erzwingt das uneingeschränkte Ja zur allgemeinen Wehrpflicht.

Die NATO

Das erklärte Ziel der Sowjetunion, ganz Deutschland für den Kommunismus zu gewinnen, schließt angesichts der geographischen Lage der Bundesrepublik eine Neutralität aus. Ihre industrielle Bedeutung stellt zusätzlich eine starke Verlockung für die sowjetische Politik dar. Es ist wirklichkeitsfremd, zu hoffen, die freie Welt werde die Bundesrepublik mit ganzem Einsatz verteidigen, ohne daß diese dazu einen angemessenen militärischen Beitrag leistet. Noch utopischer ist der Glaube, die Bundesrepublik könne eine Neutralität zwischen den beiden Machtblöcken wahren. Sie wird bei einem Krieg in Europa in jedem Falle Kampfgebiet.

Angesichts des Potentials des einzig möglichen Gegners ist eine ausschließlich nationale Verteidigung eine Illusion. Keine Nation Europas verfügt heute über genügend Menschen und über die notwendigen wirtschaftlichen und technischen Hilfsquellen, um sich allein wirksam verteidigen zu können. Durch den Zusammenschluß der freien Völker in der NATO ergänzen sich die nationalen Wehrpotentiale zu einer machtvollen Verteidigungskraft.

Die Versorgung der Bundeswehr mit Waffen, Munition, Gerät und Verpflegung ist nur noch im Rückhalt an die atlantische Gemeinschaft möglich. Auch die gleich bedeutsame Versorgung der Zivilbevölkerung, vor allem mit Nahrungsmit-

are inadequate for the needs of localised defensive actions as well as for limited and general warfare, the only alternative would be 'All or Nothing'.

However, conventional and nuclear weapons for the defence forces would present a serious risk to the enemy, even in a limited local conflict. Tactical nuclear weapons for the defence forces are clearly an indispensable element of an effective deterrent.

Atomic weapons are a 'must' for the effective defence of air-space. Only rockets with atomic warheads would give maximum effectiveness to the defending forces in an enemy air attack.

In order to deter all attempts to attack Germany, the defence force must be prepared to face any kind of military conflict. An Army whose inferiority in weaponry is known does not present any deterrent to an enemy; professionally, such an Army has no credibility. Never in the history of conflict has an archer been able to deter an attack by a gun-laden cowboy.

If the need for an effective defensive system has been recognised, atomic arms for the defence ('Shield') force becomes a necessity.

Document 2
Letter from SA Embassy, Cologne, to Sole   *19 December 1968*

SOUTH AFRICAN EMBASSY

ᴿᴱꜰ. 6/1/1

1 HEUMARKT
5 COLOGNE

ᴛᴇʟ. ADD. SALEG KOELN
TELEPHONE ʼ23 68 71

<u>AIRMAIL</u>

19th December, 1968

Dear Don,

I wish to thank you for your letter 4/2/2/1 of 5th December, enclosing copies of your Letters of Credence and copies of the speech. The speech has now been translated in the Embassy and I enclose herewith a copy of the translation. Please let me know whether you find the translation in order.

Regarding the speech, I should be grateful if you would permit me to comment on the inclusion in the speech of a

reference to nuclear energy and the production of uranium
- vide the penultimate paragraph. As you know, the East
Germans have for many years accused the Federal Republic and
South Africa of close co-operation in this particular field
and of secretly producing atomic weapons. I fear that the
reference to nuclear energy – even though you specifically
mention the peaceful uses of such energy – and South Africa
as a major uranium producer, and the fact that you
specifically express the hope, as South Africa's Governor
on the IAEA, to be able to give special attention to this
aspect of the relations between us, could be seized upon by
our enemies as further proof of the collaboration of which we
have been accused for so long. This we should avoid.
Moreover, from the German side it may prove difficult to
prepare a proper reply in this connexion for inclusion in the
Federal President's answer at the presentation of credentials
ceremony, especially as both your speech and the President's
reply will be published in the official bulletin which
enjoys wide circulation. I feel that the less said in public

Mr. D.B. Sole,
Department of Forein Affairs,
Pretoria.

at this stage about this aspect of our relations with the
Federal Republic, the more success we shall be able to
achieve behind the scenes. It is therefore strongly recommended for your consideration that the particular
paragraph in the speech be omitted.

I should be grateful to receive your views on the
above comments, before handing a copy of the speech and
translation to Dr. von Rhamm in Protocol.

yours sincerely,

353

Document 3

Instructions for SA Defence Force visitors to West Germany

Telefoon : 23 68 71
Telephone

Bylyn : 14
Extension

Militêre, Lug & Vlootattaché,
Military, Air & Naval Attaché,

Suid-Afrikaanse Ambassade,
South African Embassy,

5 KEULEN 1,
5 COLOGNE 1,

Heumarkt 1.
1 Heumarkt.

⌐ November
  November, 1969.

INSTRUKSIES EN RIGLYNE VIR NAKOMING DEUR LEDE VAN DIE SAW
INSTRUCTIONS AND GUIDELINES FOR COMPLIANCE BY MEMBERS OF

WAT DIENSBESOEKE/KURSUSSE VIR 6 MAANDE EN LANGER IN DUITS-
THE SADF VISITING GERMANY TO ATTEND COURSES OR ON OTHER

LAND MEEMAAK
DUTY VISITS

1. Instruksies en riglyne in bovermelde verband is saamgevat in
   Attached as Appendix A are instructions and guidelines in the

Aanhangsel A en word hiermee aan u uitgereik ter insae en nakoming.
above connection for your perusal and compliance.

2. Lede moet nie aarsel om hierdie kantoor te nader indien hul
   Members should not hesitate to approach this office in connec-

enige verdere informasie en/of advies sou verlang nie.
tion with any further information and/or advice which they may require.

                                          Brig
Militêre, Lug & Vlootattaché
Military, Air & Naval Attaché

Document 4

Letter from CSIR to SA Embassy, Cologne   5 November 1969

South African Council for Scientific & Industrial Research   **C S I R**

S E C R E T

Our file   Your file   Telegrams NAVORS
RVN 7/3/3/Engelter   Telephone 74-601:
                                      P.O. Box 395, Pretoria

/DIPLOMATIC BAG/

5th November, 1969.

Dr. P. le R. Malherbe,
Scientific Counsellor,
GERMANY   u 3/3-)

Dear Dr. Malherbe,

<u>Overseas Visit : Dr. A.G. Engelter</u>

Dr. A.G. Engelter, a Senior Chief Research Officer of the National
Research Institute for Mathematical Sciences will shortly be visiting various
establishments in Germany and Italy on behalf of the S.A. Navy.

Following a decision of the Chief of Defence Staff, at highest level, the
visit, including the classified visits, must be arranged by the Scientific Counsellors,
although naturally in close consultation with the Military Attachés.   I shall, therefore, be grateful if you would make arrangements to obtain clearances for Dr. Engelter

Document 5

Appendix to Engelter's itinerary: specialised interests

SECRET
S E C R E T

NATIONAL RESEARCH INSTITUTE FOR MATHEMATICAL SCIENCES

NRIMS/G/E/K/67/7
W/69/274

<u>Appendices to Itinerary of overseas tour by Dr. A.G. Engelter</u>

<u>General</u>

Dr. Engelter, who is in the Solid State Electronics Division, is stationed at the Simonstown Naval Base, where he is engaged in the instrumentation side of project Tyrant. His particular interests are :

   (i) Transducers (pressure, magnetic, sonic and ultrasonic)

   (ii) Signal transmission and processing, including the usage of such equipment, for example analogue and digital data logging systems.

(iii) The electronics only of modern mines, i.e. sensors, decision-making circuitry and power sources.

   (iv) Degaussing of ships : mainly in flux measurements (0.5 m Oe to 50 m Oe) and numerical mathematical methods for extrapolating these measurements.

       Instrumentation is required for monitoring the signatures of S.A. Navy ships, namely pressure, magnetic and low and audio frequency acoustic signals. This would include ultrasonic aspects for auxiliar tasks, for example the position determination of S.A. Navy ships while their signatures are being measured during checking the effectivity of their degaussing.

Numbered appendices

Document 6
Certificate of security clearance for Dr Engelter.

SCHEDA PERSONALE

CERTIFICATE OF SECURITY CLEARANCE

a) Cognome e nome:
    Last name, first name, middle name:

       ENGELTER, Adolf G. (Dr.)

b) Data di nascita:
    Date of birth: 18.8.1927

c) Luogo di nascita:
    Place of birth: Marburg, Germany

d) Nazionalitá:
    Nationality: German

e) Residenza di lavoro indirizzo attuale, Via e N, Cittá:
    Residence, present address, street number, city:

    c/o South African Council for Scientific and Industrial Research, P.O. Box 395,
                                   PRETORIA.

f) Nome e localitá Societá/Impresa ove è impiegato:
    Name and location of plant or agency where employed:

    South African Council for Scientific and Industrial Research, Pretoria

g) Incarico:
Title or position: **Dr. Phil., Senior Chief Research Officer**

h) Certificato di sicurezza:
Security clearance: **Top Secret**

i) Rilasciato da:
Cleared by: **Department of Military Intelligence**

j) Data del certificato di sicurezza:
Date of security clearance: **28th August, 1969**

k) Numero del documento d'identità o del passaporto:
Identity card or passport number: **B.3961837**

Document 7

Extracts from a secret report to Secretary for Foreign Affairs, Pretoria (probably from Sole), 4 February 1970.

> 9/2/8/3
> 9/2/8/17/1
> 9/2/8/16/10
> 9/2/8/38
> 9/2/8/28/1/1

4th February 1970

SECRET

THE SECRETARY FOR FOREIGN AFFAIRS,
P R E T O R I A.

DISCUSSIONS WITH THE GERMAN MINISTER OF EDUCATION AND SCIENCE

357

During the week I had my first detailed discussion with the new German Minister of Education and Science, Professor Leussink . . . we talked almost entirely about atomic energy matters. The main points covered are dealt with below:

Enrichment: Gas Centrifuges (British/Netherlands/German Project)

I referred to the interest shown in participation in this tripartite project by the Belgians and the Italians, also to the fact that Mr. Wedgwood Benn had made it clear in his statement that collaboration with other countries interested could include states outside of Europe as well as in that continent. South Africa, as a major producer of uranium, was naturally interested in the success of such a project, not only from the point of view of providing an additional market for uranium producers, but also because it was the natural trend in any important uranium producing country to improve and expand its own technology – where possible also in collaboration with other countries. I outlined to the Minister our interest in promoting sales of uranium to the Federal Republic of Germany, referred to our traditional sales relationship in this field with the United Kingdom, sketched the thinking now going on in South Africa about our own future atomic power programme and expressed to him the personal view that there would be considerable interest in South Africa in exploring the possibilities of future collaboration with the Federal Republic and its partners in the production of enriched uranium by centrifuges techniques. I enquired how the Minister himself would view the possibility of South African participation, emphasising that I was well aware that much more development work was necessary among the three countries concerned before the prospect of participation by an extra European state would be ripe for consideration. The Minister was, as expected, cautious in his reply. He said that at the present time the Federal Republic and the Netherlands were discussing the whole project in Euratom. The Euratom hurdle had first to be cleared and that would take some time. There was still a lot of work to be done on the development side and he stressed that as far as enrichment was concerned West Germany was keeping all her options open. Work was proceeding for example also on the "nozzle" process. But he agreed that participation in the tri-partite project might well be extended to a country or countries outside of Europe.

I had discussed before-hand with Dr Pretsch whether I should raise this topic with his Minister and it was on his advice that I did so. The purpose was to "keep the subject warm". I shall have an opportunity to return to the matter at a later date. Probably in about twelve months' time would be appropriate.

Visit to South African ports of the nuclear vessel "Otto Hahn":

The Minister himself took the initiative in raising this topic. He said that the German government would welcome it very much if such a visit could be arranged. I informed him that I would do my best to arrange such a visit and that I had already been in contact with you on this subject. I said that there was a difference between the amount of a guarantee against a nuclear accident required under our own regulations and the amount which had been mentioned in German government statements on the subject, but that I felt sure that this amount was negotiable. I would get in touch with the Ministry as soon as I had received a reply from you.

General:

The Minister is not so well up on his subject as his predecessor but this is understandable in the circumstances. What is interesting is that he went out of his way to seek South African assistance and support in respect of three matters to which his Government attach importance, viz. the visit of the "Otto Hahn", creation of a safeguards committee and the securement of a permanent seat for West Germany on the IAEA Board of Governors. This strengthens our bargaining position in respect of those matters in the atomic energy field where we seek understanding and assistance from West Germany.

Copy to Vienna.

GOVERNOR

Document 8
Letter to Karlsruhe from Hugo, 13 February 1970

TELEX 44-0253 Pr

O/PJW

The Head: Institute for Nuclear Technology,
Technische Hochschule Karlsruhe,
Reaktorstation, Leopoldshafen,
KARLSRUHE
Federal Republic of Germany

1 3 FEB ...

Dear Sir,

STUDY : MR. P.J. WILMOT

In the latter half of this year, Mr. P.J. Wilmot, a senior scientist
of the Board's Instrumentation Division, will be seconded for a minimum period
of approximately 2 years to the Gesellschaft für Kernforschung in Karlsruhe, in
order to gain further experience in the field of electronics, while at the same
time studying for a doctorate in Electrical Engineering at the Technische
Hochschule.

Mr. Wilmot obtained his B.Sc. in Electrical Engineering [cum laude]
at the University of Pretoria in 1967 with Electronics, Control Theory, Computer
Theory, Machines and Telecommunication as major subjects. Since 1967 he has been
studying on a part-time basis for the M.Sc. degree in Electrical Engineering
his main subjects being Network Synthesis, Control Systems, Computer Techniques
and Advanced Engineering Mathematics. Within the next few weeks he hopes to
obtain his M.Sc. degree on a thesis entitled "Analog-to-Digital Conversion using
the Ramp Method".

Since he joined the Board's Instrumentation Division in 1967, his
work has covered reactor experiment instrumentation including tie-in to reactor
control system, interfacing between existing and newly acquired apparatus, and
development work on specialised electronic modules and general electronic design.

At present he is working on the design and installation of computer-
based data acquisition and a reduction system for nuclear physics experiments.
His main interests lie in the fields of logic design and computer application
to instrumentation and control, and he wishes to pursue his studies towards a
doctor's degree on a subject related to the above.

Dr. Greifeld has agreed to Mr. Wilmot's joining the staff of the
Karlsruhe Research Centre in the second half of this year and has suggested
that - in view of Mr. Wilmot's experience and scientific interests - a secondment
to the Institut für Reaktorbauelemente would be particularly appropriate. In
this connection Dr. Greifeld has in fact written to us as follows:

"Eine der gegenwärtig zur Bearbeitung anstehenden Aufgaben
dieses Institutes beinhaltet die Simulation eines Reaktor-
kreislaufes. Zu diesem Zweck wurde eine Anlage geschaffen,
bei der die nuklearen Brennelemente durch elektrisch beheizte
, Brennstäbe ersetzt sind. Die übrigen Kreislaufkomponenten
entsprechen denjenigen eines Kernreaktors. Um das dynamische
Verhalten des Simulationskreislaufes denjenigen einer Reaktoranlage

360

Documents

anzupassen, ist vorgesehen, einen Analogrechner einzusetzen.
Die Kopplung von Rechenanlage und Simulationskreislauf dürfte ein
Problem sein, für dessen Lösung Kenntnisse auf dem Gebiet Elektronik,
Reaktorregelung, Reaktorinstrumentierung und des Einsatzes von
Rechnern, wie sie Herr Wilmot nach seinem bisherigen Werdegang
besitzt, von grossem vorteil sind. Diese Problemstellung dürfte
daher als Thema einer Promotion geeignet sein".

In view of the above, we would be most grateful if you could advise
us of the proper procedure to have Mr. Wilmot registered at the Technische
Hochschule so that he may commence his studies in the next academic year, and
if you could let us know which of the University's professors would be prepared
to act as promotor for Mr. Wilmot's thesis.

If it all possible, we would be glad if you could also let us know
if his studies would necessitate regular attendance of lectures at the University,
or if the work for his doctorate would be mainly covered by practical research
work at the Karlsruhe Nuclear Research Centre.

Yours faithfully,

J. F. B. HUGO

J.P.B. Hugo
DEPUTY DIRECTOR GENERAL

GKO/IB/11/2/1970

Document 9
Telegram to Pretoria from SA Embassy, Cologne, 28 October 1970.

8/19
9/2/8/28/1/1
8/5/1

OUTGOING PRO CYPHER TELEGRAM

TO: THE SECRETARY FOR FOREIGN AFFAIRS, P R E T O R I A.
FROM: THE SOUTH AFRICAN EMBASSY, C O L O G N E.

361

DATE: 28th October, 1970 (18.00)

- -

No. 118

Your 105. I have already reported on this trend brought to your notice by German Ambassador. See my 8/19 of 9th October entitled "Is there a threat of a second Cabora Bassa" also my 9/2/8/28/1/1 of 20th October entitled "German participation in the Rossing Project" and my report on Kaunda's visit 8/5/1/ of 20th October.

It is for this reason that I expressed reservations about Rossing in my 115.

As regards the Uran Gesellschaft it is necessary for me to talk to one of their top men and this has now been arranged for November 6th.

I must defer my final recommendation until I have spoken to Uran Gesellschaft and at this stage would merely remark that the German Government was successfully persuaded to stand firm on Cabora Bassa and that it is important that it should stand equally firm on Rossing, although it may be more difficult to do so. We should therefore not shrink from making the necessary representations if it is our best judgment that these representations could be worthwhile.

It is an open question whether publicity at this stage would be wise or unwise and it is here that I consider that we should be guided by Uran Gesellschaft.

Document 10

Extracts from a letter from Sole to Pretoria, 6 December 1971.

9/2/8/28/1/1

6 December 1971

Dear Andries,

URANIUM DEPOSITS

It is not clear from your 137/23(5) of 2 December 1971 whether you are interested in the reference to South West Africa or whether you are interested in the overall programme of the Federal Republic.

As far as South West Africa is concerned I am sure that Mr. Fourie is aware of the developments with respect to Urangesellschaft's participation in Rossing, although this is being kept very hush-hush. As to other German interests in South West Africa I am aware of other negotiations proceeding in respect of the possible exploitation of an area not far from Rossing but here the interested South West African partners are still awaiting clearance from Pretoria. I had discussions with Mr. Klein of Swakopmund when he was in this country earlier this year.

As regards Dr. Klaus von Dohnanyi's reference to South Africa in his reply to a parliamentary question, I am a little surprised that this reference slipped through because since the earlier propaganda campaign against German participation in Rossing, the practice has been to keep this kind of activity in wraps.

If you can pinpoint precisely what information you require I shall do my best to obtain it.

Mr. A. Mare,
Department of Foreign Affairs,
P R E T O R I A .

DBS/CL

Document 11
Letter from Sole to Pretoria, 24 February 1972

7/3/3

24 February 1972

A I R M A I L.

THE SECRETARY FOR FOREIGN AFFAIRS,
P R E T O R I A.

(In duplicate.)

VISIT OF SECRETARY OF STATE MAUNSCHILD :
16 To 23 April 1972.

Further to previous correspondence regarding the above, I
— attach for information a photostat of a telex sent to Dr Roux
giving details of the persons who will make up the party in
addition to himself.

It is quite clear from my discussions with Dr Haunschild
that there is a great deal of official interest in this visit.
It is not simply a courtesy gesture in response to an invitation from Dr Roux. Hence the inclusion of other experts in the
party, although the invitation was issued originally only to
Dr Haunschild and his wife.

I should be grateful if any impressions which the Department may receive from Dr Haunschild's visit might be conveyed
to me in due course.

D :

A M B A S S A D O R.

PBS/UB

Document 12

Letter from Sole to Pretoria, 28 March 1972

9/2/8/30
9/2/8/28/1/1

AIRMAIL   28 March 1972

CONFIDENTIAL

THE SECRETARY FOR FOREIGN AFFAIRS,
P R E T O R I A.

(In duplicate)

SALES OF SOUTH AFRICAN URANIUM: RÖSSING.

More publicity has appeared on the question of the withdrawal of
German Government sponsorship of uranium prospection in South
West Africa – see for example the attached article from "German

International" the English language monthly published in Bonn. According to Secretary of State Haunschild of the Ministry of Science his Ministry which normally contributes 50% of preliminary prospection surveys, the issue with respect to Rössing has been somewhat eased for Urangesellschaft by the fact that in the case of a Central African State which he did not identify but which was obviously Niger the Ministry will contribute 75%. Herr Haunschild's version was that Urangesellschaft would be in receipt of a subsidy in respect of all its approved prospecting activities overseas and the Ministry would not enquire too closely into how precisely the Urangesellschaft allocated its money. The inference was that Urangesellschaft could use some of the money for Rössing.

However when I spoke to Dr. von Kienlin of the Urangesellschaft in Frankfurt last week, he denied that the extra payments to the Central African State were being received and said that this was no more than a hope and a possibility, on which he did not place much reliance. My impression from my talk with Dr. von Kienlin was nevertheless that the financial problem was not the principal cause of concern on the part of Urangesellschaft. The real issue was the present uncertainty with respect to supplies from South West Africa and their availability over a long-term period in the future. Neither he nor his company doubted our ability to deliver – the doubts lay with the consumer. Since the uranium would be required for use ultimately in nuclear power stations operated by public utilities or under the supervision of Länder Governments, Urangesellschaft had to take into account that in the minds of many of responsible persons concerned in this particular sector there was a fear that delivery of the uranium would be banned or become impossible because of resolutions adopted by the United Nations and unwillingness on the part of the German Government to act counter to those resolutions. These fears are being played upon by the radical anti-South African elements in this country who always have a spokesman in the Cabinet on this question in the person of Herr Eppler. This was a problem about which very little could be done at the moment (although he did not say so he was obviously inferring that a change of Government would lead to an easing of the situation).

The question of Rössing will certainly come up in one form or another in the course of Secretary of State Haunschild's visit and I should be grateful to be informed in due course of any discussion which may take place on this topic.

AMBASSADOR

Document 13

Letter from Sole to SA Foundation, 14 June 1972

8/24/4

S/12

AIRMAIL
14 June 1972

C O N F I D E N T I A L

For some time I have been trying to arrange a
visit to South Africa by General Rall, Chief of the German
Air-Force, who was appointed to this position last year. I
know General Rall and his wife quite well and he has always
been interested in South and South West Africa but because
of his position and the policy of the German Government
with respect to any form of defence cooperation with South
Africa, there can naturally be no question of his being
able to accept officially an invitation either of a South
African government body or of even a private organisation
such as the South Africa Foundation. However he has in
Windhoek a very good friend in the person of Kurt Dahlmann,
Editor of the Windhoek Allgemeine who fought in the same
squadron with him during the last war. With the assistance
of Kurt Dahlmann, whom I have evoked as a kind of intermediary, General Rall has now been persuaded to agree to
set aside three weeks in May next year for a private visit
to South Africa. Unfortunately his commitments this year
do not permit of his making the visit earlier. As it would
be a private visit he would travel with his wife.

Mr. L.E. Gerber,
Director-General of the South Africa Foundation,
P.O.Box 7006,
Johannesburg.

/...

I attach importance to Rall making this visit
not only because of the position he holds in the top Command
of the Bundeswehr but because he may well succeed his predecessor Steinhoff in the top Command of NATO in Brussels.
I believe therefore a special effort should be made and
should be glad if you would consult on a confidential basis
with your Executive and members of the German Committee of
the Foundation.

There is of course a lot of time yet and it may
well be, although I personally think it is a very open
question, that by the time next May comes round there will
be another Government in power in Bonn which would not
object to a private visit. Indeed under this Government I
was able to arrange for a private visit by the Head of the
Grenzschutz, General Grüner, but in his case his fare was
met by SAA since I was able to include him amongst the six
guests which SAA allowed me to nominate for their Jumbo
inaugural flight.

Yours sincerely,

Document 14

Translation of a letter from Haunschild (in German) to Roux, 12 July 1972.

Translation

Federal Ministry
of Education and Science
The State Secretary
Dr. A.I.A. Roux, Esq.
President of the
Atomic Energy Board
Private Bag 256
PRETORIA
South Africa

53 Bonn 9, July 12th, 1972
Heussallee 2 - 10
telephone 1081

Dear Dr. Roux,
During our talk at Skukuza I had promised you to have – after my

return – investigated, how the secrecy of information could be safeguarded. This refers to the know-how of details of the enrichment process during the exploratory phase being learnt by collaborators of an industrial firm, before the actual decision on a cooperation is taken.

The result of the investigation is at hand. So far as it concerns the penal secrecy, the legal position is as follows:

1. Material Protection of Secrecy

The protection of the know-how in question as a German state secret within the meaning of the articles 94, 95, 97 of the penal code presupposes (cumulatively),

– the know-how has been confided to German government agencies;

– there is an agreement – somehow or other – on the secret treatment between the Federal government and the government of South Africa;

– the know-how is in need of secrecy materially, that is, its abandonment would be gravely detrimental to the interior security of the Federal Republic;

– the know-how is kept secret by an official German agency or at its suggestion (factual secrecy).

2. Preventive Technical Secrecy

In this case it is a question of the formal treatment as a secret knowledge without judging the value of the secret.

First of all, a binding assent by the government ought to be given so far to apply the procedures of secrecy being in force, or, to render the expert – having received the know-how for testing – liable to treating it according to the scheduled procedure. (This presupposes that the receiver incurs such liabilities).

He who imparts such secret material either completely or in part to someone else or renders it public thereby endangering vital public interests, makes himself liable to prosecution regardless of the material quality of the knowledge as a German state secret according to the article 353 c of the penal code. The prosecution presupposes an authorization of the Federal government.

I hope this short representation of the state of affairs and of the legal position does answer your question sufficiently.

During our talk we agreed that an arrangement on the secrecy of the process between our countries would not be opportune at present. I should like to stress, however, that private German industrial groups are completely at liberty to decide upon participation in foreign processes including upon agreements on the confidential treatment of the then obtained information. This also does apply to

the disposal of non-secret knowledge conversely.
I have informed STEAG about this statement.

yours sincerely,
Haunschild

Document 15

Translation of a letter (in German) from Admiral Steinhaus, Ministry of Defence, Bonn, to Brigadier-General Hamman of SA Embassy, 30 November, 1972.

Translation

Fleet Admiral Rolf Steinhaus
in the Federal Ministry of Defence

53 Bonn, Nov. 30th, 1972
P.O.B. 161
Tel.: 20101
Tlx.: 0 886 575, 0 886 576

Brigadier General D.J. Hamman
Embassy of the Republic of South Africa

5 Köln
Heumarkt 1

Dear General Hamman
As enclosure I'm sending to you – as we have arranged today on the phone – the documents in hand concerning the question of the "expansion of NATO in the South Atlantic". As I had told you, the final formulation of the controversially discussed recommendation has not been finished yet, but, it was adopted, however. The final text will be issued at Brussels.

Best regards
sd. R. Steinhaus

Document 16

Letter from M.E.Beyers to SA Embassy, Cologne 10 January 1973

NATIONAL MECHANICAL ENGINEERING RESEARCH INSTITUTE

OF THE SOUTH AFRICAN COUNCIL FOR SCIENTIFIC AND INDUSTRIAL RESEARCH

AERONAUTICS RESEARCH UNIT

Telephone 74-6011   Telegrams NAVORSING

P.O. Box 395,

OUR FILE

YOUR FILE

NEA/T/99

W 3/3
10 January 1973

S. A. S. L. O. **PRETORIA**
Cologne
RECEIVED 1973
2 2. JAN 1973

Miss I. Golowitsch,
Office of the Scientific Counsellor,
South African Embassy,
5 - KÖLN,
Heumarkt 1,
Germany.

Action: _____

File: _____

Pend: _____

Dear Miss Golowitsch,

OVERSEAS VISIT : M.E. BEYERS

Thank you for your letter referenced above. My overseas visit has been officially approved.

On the basis of your information on the DFVLR it is now clear to me that my main interest lies in a visit to the facilities in Göttingen, but that a visit to Porz could also be valuable.

Basically, I am interested in experimental investigations of the flight dynamics of rigid bodies, and in particular, techniques for free-flight testing in conventional wind tunnels and in shock or gun tunnels. The following topics are of special interest :

    (1) models and launching equipment

    (2) optical data acquisition systems

    (3) data reduction techniques

At present we are conducting free-flight tests in the CSIR supersonic wind tunnel using a pneumatic launch gun. Data reduction is fully computerized.

I would therefore like to visit the DFVLR at Göttingen and if it can be conveniently accomplished, also at Porz. I will welcome your suggestions and will be in touch with you in connection with the other arrangements.

Yours sincerely,

M.E. BEYERS

M.E. BEYERS

MEB/JCC

Director of the Institute H G Denkhaus Dr Ing   Head of Unit C G van Niekerk D Sc (Eng)

Document 17

Letter from Chief of SA Navy   18 May 1973

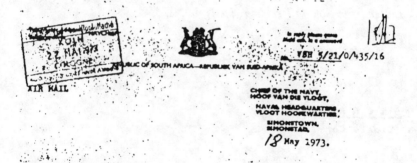

AIR MAIL

REPUBLIC OF SOUTH AFRICA—REPUBLIEK VAN SUID-AFRIKA

VBH 5/21/0/435/16

CHIEF OF THE NAVY,
HOOF VAN DIE VLOOT,
NAVAL HEADQUARTERS,
VLOOT HOOFKWARTIER,
SIMONSTOWN,
SIMONSTAD.

18 May 1973.

Military, Air and Naval Attaché,
South African Embassy,
1 Heumarkt,
5 Cologne 1,
West Germany.

Director General Management Systems,   -   For Information
Directorate Codification and
Cataloguing,
Private Bag X414,
Pretoria.

CODIFICATION : ADVOKAAT SPARES

1.   It would be appreciated if you could advise what progress has
been made with the codification of the spares reflected in the
six volumes of the Consolidated Rationalised Spares Recommendation
delivered to you by Messrs. A.E.G. Telefunken of Germany.

371

2. Because much difficulty is experienced in coupling the items,
codified by your Codification Department, and the corresponding
item in the Rationalised Spares Recommendation, it is suggested that
you investigate the possibility of supplying this Headquarters with
a cross-reference list reflecting the NSN against the specific item
in the Consolidated Rationalised Spares Recommendation. Data
arranged in the following sequence will suffice:-

Rationalised Spares
Recommendation Identity   NATO Stock No.

Item 09 - Page 0240 - Book 2   5905-12-154-8633

Vice-Admiral,
Chief of the Navy.

Document 18

Letter from Sole to Pretoria, 15 June 1973

9/2/8/28/1/1

15 June 1973

S E C R E T
(Single Copy)

THE SECRETARY FOR FOREIGN AFFAIRS,
P R E T O R I A.

VISIT OF DR VOELCKER

With reference to your 137/13/1(23) of 7 June the
envelope in question was handed by me personally
to Dr Voelcker yesterday evening, 14 June. Because
of his other commitments Dr Voelcker was not able
to meet me earlier.

372

Dr Voelcker also gave me a copy of a telex which
he had sent to Dr Roux the previous day. A photostat
of this telex is attached. I pointed out to
Dr Voelcker that we operate on the basis that both
telex and telephone communication to South Africa
is periodically monitored. I added that in this
context we had warned other German firms engaged
in undertakings of a confidential nature to
exercise care in the use of open telex and
telephone lines. These firms had been told
that confidential communications to, for example,
the Armaments Board could, if they wish, to be sent
through this Embassy utilising our cipher facilities
for this purpose. The same facilities were available
to STEAG for communications with the Atomic Energy Board.

The preceding is for your own background only.

A M B A S S A D O R.

DBS/svdh

Document 19

Letter from Fourie, Pretoria, to Sole, 27 August 1973.

REPUBLIEK VAN SUID-AFRIKA   REPUBLIC OF SOUTH AFRICA

Ref. No. 137/23/1   DEPARTEMENT VAN BUITELANDSE SAKE
DEPARTMENT OF FOREIGN AFFAIRS
Privaatsak/Private Bag X141
PRETORIA

27 August 1973

The Nuclear Axis

Mr D. B. Sole,
Ambassador,
COLOGNE.

Dear Don,

GERMAN INTEREST IN SOUTH AFRICAN URANIUM DEVELOPMENTS

In response to your minute 9/2/8/30, 9/2/8/28/1 of 24 July 1973, the Atomic Energy Board has reminded us that matters such as uranium sales and the allocation of equity in the share capital of Rössing, although done with the approval of the Minister and/or the Board, are the sole functions and responsibility of that company in the normal course of business practice, and information pertaining thereto must be treated as commercial secrets unless the buyer wishes to make a public or press announcement at the appropriate time, as in the case of Total, to which you refer.

For these reasons it is not possible for the Board to divulge secret information to outside persons who are not directly concerned with such negotiations – even if they were to make enquiries through one of our Ambassadors.

However, for your personal information I may mention that Total, having concluded an agreement with Rössing for a substantial supply of uranium U_3O_8, has acquired an option to take up 10 per cent in the equity of Rössing. This deal has nothing to do with Urangesellschaft and "the German interests concerned" have therefore not surrendered the option which they have had with respect to Rössing to the French company.

I may also mention that Urangesellschaft has in fact re-negotiated a somewhat smaller quantity of U_3O_8 than originally planned,* and as a result the percentage equity to which they are entitled has been reduced.

Whatever the facts, I fully agree that the course you took with the Member of Parliament who spoke to you, is the only proper and advisable one.

Yours sincerely,
Brand

* The total contract is far in excess of this reduction – it is even larger than the quantity originally contracted for by Urangesellschaft.

Document 20

Part of letter from Steag to Roux   2 October 1973

STEAG Aktiengesellschaft 4300 Essen, Postfach 7020

S E C R E T

Dr. A.J.A. Roux
President of
Atomic Energy Board
of South Africa

Private Bag 256

P r e t o r i a /Südafrika

4300 Essen
Bismarckstraße 54
Tel.: (021 41) 7 99 41
Telex: 08 57 693

Ihre Zeichen	Ihre Nachricht vom	Unsere Zeichen	Sachbearbeiter	Durchwahl	Datum
		VÖ/G1		7994 2339	Oct.2, 1973

Dear Dr. Roux,

Referring to article 3 of our agreement we have requested
formal approval for sublicensing our rights according to
our contract with Gesellschaft für Kernforschung Karlsruhe
(GfK). GfK has agreed to our request in principle but needs
approval of Staatssekretär Haunschild as chairman of GfK
supervisory board.

We have unofficially been informed that on request of
Mr. Haunschild Staatssekretäre of Ministery of Economy,
Ministery of Foreign Affaires and of Chancellor Brand's
office have met on 27th of September to discuss this
matter. They have given a positive reaction to the GfK
position. However on request of the Ministery of Foreign
Affaires a legal investigation to find out whether the
Außenwirtschaftsgesetz is applicable in this case is still
necessary. The expert of the Ministery of Economy, who has
meanwhile contacted us has already unofficially confirmed

that this law is not applicable and that he cannot see a
reason to withhold Mr. Haunschild's approval.

- 2 -

Vorsitzender des Aufsichtsrats Bergassessor a. D. Karl-Heinz Hawner
Vorstand: [...] po Karlheinz Bund Vorsitzender, Dr rer pol Hans Krämer, Dipl.-Ing. Hans Puhr-Westerheide, Dipl.-Ing Rolf Ravoth, Dr rer pol Ernst Schadeberg
Sitz der Gesellschaft Essen Registergericht Essen Handelsregister-Nr : B 13

Document 21

Extracts from letter from Chief of SA Navy, March 1974

CONFIDENTIAL

the German Navy has lost none of its efficiency or traditions.
I sincerely hope that we have another such visit within the
not too distant future.

I am very aware that this very successful visit would not have
taken place at all were it not for the personal efforts made
by your good self and I want to express my thanks and that of
all my officers to you. I have not forgotten either how you
smoothed the way for PRESIDENT STEYN's visit to Hamburg, truly
we have a good friend in Cologne.

Yours most sincerely,

P.S.

You may be interested in Deutschland's unscheduled stop off
Durban on 21 March. The following extracts from my signal log
tell the story:

376

Documents

From Deutschland to Commander Naval Operations (SA):

"REQUEST PERMISSION BY SOUTH AFRICAN AUTHORITIES TO
ENTER TERRITORIAL WATERS FOR SAILING TO DURBAN DUE
TO ACUTE APENDICITIS.

2.   REQUEST FACILITIES TO DISEMBARK PATIENT AT
DURBAN ROADS ETA 211615B MAR LOCAL TIME.

3.   PERSONALITIES OF PATIENT, NAME, OKUBASELASSIE
FIRST NAME, SAILON BIRTHDATE, 21.07.53 RANK,
MIDSHIPMAN NATIONALITY ETHIOPIA.

4.   PLEASE KEEP INFORMED GERMAN AND ETHIOPIAN
AUTHORITIES AT OWN DISCRETION

210728Z MAR"

From Commander Naval Operations (SA) to Naval Officer-
in-Command, Durban:

DEUTSCHLAND ETA 211615B TO LAND ACUTE APENDICITIS
PATIENT ETHIOPIAN MIDSHIPMAN

2.   RENDER ALL ASSISTANCE POSSIBLE

210925Z/MAR 74"

/3....

CONFIDENTIAL

CONFIDENTIAL

-3-

From Deutschland to Commander Naval Operations (SA):

"1.   ORIGINATOR APPRECIATES PROMPT SUPPORT AND
EXCELLENT COOPERATION

2.   ALL FACILITIES WERE WELL PREPARED

3.   THANK YOU AND AF WIEDERSEHEN

211808Z MAR"

From: VICE-ADMIRAL J. JOHNSON, CHIEF OF THE NAVY

CONFIDENTIAL

Document 22

Partial translation of letter (in German) from Bellingan, SA Embassy, to Bundesgrenzschutz, 17 April 1975.

Translation

1st.–Lt. Ulrich Wegener   17 April 1975
Federal Border Guards
GSG 9

My dear First Lieutenant,
   Since our Ambassador is away at the moment, I should like to take the opportunity of thanking you again for your help in looking after us during the transport of our sensitive items from Cologne to our new Embassy in Bad Godesberg. I am sure our Ambassador would want to thank you personally, if he were here.

Document 23

Telex from Stoker, Bonn, to Pretoria, 15 August 1975

```
–          15.a
X
–          1
T

E          bonn    15/08/75    tlx948
L
           csir   pretoria
E

X          professor stoker requests to inform dr. hewitt of the south
–          african c.s.i.r. in pretoria that the gesellschaft feur
T          kernenergienutzung in schiffbau und schiffahrt mbh who will
E          transport the lindau neutron monitor bz the otto hahn free of
           charges from rotterdam to an harbour in south africa for tsumeb,
L          was advised not to unload in walvis bay because of possible
E          political demonstrations against any foreign vessel entering the
X          har our. c nsequently they decided to unload in durban, which
```

T E L E X T E L E X - T =

harbour the otto hahn has to visit on this trip.

the monitor has to be dismantled and packed virtually immediatelly
for delivering in rotterdam in time for transportation to south
africa in september. we have to meet the costs of unloading the
monitor at harbour and of the transportation to tsumeb. the
monitor comprises of 29 tons of lead, 3,5 tons of paraffin wax
in boxes, 18 large neutron counter tubes and electronic
recording equipment. please inform

  professor stoker at the

  conference office

  14th international cosmic ray conference

penta hotel

muenchen

telex no. 5-29046

whether the additional costs of transportation from durban to
tsumeb are acceptable

n.h. stoker

c/o saslo, bad godesberg

saleg

Document 24

Translation of a memo (in German) from Wenzel, Steag, 25 November 1975.

Translation

STEAG   Essen, 25th November 1975
Nuclear Installations   Wen/B1

Memo

RE: Export permission for nuclear components

Prof. Fiedler informed me by telephone that he participated in a meeting at the Federal Ministry of Economic Affairs in Bonn on November 24th, 1975 with the aim to achieve exemption from the export regulations for the compressors for uranium enrichment plants. Mr. Fiedler is of the opinion that the machines will in future no longer be subjected to these controls, since the GHH argued in this way that the machines in question are, after all, entirely normal compressors.

However, the point of importance for all those interested in the spreading of the jet-nozzle system was the announcement of the likewise present Mr. Heil of the Federal Ministry for Research and Technology that he would use all means at his disposal to see to it that the entire jet-nozzle system is placed under export permission regulations. It is therefore recommended that Prof. Becker be informed about this trend as quickly as possible in order that a credible description of the unimportant military significance of the jet-nozzle system be produced without delay.

> sign. Wenzel
> (Head of Dept. on Nuclear Installations)

cc: Dr. Völcker (Director, STEAG Nuclear Energy)
    Mr. Geppert (STEAG)

Document 25
Part of letter from Steag to UCOR   *12 March 1976*

Uranium Enrichment Corporation
of South Africa Ltd.

P.O. Box 4587

P r e t o r i a
Republic of South Africa

Iw/Bl   Iwand   3230 March 12, 1976

Ref.: <u>Minizet-Project</u>

Dear Sirs,

Pursuant to our proposal from 20th January 1975 and your letter from February 1975 we have acted as technical agent for UCOR!

Because of increase of wages since 1th July 1975 we have to charge higher hourly rates.

The now valid hourly rates are:

    DM 78,90   for mechanical engineers
    DM 90,90   for instrumentation engineers
    DM 66,60   for draughtsmen.

Assuming that you will agree with the above mentioned hourly rates we beg you to settle the following amount:

- 2 -

Document 26
Letter from Völcker, Steag, to Roux   *(n.d.)*

253 sa

857826 steha d

dr. a.j.a. roux
president atomic energy board
pretoria
republic of south africa

dear dr. roux,

with reference to article 14 of the memorandum of understanding between ucor/aeb and steag we have the pleasure
to inform you as follows:

1. steag board of management has agreed to the memorandum.

2. state secretary haunschild was informed by dr. bund
   about the memorandum and has agreed to proceed as
   planned.

3. gfk has given approval to the memorandum in principle.
   gfk has recommended an aquivalent wording with respect
   to point 2 a and 2 b in the final agreement.

4. with respect to the political situation we refer to the
   recent letter exchange between prime minister vorster
   and chancellor brandt.

5. we kindly ask you to proceed in this matter as agreed
   during our meeting.

                                               44-0???

best regards,
steag - dr. v?oelcker   ?57?25 stoha d

Document 27
United States: NRO Armies Inventory, South Africa

NRO RESTRICTED TRQ139906   ARMIES INVENTORY S AFRICA
9C5 NSA 906 NRO 906 NSD 905 DIA 906 DCA 906 DNA
906 DSA 906 DIS 906 DIA 906 USSUMS 906 DCAA 906
DOS 12 906 JCS 906 NSC**14 CC
906A RESTRICTED CIA 4****1 CC

                                   100176

*** FV4017 MK 10 CENTURION 224 DEL 68 UK 70% IN SVCE
***FV4007 MK 7 CENTURION 162 DEL 57 UK MANU 50% IN
SVCE RES NO SPARES
***COMET 66 DEL 55 MANU 30% IN SVCE RES NO SPARES
***BULLDOG M41 96 DEL US MANU CHC LIC #25658 90 51
90% IN SVCE RES + SPARES
***PATTON TYPE M47 REF 105 TG 104 DEL 71 ITAL GOV
LIC CHC COMM CMC APPR 050266 60% IN SVCE RES NO
SPARES
***AML 245 H60 SA PROD LIC PANHARD 880 COMPL 90%
IN SVCE + SPARES PROD BAL COMPL 1930 1400 VEH
***AML 245 H90 SA PROD LIC PANHARD 160 COMPL 90%
IN SVCE + SPARES PROD BAL COMPL 1978 400 VEH
***M3A1 350+ DEL US MANU CHC LIC #16544 54 49 20%
IN SVCE RES NO SPARES
***FV703 MK 2 FERRET 460 DEL 68 69 UK 38% IN SVCE
RA + 20% IN SVCE RES + SPARES
***FV603 SARACEN 676 DEL UK MANU 62 63 64 66 36%
IN SVCE RA + 50% IN SVCE RES + FEW SPARES
***FV610 SARACEN 24 DEL UK MANU 64 90% IN SVCE +
FEW SPARES
***SHORLAND MK 3 208 ORD MANU 74 30 DEL BAL PRE 80
***SHORT SB301 312 ORD MANU 75 101 DEL BAL PRE 78
90% IN SVCE MILPOL
***STAGHOUND T17E1 448 DEL US GOV MAP 51 40% IN
SVCE RES NO SPARES
***COMMANDO V150 320 ORD PORT PROD LIC CADGAGE
COMM CMC APPR 399428 86 DEL BAL PRE 80 NO SPARES
***PIRANHA 110 ORD SWISS MANU NO DEL
***M113A1 414 ORD 73 ITAL GOV LIC FMC COMM CHC APPR
325541 110 DEL BAL PRE 83
***25 PDR HOW 112 DEL 51 UK GOV 80% IN SVCE + SPARES
***105 HOW M7 214 DEL US GOV MAP 51 43% IN SVCE RES
NO SPARES
***25 PDR SP SEXTON 180 DEL CAN MANU 46 80% IN SVCE
NO SPARES
***17 PDR AT 234 DEL 56 UK GOV 65% IN SVCE NO SPARES
***90MM AT SA PROD LIC FR GOV 160 COMPL 90% in SVCE
PROD OPEN

```
***155MM SP M109 FH70 52 ORD 71 ITAL GOV LIC GMC
COMM GMC APPR 0963422 12 DEL BAL PRE 80
***ENTAC ATGW 92 UNITS 55 46 UNITS 56 FR MANU 56
90% IN SVCE RA NO SPARES
***CROTALE AAM 160 UNITS DEL 69 FR MANU 10S IN SVCE
RA + SPARES
***TIGERCAT AAM 162 UNITS DEL 72 UK MANU 100% IN
SVCE NO SPARES

NRO RESTRICTED TPQ189907 ARMIES INVENTORY TANZANIA
NEXT***
```

Document 28
Title page, IMF minutes of 21 January 1976

NOT FOR PUBLIC USE

INTERNATIONAL MONETARY FUND

Minutes of Executive Board Meeting 76/6

10:00 a.m., January 21, 1976

W. B. Dale, Acting Chairman

Executive Directors	Alternate Executive Directors
J. Amuzegar	
P. Åsbrink	J. H. Kjaer
	T. Leddy
J. de Groote	
	M. Finaish
	D. Lynch
	S. Sevilla
S. Jagannathan	W. M. Tilakaratna
	W. Temple-Seminario
K. Kawaguchi	M. Wakatsuki
	Sein Maung
P. Lieftinck	T. de Vries
H. R. Monday	J. B. Zulu
F. Palamenghi-Crispi	
E. Pieske	
	G. Laske
	P. J. Bull
	G. Heyden Q., Temporary

384

Documents

J. H. Wahl   J. Foglizzo
R. J. Whitelaw
A. W. Yaméogo

              R. V. Anderson, Acting Secretary
                K. S. Friedman, Assistant

Extracts from IMF Minutes of Executive Board Meeting, 21 January 1976.

Decision No. 4943-(76/6),
adopted January 21, 1976

3. SOUTH AFRICA – 1975 ARTICLE VIII CONSULTATION AND STAND-BY ARRANGEMENT

The Executive Directors considered the staff report for the 1975 Article VIII consultation with South Africa (SM/76/7, 1/6/76). In addition, they considered South Africa's request for a stand-by arrangement for one year in an amount equivalent to SDR 80 million (EBS/76/6, 1/5/76). They also had before them a report on recent economic developments in South Africa (SM/76/8, 1/12/76). [p.6]

The Principal Resident Representative, South Africa, made the following statement:

At the outset I should like to convey the appreciation of my South African authorities to the staff for the very thorough and comprehensive reports for discussion today. These studies not only display a high professional standard but reveal all pertinent economic data on South Africa and analyze in detail recent developments as well as short-term prospects. [p.6]

Monetary policy, Mr. Bull considered, had been fairly flexible, and the progress with the incomes policy was highly commendable. On the external side, however, the outlook did not seem as promising, and the authorities would probably face a number of risks in the coming months. He was therefore pleased that South Africa had requested a stand-by arrangement which would give the authorities some additional room for manoeuvre and some feeling of international support, which they deserved; the Government was making considerable efforts to deal with its balance of payments problems, and he fully supported the proposed decision. [p.13]

Document 29

Extracts from IMF Minutes of Executive Board Meeting, November 1976.

IMF Minutes of Executive
Board Meeting 76/156   10am, November 10, 1976.

3. SOUTH AFRICA – PURCHASE TRANSACTION – COMPENSATORY FINANCING

The Executive Directors considered the staff's analysis and recommendation with respect to a request from South Africa for a purchase under paragraphs 2, 3 and 4 of the Decision on Compensatory Financing of Export Fluctuations (EBS/76/468, 10/19/76; and Supplement 1, 11/5/76).

Mr. Dini commented that the South Africa economy was passing through a difficult period due to a number of factors including the recent recession, political uncertainties at home, and the fall in the price of gold that had hurt not only gold producers but also countries that had gold in their reserves ... So far as the shortfall was concerned, the staff had departed from current practice, since it had not arrived at an estimate of the possible shortfall in the leading export commodity, namely, gold. Naturally, forecasting the future of gold prices was difficult, but if the Fund intended to allow purchases under the compensatory financing facility, it should be prepared to make forecasts of market developments in all relevant commodities, including gold. He would therefore like to see the staff intensify its analytical work on the gold market, and present the result to the Executive Directors when appropriate.

Regarding the nongold merchandise exports, Mr. Dini stated, he had some difficulties with the group entitled "exports other than primary commodities," which included manufacturers and semi-processed products. The staff argued that the shortfall in that category was due mainly to the world recession; and it supported its conclusion by identifying business cycle conditions and the export price index in the industrial countries as the explanatory variables for South Africa's merchandise exports. However, another factor with some effect on merchandise exports must be the level of domestic economic activity, if only because in a country like South Africa goods produced for export were also in demand at home. In 1975 South Africa goods had pursued relatively expansionary domestic policies, and consumption and other expenditures had remained at a high level during the first half of 1976. He could therefore not accept the staff view that the shortfall in exports other than primary commodities was due largely to external circumstances, and he would like the staff to provide some explanation. The staff's calculations were the more suspect because the shortfall calculated by the staff appeared to be exactly equal to the amount – 50 per cent of quota – requested by South Africa plus the amount of calculated double compensation under the previous drawings from the Fund. [p. 12-13].

Mr. Leddy agreed with Mr. de Vries that the existence of a balance of payments need in South Africa was clearly established. One of the primary causes of the current difficulties had been the authorities' excessive optimism about an export revival and changes in the terms of trade, which had made them reluctant to adopt the needed adjustment measures. He concluded that South Africa was cooperating with the Fund by virtue of the stand-by agreement recently negotiated, and he wondered whether the staff could report on progress under that arrangement. South Africa's balance of payments seemed to have improved in the third quarter of 1976, and more recent data might be interesting. He would also be interested to know whether the import deposit requirement had had

387

any effect on the current account position and, in that connection, he expected that South Africa would fulfill the commitment under the stand-by arrangement to eliminate the requirement in February 1977. South Africa's export shortfall was probably due to factors largely beyond its control. The calculation of the export shortfall had apparently not been dependent on any assumptions with respect to gold, and the staff's approach in that respect was reasonable. He was therefore prepared to support the request.[p.14]
►

The shortfall had been calculated largely on the basis of the projected annual growth rate of 16 per cent for nongold exports and of 17 per cent for the so-called "other merchandise exports," Mr. Pieske noted. He agreed with Mr. Dini that the shortfall appeared to have been tailored to the size of the requested drawing, and the assumed annual growth rates used for the purpose appeared rather high. Table 8 of EBS/76/468, for instance, showed that the rate of growth of the "other merchandise exports" had exceeded 17 per cent only once in recent years. One of the explicit assumptions underlying the forecast was that the restrictive monetary, fiscal, and incomes policies recently instituted would succeed in keeping the inflation rate in South Africa at a level commensurate with that of its trading partners. It might prove difficult to fulfill that assumption. In the past, inflation in South Africa had usually run well ahead of that in industrialized countries; the most recent figures were 11 per cent and 7 per cent respectively. On the other hand, the uncertainties regarding future export developments were counter-balanced by the fact that the judgmental forecast did not take into account any possible shortfall in gold exports. He was therefore satisfied that, taking the gold portion of South Africa's exports into account, the requirements both for the shortfall and for need had been met. [p.15]

Document 30
Extracts from IMF Monthly Report on gold prices

DOCUMENT OF INTERNATIONAL MONETARY FUND AND NOT FOR PUBLIC USE

Any views expressed in the Departmental Memoranda (DM) Series represent the opinions of the authors and, unless otherwise indicated, should not be interpreted as official Fund views.

DM/77/44   INTERNATIONAL MONETARY FUND

Treasurer's Department

Monthly Report of Gold Prices in World Markets

Documents

Prepared by the Financial Relations Division[1]/

Approved by Paul M. Dickie

May 23, 1977

The price of gold in leading international markets fluctuated within a narrow trading band of $147 and $152.50 an ounce in generally quiet trading during April. In London the price eased by $1.65 an ounce, falling from $148.90 an ounce at the end of March to $147.25 an ounce at month-end April. Similarly, in Zürich the price fell by $1.50 an ounce to end the month at $147.38 an ounce. The price of gold in Paris moved in parallel easing by $1.26 an ounce to $148.65 an ounce, although the premium over prices in the London and Zürich markets rose somewhat. However, while gold prices eased on an end-of-month basis, monthly average prices in London, Zürich, and Paris rose somewhat from the previous month. As regards the futures markets, the average premium of futures prices over spot prices, adjusting for interest costs, declined from an average premium of about one and a half per cent in March to just slightly above par in April. This, along with a decline in futures trading volume, reflected an abatement in speculative interest in the metal.

The steadiness in bullion prices in April was basically attributable to a lack of change in the factors that underpin the market. Demand from industrial users remained strong, with support increasing below the $150 an ounce level, while investors cautiously appraised the market, doubtful whether a price above $150 an ounce was sustainable at this time. Essentially the market appeared to be going through a period of consolidation, following recent advances which resulted in a price rise from the $135 level which prevailed in late 1976 and early 1977.

---

[1]/ Major contributor was Michael Martin.

- 4 -

South Africa effected a new swap arrangement, selling probably just under 3 million ounces of gold or about one quarter of official gold holdings on a spot basis from its reserves during the week ending April 29 and repurchasing it forward. Previously in March 1976 the Bank effected a similar transaction involving about 5 million ounces which has yet to be unwound. According to the Reserve Bank's balance sheets, official gold holdings declined by 2.986 million ounces in the week ending the twenty-ninth. However, since sales or additions to stocks might have been made during that week, it is not possible to ascertain the precise size of the swap. Assuming the transaction was based on a price of $130 an ounce, approximately 85 to 90 per cent of the prevailing market price, the swap would have yielded about R 340 million (SDR 336 million) in liquid funds. However, total gold and foreign exchange holdings increased by only R 94.4 million (SDR 93 million) in the week ending the twenty-ninth due to the use of some of the proceeds to repay outstanding foreign loans. The official reason for the swap advanced by Governor De Jongh was the need "to make advance provision for any possible adverse effect on foreign capital movements of the anticipated seasonal increase during May and June in net bank credit to the government sector."

Essentially the swap is similar to a loan collateralized by gold, for it allowed the Reserve Bank to increase the liquidity of its reserves without adversely affecting the price of gold through an increase in supplies on the private market. In this light when the announcement of the transaction was made on May 2, the market responded positively, bidding prices up, as the swap was interpreted to mean that the amount of gold coming onto the market would probably be less than otherwise. Of course, the swap, like a collateralized loan, entails interest cost; only it is in a different form, i.e. the differential between the spot and futures prices at which the transaction was contracted.

The Nuclear Axis

In part to enable the swap, the South African Reserve Bank announced on April 25 suspension of the legal requirement to maintain minimum gold reserves equal to 25 per cent of the domestic liabilities of the Central Bank. The ratio which stood at 45.0 per cent at the end of 1974 had fallen to 25.1 per cent by February of this year, and the swap would have brought it to considerably less than the 25 per cent requirement. In addition, according to Finance Minister Owen Horwood, the Bank's practice of valuing its gold reserves at the statutory price of R 29.55 per ounce when the market price is around R 130 made it difficult for the bank to comply with the requirement.[1]

1/ As mentioned in the monthly report issued March 22, 1977 (DM/77/26), the South Africans have announced that the statutory price of R 29.55 will be abolished and replaced with a valuation in line with market prices soon after ratification of the Amendment of the Fund's Articles.

Document 31
Title page, IAEA Report on Safeguards

International Atomic Energy Agency

BOARD OF GOVERNORS

GOV/1842.
6 June 1977
RESTRICTED Distr.
Original: ENGLISH

For official use only
Item 2(a) of the provisional agenda
(GOV/1832)

SAFEGUARDS

(a) SPECIAL SAFEGUARDS IMPLEMENTATION REPORT

Note by the Director General

1. Governors will recall that in February 1977 the Board considered the proposed structure and format of the Special Safeguards Implementation Report (SSIR). The first of these reports, covering the Agency's safeguards activities in 1976, h' now been prepared by the Secretariat and is annexed hereto. Its structure and format generally follow the lines recommended by the Standing Advisory Group on Safeguards Implementation (SACSI).

2. The Director General proposes that the main recommendations contained in the report concerning changes in safeguards approaches should be referred to SACSI for

Documents

evaluation and that, taking into account SAGSE's advice, they should be reflected in the programme and budget for 1979 and subsequent years and, to the extent possi in carrying out the 1978 programme.

3. The recommendations set forth in paragraph 2.1 of the report are addressed t Member States. In view of their importance and the need for early action, the Director General proposes that the Board authorize him to address a communication conveying the recommendations to the States concerned.

4. The review of the Agency's safeguards operation has also confirmed the need reinforce the Agency's capacity for evaluating the effectiveness of the safeguard by establishing suitable machinery in the Secretariat, as recommended in the rep of the Group of Experts to Advise on the Organization of Work in the Department Safeguards and Inspection.[1]

---

[1] See GOV/INF/322.

Extracts from IAEA Report on Safeguards

3.1 In each safeguards agreement concluded in connection with NPT the technical safeguards objective is clearly defined as "the timely detection of diversion of significant quantities of nuclear material . . . ." Quantification of the notions "timely detection" and "significant quantities" has been developed along with the implementation of NPT safeguards agreements.

3.2 To judge timeliness of detection it is necessary to know the period of time required to manufacture a single nuclear explosive device from diverted material. Depending on the chemical composition and the physical form of such material, the estimates used at present by the Secretariat vary between ten days and six months. These estimates are based on advice given to the Secretariat by various expert groups. [p.8]

3.6 Further important factors involved in evaluating safeguards effectiveness are the physical accessibility of nuclear material to be verified, the degree of verifiable containment in the design of facilities and the co-ordination arrangements with States facilitating verification of the flow and inventory of nuclear material. In other words, the actual effectiveness of safeguards is closely related to the effectiveness of national systems of accounting and control of nuclear material and the extent to which States' legislation and operators' practices enable the Agency to carry out the planned verification activies in a timely fashion. [p.10]

391

Appendices

Appendix 1

The failure of international safeguards systems

At 11.30 am on Tuesday 9 August 1949, at Lake Success in New York, the first meeting of the six permanent members of the UN Atomic Energy Commission began. Canada was represented by General A. G. L. McNaughton, France by Jean Chauvel, Britain by Sir Alexander Cadogan, the USA by John D. Hickerson, the Soviet Union by S, K. Tsarapkin and China by Dr T. F. Tsian.

The meeting was convened in response to the General Assembly Resolution of 24 January 1946 asking the six powers 'to meet together and consult in order to determine if there exists a basis for agreement on the international control of atomic energy to ensure its use for peaceful purposes and for the elimination from national armaments of atomic weapons.' At a second meeting of the six powers Sir Alexander Cadogan introduced a list of topics which his delegation offered as a basis for the discussion which included the fundamental points finally agreed upon by the Western powers. Two of the points deserve special attention:

> *'International system of control*
> (a) There should be a strong and comprehensive international system for the control of atomic energy and the prohibition of atomic weapons, aimed at attaining the objectives set forth in the resolution of the General Assembly of 24 January 1946. Such an international system should be established, and its scope and function defined by an enforceable multilateral treaty in which all nations should participate on fair and equitable terms.

(b) Policies concerning the production and use of atomic energy which substantially affect world security should be governed by principles established in the treaty. Production and other dangerous facilities should be distributed in accordance with quotas and provisions laid down in the treaty.

Prohibition of atomic weapons
(a) International agreement to outlaw the national production and use of atomic weapons is an essential part of this international system of control.
(b) The manufacture, possession and use of atomic weapons by all nations and by all persons under their jurisdiction should be forbidden.
(c) Any existing stocks of atomic weapons should be disposed of, and proper use should be made of nuclear fuel for peaceful purposes.'

The British proposal provided also for the establishment within the framework of the Security Council of an International Control Agency.[1]

Today, more than thirty years later, these early UN efforts to control atomic energy are forgotten history. The current situation is described in the introduction of the SIPRI book *World Armaments: The Nuclear Threat:*[2]

'The probability of a nuclear world war is steadily increasing. This conclusion is virtually inescapable if only the consequences of advances in and the worldwide spread of military technology are considered. Given the catastrophic nature of a general nuclear war, an increasing probability of its occurrence is, to say the least, alarming. The main reasons for pessimism are:
— The Soviet-US arms race may lead to a first-strike capability.
— The spread of peaceful nuclear technology is proliferating worldwide the capability to produce nuclear weapons.
— The international trade in arms is rapidly militarizing the entire globe.
— The arms control approach has failed to restrain the nuclear arms race, prevent the proliferation of nuclear exposives or control the arms trade, let alone lead to

393

nuclear disarmament.'

The current developments in the field of atomic energy, notably the pace of nuclear proliferation, certainly support SIPRI's pessimism. Within the context of these developments, the acquisition of a nuclear bomb by South Africa represents perhaps the most serious breakdown ever of the system of international safeguards.

One of the most striking aspects of the South African nuclear programme had been its orientation to military uses. South Africa has followed the path chosen by the United States, the Soviet Union, Great Britain and France, where commercial uses of atomic energy were merely a by-product of the military programme. South Africa has the bomb several years before its first nuclear power reactor is due to come into operation. It is a matter of record, documented in this book, that South Africa has been helped all along the way to satisfy her nuclear ambitions by the same powers which protested so loudly about South Africa testing its nuclear device in August 1977.

In 1954 South Africa was invited to join the talks leading to the establishment of the International Atomic Energy Agency two years later. At the IAEA South Africa held a privileged seat on the Agency's Board of Governors with Ambassador Donald Bell Sole representing them for several years. Western support was not affected by South Africa's failure to sign the Nuclear Non-Proliferation Treaty (NPT). In June 1977 when the truth about South Africa's nuclear programme had been exposed, all the Western countries represented on the Board of Governors of the IAEA still voted against the African proposal to remove South Africa from it.

The 'official death' of the proposed United Nations Atomic Energy Commission was formally confirmed by the resolution adopted at the last (24th) meeting:

> '. . . that further discussions in the Atomic Energy Com-
> mission would tend to harden these differences and
> would serve no practical or useful purpose until such

time as the sponsoring Powers have reported that there exists a basis for agreement.'3

The deadlock between the two super-powers which had existed since the First Report of the Atomic Energy Commission to the Security Council turned out to be absolutely insoluble.

The UN Atomic Energy Commission was officially dissolved on 1 January 1956. By that time the US atomic bomb monopoly had been broken by both the Soviet Union and Great Britain, with France feverishly trying to catch up. The American policy of secrecy gave way to salesmanship, and the so-called 'Atoms for Peace' programme was available to any country in the Western world. South Africa was one of the first to get into the picture and has derived enormous benefits from it ever since.

Efforts to ensure that nuclear energy would be used for commercial purposes only were revived through the International Atomic Energy Agency (IAEA) established in Vienna on 26 October 1956. It came legally into existence on 12 July 1957. The Agency is a largely autonomous international organisation, which reports annually to the UN General Assembly and, when appropriate, to the Security Council. It has a membership of 109 countries, amongst which are all the major nuclear powers, except China. The objective of the IAEA is to 'accelerate and enlarge the contribution of atomic energy to peace, health and prosperity throughout the world.' Towards achieving this aim, it was intended initially that the Agency would accumulate nuclear materials for supply to member States; act as an intermediary for securing materials, equipment or facilities; encourage the exchange and training of scientific personnel; and promote the exchange of scientific and technical information. The IAEA is performing these functions, except that it has not become a store-keeper of nuclear materials.

In order to fulfil its role in encouraging the peaceful use of nuclear technology, the IAEA developed two systems of safeguarding nuclear power industries. The initial safeguards were designed to apply to projects involving IAEA assistance, to meet requests by parties to any bilateral or multilateral agreements, and to meet requests by individual countries with nuclear commitments. The more recent safeguards stem from obligations undertaken by states which are party to the NPT.

IAEA safeguards have been shaped by the nature of specific problems and by the degree to which countries will permit their nuclear industries to be regulated. These safeguards normally apply to particular facilities rather than to all facilities in a country. Initial safeguards were developed in 1958 in response to a Japanese request for assistance in obtaining fuel for a heavy-water research reactor. Subsequently, safeguards were expanded in a series of steps to become more generally applicable to a wide range of nuclear processes. The first of these general safeguards systems appeared in 1961 and related to reactors of less than 100 megawatts thermal capacity. In 1965, safeguards were extended to cover reactors with greater than 100 megawatts thermal capacity. This system was revised in 1965, and was extended in 1966 to incorporate provisions relating to reprocessing plants. In 1968 the system was further extended to include provisions for the safeguarding of nuclear material in conversion and fuel fabrication plants. This system, known as 'The Agency's Safeguards System (1965, as provisionally extended in 1966 and 1968)' and reproduced in IAEA document INFCIRC/66/Rev.2, constituted the extent of safeguards development at the date on which the NPT entered into force.

In February 1974, the IAEA Board of Governors decided that subsequent IAEA safeguards agreements should normally contain provisions which relate the duration of the agreements to the period of actual use of the safeguarded items. The Board also decided that the agreements should normally confirm the right of the IAEA to continue to apply safeguards to special fissionable materials produced until the provisions for termination of safeguards contained in INF-CIRC/66/Rev.2 have been satisfied.

Experience with IAEA safeguards demonstrate that countries have not been prepared to accept continuous surveillance of nuclear activities by an external authority. The control system established by the Agency involves accounting methods augmented by regular 'on the spot' inspections. The inspections are carried out by a team of skilled personnel within the IAEA, and as of 30 June 1976 there were 79 inspectors, of whom about 50 regularly carried out inspections. Many observers believe that this force is far too small to maintain effective surveillance of existing installations covered by IAEA safeguards.

In brief, the accounting procedure is based on a system of records and reports which are maintained by a country with respect to facilities and nuclear materials in its territory. It is then for the inspectorate to carry out regular audits of records and reports, to check the amount of safeguarded material, and to scrutinise the operation of facilities subject to safeguards.

The most serious attempt to introduce international controls of atomic power was the Americans' Baruch Plan. However, it is extremely doubtful whether even the authors of the Plan had any expectation that its provisions for stringent regulation and internationalisation would ever be put into effect.

J. R. Oppenheimer, who himself claimed authorship of the Acheson-Lilienthal Plan which later became the Baruch Plan, states that at the time he did not expect the Soviet Union to accept the plan, because he thought that if they did so and opened their frontiers and freely admitted Western inspection, the Soviet system would collapse.[4]

Similarly an American author, R. E. Lapp, a physicist who at the time of the Atomic Energy Commission discussions held a post at the Pentagon and was thus presumably familiar with the nuclear diplomacy of the United States, says:

'In retrospect, one shudders at what might and probably would have happened, had the Soviets elected to accept our United Nations proposals. The United States Senate would probably have refused to agree to the plan, and the United States would then have been in the position of a warmonger, or, at the very least, not a seeker of peace.[5]

The problem has been exacerbated by rising fuel costs and fears of a world energy shortage. The number of countries which are now clamouring for nuclear power has dramatically increased, encouraged by reactor salesmen trying to compensate for the dwindling reactor market in the United States.

Provision for the IAEA to 'exercise strict control over the operations connected with the production of power where diversion of fissionable materials to weapons can most readily take place, and to approve means to be used for chemical processing of spent fuel elements' (Article XII of the IAEA Statute) has not prevented countries such as India, Israel and most recently South Africa from joining the nuclear 'club'. The Non-Proliferation Treaty (NPT), signed in London, Washington and Moscow on 1 July 1968 and entering into force on 5 March 1970, also failed to achieve its primary objective: to freeze the number of States possessing nuclear weapons.[6]

The NPT is based on the questionable assumption that at least some degree of separation is possible between commercial and military nuclear technologies. It prohibits signatories from transferring nuclear weapons to other states and from manufacturing them if they do not have them already, but it leaves everyone free to develop a 'peaceful' nuclear capability which can in fact be converted into a military capability.

The most alarming aspect of nuclear proliferation is that it has been spurred primarily by private business rather than by the policies of the governments, although these are involved in nuclear deals by virtue of their control over atomic energy establishments in their own countries. The West German-South African nuclear deal is an example of how this can happen.

About ninety nuclear power reactors have been, or are being, constructed abroad by the major nuclear exporters, with a total electrical generating capacity of about 50,000

MWe. These exports are worth some $20,000 million and the fuel contracts that normally go with the reactors more than double this sum.

While reactors remain the dominant item on the international nuclear market, all other parts of the fuel cycle have now entered the commercial realm. In that respect West Germany and France have taken a lead. The West German transfer of nuclear enrichment technology to South Africa as documented in this book, together with its sale of the complete nuclear fuel cycle to Brazil and the French deal with Iran, constitute the most serious loopholes in the international safeguards system.

Although the NPT Review Conference held in May 1975 went largely unnoticed, growing public interest in nuclear proliferation issues was discernible in the autumn of that year, and has since become considerably stronger. The political struggle over nuclear power programmes in Sweden, which ended the forty-year rule of the Social Democratic Party, has spread to West Germany, France and many other countries. This has helped to draw attention to the relationship between nuclear power and nuclear weapons.

International concern about the need to check the military implications of international nuclear commerce led to the establishment of the London Club. This grew out of a secret meeting between Canada, France, West Germany, Japan, Britain, the United States and the Soviet Union held in 1975 in London.

The rules have somewhat tightened the terms of nuclear supplies and reduced the advantages that non-parties may derive from remaining outside the NPT. But they are still insufficient to guarantee that no further nuclear weapon proliferation will occur as a result of transfers. They suffer from two major omissions: (a) they fail to require full-scope safeguards, that is, safeguards on all peaceful nuclear activities in recipient states, as a condition for nuclear supplies, and (b) they do not definitively preclude exports of highly sensitive facilities.

None of the existing safeguards provided for by the International Atomic Energy Agency, the NPT or the London Club can effectively stop the spread of nuclear proliferation. What has happened, very simply, is that existing controls and treaties have proved wholly inadequate. The capacity to produce nuclear weapons is spreading all the time, and so is the demand for nuclear power stations which both contribute to this capacity and produce radioactive waste for which nobody has yet found a safe system of disposal. If no better controls can be devised a disaster of some sort may be inevitable.

1. International Control of Atomic Energy: Interim Report on the consultations of the six permanent members of the UN Atomic Energy Commission approved 24 October 1949. Doc. A/ 1045 Corr. 1.
2. Stockholm International Peace Research Institute, 1977.
3. AEC, Fourth Year, Official Records, 29 July 1949, p. 38.
4. In the matter of J. R. Oppenheimer, U.S. Govt. Printing Office, 1954, p.38.
5. R. E. Lapp, *Atoms and People* (New York 1956), p. 159.
6. The Non-Proliferation Treaty prohibits the transfer by nuclear-weapon states to any recipient whatsoever of nuclear weapons or other nuclear explosive devices or of control over them. Prohibits the receipt by non-nuclear-weapon states from any transferor whatsoever, as well as the manufacture or other acquisition by those states, of nuclear weapons or other nuclear explosive devices.

    Non-nuclear-weapon states undertake to conclude safeguards agreements with the International Atomic Energy Agency (IAEA) with a view to preventing diversion of nuclear energy from peaceful uses to nuclear weapons or other nuclear explosive devices.

    The parties undertake to facilitate the exchange of equipment, materials and scientific and technological information for the peaceful uses of nuclear energy and to ensure that potential benefits from peaceful applications of nuclear explosions will be made available to non-nuclear-weapon parties to the treaty. They also undertake to pursue negotiations on effective measures relating to cessation of the nuclear arms race and to nuclear disarmament, and on a treaty on general and complete disarmament.

    It was signed at London, Moscow and Washington, 1 July 1968, and entered into force on 5 March 1970.

Appendix 2

Production of nuclear weapons

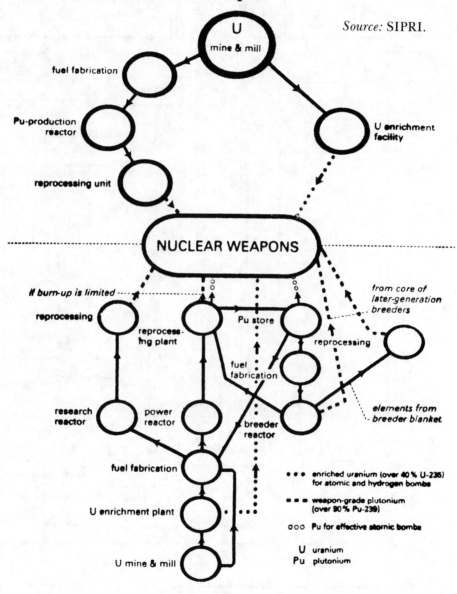

Source: SIPRI.

Appendix 3   Bonn's capital share and influence i

Source: Sechaba, November/December 1975.

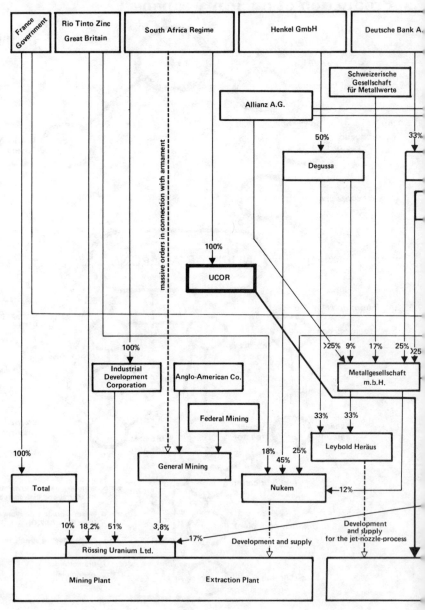

he nuclear co-operation with South Africa

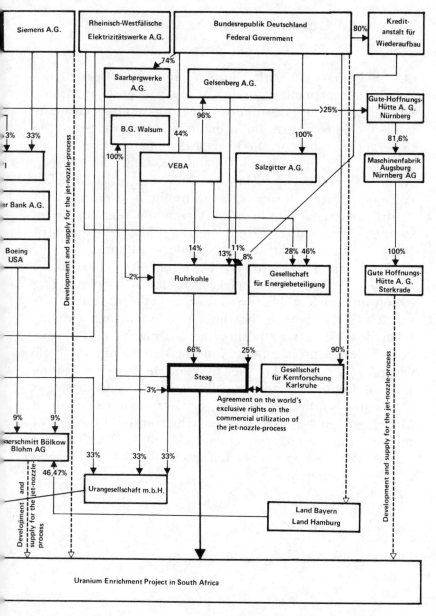

Appendix 4

The West German 'Memorandum' *

On 28 January 1977 the head of the Africa Desk of the Political Department of Bonn's Ministry of Foreign Affairs, Dr Helmut Müller, distributed to a group of nine journalists invited by him the previous evening, a 'Memorandum' of the Federal Government. The 13-page document does not carry any title, sender, author or signature. Only the German version bears the date 17 January 1977. The official English version, we are informed, was transmitted to the OAU in Addis Ababa and a number of African ambassadors there; some time later it was also sent, together with a Verbal Note of Foreign Affairs, to African ambassadors in Bonn.

In this Memorandum, the Government of the Federal Republic replied to the 'allegations that it is cooperating closely with the Republic of South Africa in the nuclear and the military fields', claiming that the charges originated 'evidently from the same source' (the implication being that the information was hostile propaganda). While the Memorandum refers to respective accusations of the OAU and the statements of the Federal Republic before the UN General Assembly, it is obviously also directed against the German Anti-Apartheid Movement and a press release which listed 14 items of military and nuclear cooperation as well as other cases of cooperation between the Federal Republic of Germany and the Pretoria regime.

* See Chapter 3.

Appendix 5

The 'jet-nozzle' system of enrichment

The 'jet-nozzle' system is based on pressure diffusion in a gaseous mixture of UF_6 and an additional light gas (He or H_2) flowing at high speed through a nozzle along curved walls. The heavier molecules are less deflected and enriched in the stream with the largest curvature. The addition of He increases the velocity of the UF_6 molecules and thus improves the separation effect. (See Appendix 9.)

Jet-nozzle technology requires low specific investment costs to become economically attractive for medium-sized plants, but it has a very high power consumption. It could be an attractive alternative for a medium-sized country with cheap electricity, for example from a hydroelectric power station, and no serious problems of cooling capacity. One could think of underdeveloped countries with rich fields of uranium ore.

Another important factor is the number of years needed to bring a new plant into full operation. Centrifuge and jet nozzle plants need about two years' less construction time than a gaseous diffusion plant. This means that decisions can be better adjusted to the market demands.

Source: P. Boskma, 'Uranium enrichment technologies and the demand for enriched uranium', in *Nuclear Proliferation Problems* (SIPRI: Stockholm, 1974).

For a detailed explanation of commercial-type jet nozzle uranium enrichment technology see Robert Gillette, 'Nuclear Proliferation: India, Germany May Accelerate the Process', in *Science*, Vol. 188, no. 4191 (30 May, 1975) pp. 911-914.

Appendix 6

Operating nuclear facilities not subject to IAEA or bilateral safeguards, as of 31 December 1977

Country	Facility	Indigenous or Imported	First year of operation
Egypt	Inshas	Imported (USSR)	1961
India	Apsara research reactor	Indigenous	1956
	Cirus research reactor	Imported (Canada/USA)	1960
	Purnima research reactor	Indigenous	1972
	Fuel fabrication plant at Trombay	Indigenous	1960
	Fuel fabrication plant at the Nuclear Fuel Cycle Complex at Hyderabad	Indigenous	1977
	Reprocessing plant at Trombay	Indigenous	1964
	Reprocessing plant at Tarapur	Indigenous	1977
Israel	Dimona research reactor	Imported (France)	1963
	Reprocessing plant	Development in collaboration with France (Saint Gobain Techniques Nouvelles)	1963
South Africa	Pilot enrichment plant	Developed in collaboration with FR Germany	1975
Spain	Vandellos power reactor	Jointly operated with France	1972

Source: SIPRI Yearbook, 1978

Appendix 7 Estimated yield limits for various evasion techniques which are considered technically feasible*

Evasion Techniques	Estimated Yield Limit to Avoid Detection**	Constraints on Tester
Tamped shot in low coupling media	1-2 KT***	Low yields; relatively few areas of low coupling media, most in undeveloped regions; evader would probably test in seismic region.
Decoupling Cavity	50 KT	Large volume of rock or salt required; long preparation time; expensive.
Detonate following nearby earthquake	100 KT	Device must be pre-positioned; evader would probably have to test in seismic regions; 1 opportunity every 1-2 years to conduct several simultaneous events in a series; decision to test must be made quickly.
Detonate following large distant earthquake	100 KT	Requires multiple emplacement holes; evader would have to test in seismic regions. Requires considerable testing experience.
Multiple shot simulation of earthquake signal	50 KT	Device must be pre-positioned; local earthquakes must be about one seismic magnitude larger than explosion; decision to test must be made very quickly.

*See Chapter 5.
** Estimates based on detection capabilities of stations remote from event.
*** Could be as high as 10 KT dependent upon the availability of sufficiently deep low coupling media.

407

Appendix 8

Israeli-South African Military Collaboration and Co-operation in Defence Production

1962   Israel supplies 32 Centurion tanks to the RSA. (SIPRI, *Arms Trade Register*, Stockholm, 1975.)

1963   Israel 'possibly' supplies spare parts to the RSA. (SIPRI, *Arms Trade with the Third World*, Stockholm 1971, p.680.)

5/1967   South African Jews go to Israel as volunteers, officially military employed. (*Rand Daily Mail*, 31 May, 1967.)

6/1967   According to *International Herald Tribune* a South African mission flew to Israel during the June war to study tactics and use of weapons. (*International Herald Tribune*, 30 April, 1971.)

10/1967   General Mordechai Hod, chief of the Israeli Air Force reports 'in detail' to the South African chief of staff about experience in the 'Blitz'. (*Rand Daily Mail*, 10 Oct, 1967.)

1/1970   The *Jewish Telegraph Agency* reports that 'the South African government is about to export tanks to Israel'. This refers to a 65 t. tank with a heavy gun built on the lines of the British 'Chieftain'. (*Jewish Telegraph Agency*, 20 Jan, 1970.) This is denied by the Israeli ambassador to the U.N., Tekoa, in a letter to the Security Council.

5/1971   Israel offers to sell three aircraft to South Africa to replace three crashed aircraft of the South African Air Force. (*Rand Daily Mail*, 11 Sept, 1971.)

1971   In a heavily disputed article in the *International Herald Tribune*, C. L. Sulzberger reports that South Africa is producing the Israeli submachine gun Uzi under a licence granted through Belgium. Further he reports rumours that Israel has handed on the blueprints of the French Mirage obtained through

408

espionage to South Africa. (*International Herald Tribune,* 30 April, 1971.) The South African aircraft industry, like the Israeli, is building a fighter plane along the lines of the Mirage.

1971 The Israeli firm Tadiran offers South Africa electronic equipment for military purposes. (*Financial Mail,* Johannesburg, 24 Sept, 1971.) Tadiran is jointly owned by Koor (50%), General Telephone and Electronics International (USA) (35%), Israel's Ministry of Defence (15%).

1971 A five man sales mission from the state-owned Israeli Elta is sent to the Republic of South Africa to sell ground-to-air communication equipment for the South African Ministry of Defence. (*Financial Mail,* Johannesburg, 24 Sept., 1971.) Elta is a subsidiary of the state-owned Israeli Israel Aircraft Industries (IAI).

1972 Major supplies of Israeli spare parts for Mirage and other weapon systems to South Africa. Israel is presumably charged with the servicing of the Mirage. (*Afrika Heute,* Bonn, June/July 1973.)

5/1973 During the Paris Air Show an Israeli 'official' points out the usefulness for South Africa of the Arava. The Arava had already been taken for trials to South Africa and there are rumours that South Africa has ordered a certain number of the aircraft.

10/1973 The Egyptian government discloses that a Mirage of unknown nationality had been shot down on the Suez front. According to the London *Daily Telegraph* South Africa had during the October war sent Mirage fighters in support of Israel via the Portuguese Atlantic islands. (*Daily Telegraph,* 31 October, 1973.)

10/1973 According to the *Rand Daily Mail* 1,500 Jews with South African connections served in the Israeli Armed Forces during the war, 800 of whom took part in the Suez crossing. Dr. C. L. Kowalsky was captured by the Syrians at Mount Hermon. After his release he visited his relatives in South Africa. (*Rand Daily Mail,* 22 Oct, 1973; *Rand Daily Mail,* 12 July, 1974.)

1974   SIPRI confirms the assumed delivery of sea-to-sea missiles of the type Gabriel to South Africa. Date of delivery: December 1974, for the equipment of seven new South African ships. (SIPRI, *Yearbook 1975*, Stockholm 1975, p. 239.)

1974   According to *Monitor*, Israel supplies South Africa with the following: light guns, heavy mortars, hand guns, electronic and mechanic warning systems for border security. (*Monitor,* 10 Sept, 1974.)

1975   The former chief of the Israeli secret services and present President of Koor industries, General Meir Amit, during a visit to South Africa discloses that Israeli officers regularly lecture before an audience of South African officers about modern warfare and anti-guerilla tactics. Asked whether Israel and South Africa have good military relations he answers: 'That is an understatement'. (*Rand Daily Mail,* 7 July, 1975; *Guardian,* 9 July, 1975; *UN, Unit on Apartheid, Notes and Documents,* No. 31/75, September 1975.) Koor industries are owned by Israeli trade-union Histadruth and one of the most important armament producers of Israel.

1976   The *Financial Times* (London) reports that Israel is offering its tactical aircraft Kfir to South Africa. (*Financial Times* 20 Feb, 1976.)

1976   South Africa is going to build six patrolboats of the Reshef class in a licence of the Israeli firm Ramta. They will be equipped with 'Gabriel' missiles. (*Les Flottes de Combat 1976,* Paris, p. 56; *Jerusalem Post,* 9 March 1976; *Financial Times,* 12 Feb, 1975.) Ramta is a subsidiary of the state-owned IAI. The shipyards in Durban, South Africa, and Haifa are being extended.

4/1976　South Africa's Premier Vorster visits IAI as well as Reshef patrolboats in Sharm-el-Sheikh; the supply of Kfir is denied. (*Jerusalem Post*, 13 April, 1976; *Guardian*, 10 April, 1976.)

Appendix 9

Jet nozzle

Light Fraction

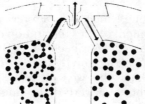

Feed Gas　　　　　　　　　　　　　　　Heavy Fraction

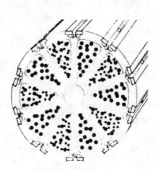

Notes

Chapter One

1. This possibility was also entertained by *Der Spiegel* (Hamburg) of 28 June 1976, as follows:

    'The diplomatic mission of South Africa happened to move from the Heumarket in Cologne to new premises in Bonn-Bad Godesberg, Auf der Hostert 3. And because the new filing-cabinets had not been delivered in time the employees of the Embassy piled up their papers in the basement garage where the entrance gate had not yet been installed: for weeks anyone interested was able to help himself free and easy and undisturbed to the contents of document boxes, waste-paper baskets and the locked safe of ambassador Donald Bell Sole – so much so that even today the diplomats are still lacking an overview of what exactly got lost.'

2. ibid.
3. IAEA, Board of Governors, GOV/OR.SO1, 24 June 1977. p.5.
4. ibid. p. 8.
5. *Der Spiegel*, no. 31, July 1977, pp. 27-28.

Chapter Two

1. Eleanor Lansing Dulles, John Foster and Allen Dulles's sister.
2. The establishment of the constitution of the Federal Republic of Germany is well described by J. F. Golay, *The Founding of the Federal Republic of Germany* (University of Chicago Press, 1958).
3. 'The French knew that we were ready to proceed in the three zones, but otherwise were determined to go ahead in two.' Lucius D. Clay, *Decisions in Germany* (New York, 1950).
4. Quoted by Lucius Clay, *op.cit.*, p. 80.
5. Hans Speier, *German Re-armament and Atomic War: The views of German military and political leaders* (Raw, Petersen; Evanston, 1957).
6. ibid, p. 7.
7. For the text of his statement see the Department of State Bulletin, 4 September 1950.
8. René Pleven, then Prime Minister of France, balking at the idea of the remilitarization of West Germany, put forward his own proposals for Western defence, called the 'Pleven Plan'.
9. Anthony Eden, then British Foreign Secretary, on 1 December 1951 wrote a memo to Prime Minister Winston Churchill which he concluded as follows:

412

'(a) We should support the Pleven Plan, though we cannot be members of it. This is what the Americans are doing, and it is the course Eisenhower wants us to take.

(b) If the Pleven Plan does collapse, we should try to work out a more modest scheme with our allies, based upon the technical military arrangements agreed upon, but without elaborate political superstructure.

(c) Any move for (b) will require careful timing. If we move too soon, the Pleven Plan will collapse, and we shall be told we have killed it.'
A. Eden, *Full Circle* (Cassel, London, 1960), p. 33.

10. H. Speier, *op.cit.*, p.50.
11. ibid.
12. The existence of the Memorandum was revealed by Professor H. Arntz in an official publication of the Federal Press Office, *Germany Reports*, published in 1964.
13. Quoted by Catherine McArdle Kelleher, *Germany and the politics of nuclear weapons* (Columbia University Press, 1975), p. 18.
14. A. Eden, *op.cit.*, p. 47. See also Note 20 below.
15. ibid.
16. *Le Monde* (Paris), 30 August 1954.
17. *New York Herald Tribune*, 1 September 1954.
18. The unilateral declaration by the Federal Chancellor made in London constitutes Annex I to Protocols II and IV of the Paris Agreements which came into force on 23 December 1954.
19. The possibility of German aggression was expressly mentioned in one of the eight intents in the Preamble to the Treaty. Article 4 of the Treaty, which provided for automatic military aid if any of the parties should be the object of an armed attack in Europe, did not specify the potential aggressor.
20. cf. Eden's remarks on p. 51 of his memoirs (*Full Circle, op.cit.*) with those on p. 43.
21. Eisenhower, *Waging Peace 1956-61* (Heinemann, London, 1966), p. 366.
22. ibid, p. 159.
23. K. Adenauer, *Erinnerungen 1953-1955* (Deutsche Verlags Anstalt, Stuttgart, 1966), pp. 308-313.
24. The vast scope of the Authority's competence was laid down in Article 15 of the Agreement, according to which:

'The Authority shall have the right to examine transport, price and trade practices, quotas, tariffs, and other governmental measures or commercial arrangements instituted or permitted by the German authorities, which affect the coal, coke or steel of the Ruhr. If the Authority determine that such practices, measures or arrangements are artificial or discriminatory and are of such nature as

1. to impede access by other countries to the coal, coke or steel of the Ruhr,
2. to distort the movements of Ruhr coal, coke or steel in international trade, or
3. otherwise to prejudice the accomplishment of the purposes of the present Agreement,

the Authority shall decide that such practices, measures or arrangements shall be appropriately modified or terminated.'

25. Powers in the following fields were specifically reserved: (a) disarmament and demilitarization, including related fields of scientific research, prohibitions and restrictions on industry, and civil aviation;
(b) controls in regard to the Ruhr, restitution, reparations, decartelization, deconcentration, nondiscrimination in trade matters, foreign interests in Germany and claims against Germany;
(c) foreign affairs, including international agreements made by or on behalf of Germany;
(f) respect for the Basic Law and the Land constitutions;
(g) control over foreign trade and exchange.

26. K. Adenauer, *op.cit.*, p. 309.
27. Both points were, at Adenauer's request, included in the official communiqué issued on 17 September 1954 which described the talks with Dulles as having been conducted 'in a friendly and cordial atmosphere which has characterized the relations between the two governments.' The text of the communiqué was published by all leading Western papers on 18 September 1954, and was reproduced in full in *Keesing's Contemporary Archives 1952-1954*, p. 13797.
28. K. Adenauer, *op.cit.*, p. 312.
29. ibid, p. 347. 'ABC weapons' is military jargon for atomic, biological and chemical weapons.
30. *Oppenheim International Law*, ed. Lt. Lauterpacht (Longmans, London, 1963), vol. I, 8th edition, p. 939.
31. These points were analysed by Leiters and de la Malève in 'Paris from EDC to WEU', in RAND MEMORANDUM 1668 Rc of 1956 quoted by C. McArdle Kelleher, *op.cit.*, p. 28.
32. Quoted by Konrad Huber in 'L'utilisation pacifique de l'énergie atomique en Allemagne', in *Politique Etrangère* (Paris), no. 6/1956, p. 678.

33. T. Sommer, 'The objectives of Germany', in *A World of Nuclear Powers?*, ed. A. Buchan (American Assembly, Columbia University, 1966), p. 41.

34. A Declaration on the Admission of the German Federal Republic to the North Atlantic Treaty Organization and on the future deployment of NATO forces in Europe of 3 October 1954.

35. *Bundestag Record*, 15 December 1954, p. 3124.

36. See Józef Garliński, *Hitler's Last Weapons* (Julian Friedmann, London, 1978).

37. For a complete review of the Nazi gas-centrifuge project, see David Irving, *The German Atomic Bomb: The History of Nuclear Research in Nazi Germany* (Simon and Schuster, New York, 1967), especially pp. 36-37, 127-29, 173 and 229-30.

38. A. Grosser, *Germany in our Time, A Political History of the Postwar Years* (Pall Mall, London, 1971), p. 57.

39. Quoted in A. Grosser, *op.cit.*, p. 23.

40. Irving, *op.cit.*, p. 230.

41. See Donald G. Avery and Airwyn Davies, *Uranium Enrichment by Gas Centrifuge* (Mills and Boon, London, 1973), p. 91.

42. Kurt Beyerle, Wilhelm Groth, Paul Harteck and Hans Jensen, *On Gas Centrifuges: The Enrichment of the Xenon, Krypton and Selenium Isotopes by the Centrifuge Process* (Verlag Chemie, Weinheim, 1950), p. 67.

43. On Metallgesellschaft, see British Intelligence Objectives Subcommittee, Final Report No. 335, Item No. 21, *Metallgesellschaft and the Lurgi Group of Chemical Engineering Companies* (H.M.S.O., London, 1946), pp. 7-8 and 15-23.

44. *Daily Telegraph*, 22 July 1972.

45. 'Ceremonial Delivery of First German Uranium', *Atomwirtschaft*, December 1956, p. 428.

46. ibid., p. 23.

47. R. E. Lapp, *Atoms and People* (New York, 1956), pp. 191-192.

48. Grosser, *op.cit.*

49. Irving, *op.cit.*

50. *Trials of War Ciminals before the Nürnberg Military Tribunals under Control Council Law No. 10*, Vol. VII and VIII: "The I.G. Farben Case" (United States Government Printing Office, Washington DC, 1953), pp. 13-14.

51. ibid., p. 19.

52. ibid., p. 30.

53. 'General Order No. 2, Pursuant to Military Government Law No. 52 – Blocking and Control of Property', Military Government of Germany, United States Zone, 5 July 1945; cited in

Richard Sasuly, *IG Farben* (Boni and Gaer, New York, 1947), pp. 264-65.

54. *Der Spiegel*, 15 March 1976.
55. *Nürnberg Military Tribunals, op.cit.*, pp. 13-14, 42, 344, 347..
56. Irving, *op.cit.*, pp. 82-83.
57. ibid., pp. 47 and 55.
58. *Atomwirtschaft*, December 1956, p. 428.
59. David Peirson, 'Centrifuge Enrichment for United Kingdom Nuclear Fuel', *Kerntechnik*, April 1973, p. 161.
60. Lapp, *op.cit.*, pp. 191-92.
61. B. F. J. Schonland, 'Nuclear Energy and Southern Africa', conference paper, published in the Conference Proceedings, vol. I, p. 1078.
62. Conference paper, *op.cit.*, p. 125.
63. Conference paper, *op.cit.*, p. 145.
64. *Süddeutsche Zeitung*, Munich, June 29 and July 12, 1955.
65. OEEC was established on 16 April 1948 by the 'Convention for Economic Co-operation' and one year later on 17 February 1949 became a permanent body. The signatories of the Convention were: Austria, Belgium, Denmark, France, Greece, Iceland, Ireland, Italy, Luxembourg, the Netherlands, Norway, Portugal, Sweden, Switzerland, Turkey and the United Kingdom. The West German occupation zones were represented by the military commanders. In October 1949, the Federal Republic of Germany became a fully-fledged member. In 1950 America and Canada became associate members.
66. Article of the Convention on the Establishment of Security Control in the Field of Nuclear Energy.
67. TASS Press Release of 13 July 1956.
68. 'The Agency shall seek to accelerate and enlarge the contribution of atomic energy to peace, health and prosperity.' (Article II of the Statute)
69. Interview with Matthöfer in *Frankfurter Rundschau* (Frankfurt), 28 April 1977.
70. Natural uranium obtained after processing in the extraction plants consists of 99.3% of the heavy isotope U-238 and only 0.7% of the lighter U-235 called 'fissionable material', which is used as a nuclear fuel for power-reactors and, significantly, also for nuclear explosives. There is, however, a fundamental difference between the degree of enrichment of the uranium used for the nuclear reactors and the degree necessary for the manufacture of atomic bombs. While 3% concentration of U-235 is sufficient for the reactors, 90% enrichment is necessary

to make the bomb. The enrichment of the isotope U-235 is technically a very difficult and extremely expensive process. The jet-nozzle process has been described by its inventor Professor E. W. Becker in 'Die Physikalischen Grundlagen der 235 U-Anreichung nach den Trenndüsenverfahren', published in *Zeitschrift für Naturforschung*, vol. 26A, no. 9, 1971, p. 1377. It is based on pressure diffusion in a gaseous mixture of UF6 and an additional light gas (He or H2) flowing at high speed through a nozzle along curved walls. According to the evaluation made by Dr H.Mohrhaner, in *Atomwirtschaft*, of the potential of the uranium enrichment methods in Europe, the jet-nozzle system was promising. ('Stand der Urananreichung in Europe', vol. 17, June 1972, p. 300.) See also Appendix 5.

71. *SIPRI Yearbook 1977*, p. 43.
72. *Der Spiegel*, 15 March 1976.
73. *Der Spiegel*, July 1977.
74. In January 1977, the Federal Ministry of Foreign Affairs produced a memorandum strongly refuting all accusations of its nuclear and military co-operation with South Africa. See Appendix 4.
75. *International Proliferation Problems* (SIPRI, Stockholm, 1973), p. 56.
76. 'World Armament and Disarmament', *SIPRI Yearbook 1977*, p. 10.
77. *Der Spiegel*, 15 March 1976.
78. The interview was picked up by international news agencies, and extracts were published by the Swedish paper *Dagens Nyheter* of 3 May 1977. However, when the author asked for the official version from the Federal Press office, they denied its existence. Its text is therefore reproduced in the form in which it appeared on the telex.
79. ibid.
80. Negotiations with the Dutch Government over a joint German-Dutch establishment for the commercial enrichment of uranium based on the centrifuge technology started in 1967. This was revealed during the parliamentary debate in Holland by deputy H. De Goede, who said that the reasons for which Germans were seeking partners for the enrichment venture were political. (Proceedings of Second Chamber 1967-1968 on 3 April 1968, p. 1719.)
81. 'It is noteworthy that even the United States with its 62 nuclear power plants has decided to defer the commercialization of reprocessing activities, and has thereby made uncertain the future of the "fast breeder" reactor which uses plutonium as a

fuel and which is being used as a replacement for existing reactors.' *SIPRI Yearbook 1977*, p. 363.

82. For a critical assessment of the size of the West German nuclear industry, see Wolfgang Bergmann, 'Verfilzte Kernkraftsippe', in *Forum, Zeitschrift für transnationale Politik* (Bonn, no. 3/4, 1976), pp. 52-55.

Chapter Three

1. Heinrich Böll, *Der Spiegel*, No. 46, 1976.
2. Rudel's interview with *Das Bild*, quoted in *Der Spiegel*, no. 46, 1976, p. 36.
3. They were dismissed by Defence Minister Leber after a protest by 40 SPD Members of the Bundestag led by Manfred Coppik, Karl-Heinz Hansen, Erich Meinike and Ernest Waltemathe who requested disciplinary measures on the grounds that the behaviour of these Luftwaffe officers was not in accordance with the political aims of a democratic society of which the Army is an integral part. (*Frankfurter Rundschau*, 30 October 1976.)
4. Rall in his letter to Ambassador Sole of 5 June 1973.
5. Confidential source.
6. *Frankfurter Rundschau*, 25 September 1975.
7. *Der Spiegel*, no. 46, 1976.
8. Interview in *Capital*, no. 3, 1974.
9. The group of western states was enlarged in 1956 when the Soviet Union, Brazil, Czechoslovakia and India took part in drafting the statute of the IAEA.
10. Article III of the Agreement Governing the Relationship between the United Nations and the International Atomic Energy Agency.
11. The text of the Treaty was reproduced in *Nuclear Proliferation Problems* (SIPRI, Stockholm, 1974).
12. 'Fissionable material' in this context means enriched uranium.
13. *Report from South Africa*, April 1975.
14. *Der Spiegel*, no. 41, 1975, p. 27.
15. *Financial Times* (London), 18 February 1970.
16. The dinner was held on 13 March 1973 at 6 pm in a restaurant called 'Haus Wohnung' in Essen.
17. Memorandum from Ambassador Sole to the Secretary for Foreign Affairs, Pretoria, no. 9/2/8/28/1/1/, 15 June 1973.
18. So far Steag has not revealed the content of the 'memorandum of understanding', Herr Haunschild has not explained what exactly was planned and Herr Willy Brandt according to the SPD spokesman Heye 'could not remember' the

exchange of letters with Vorster. Heye was quoted in the German radio series *Anhang zum Nachrichtenspiegel I*, in a programme on 6 October 1975 entitled 'Doch Kernenergie-Zusammenarbeit Bundesrepublik-Südafrika?' ('Nuclear Cooperation between the Federal Republic and South Africa after all?').

19. Law governing foreign trade and foreign economic relations.
20. Interview with Anthony Terry of the *Sunday Times* (London), 7 October 1977.
21. ibid.
22. *The Cape Times* (Cape Town), 22 February 1977.
23. Letter from Rohwedder of 5 May 1975.
24. Shorthand report on the Federal Parliament, 7th period of legislation, 167th session.
25. The $184m order for compressors designed for the acceleration of the gaseous uranium mixture was placed by the South African Electricity Supply Commission (ESCOM) with Maschinenfabrik Augsburg Nürnberg AG (MAN) on 11 September 1974. The compressors eventually found their way to the original destination through a different channel, described on p. 46.
26. The contract for delivery of two nuclear power stations of 1000 Megawatts each to be built at Koeberg (Cape Town) – a contract worth Rand 800m – was signed in Pretoria on 25 May 1976 between ESCOM and the French consortium Framatome. The loss of the order, for which the West German company KWU was one of the bidders (others were US General Electric and the US-Swiss-Dutch consortium General Electric/Brown Boveri/Rijnschelde), caused fury in the right-wing West German press which blamed the Federal Government's anti-apartheid policy for the loss of this golden opportunity.
27. Egon Bahr in an interview with *Deutsche Allgemeine Sonntagsblatt* of 3 July 1977.
28. P. Ripken and von Hein Anhold, 'Hermes-Garantien geraten immer stärker ins Zwielicht', in *Frankfurter Rundschau*, 10 August 1977.
29. 'West Germany's political arm in South Africa', *Daily News* (Dar-es-Salaam), 12 July 1977, p. 4.
30. Quoted in Stephan Schmidt (ed.), *Schwarze Politik aus Bayern – ein Lesebuch zur CSU* (Herman Darmstadt, Luchterhand, 1974), p. 149.
31. *South African Digest*, January 1974.
32. 'Warum darf Südafrika nicht fallen', in *Deutschland Magazin*, no. 4, 1977.

33. *Der Spiegel*, 11 April 1977.
34. Dieter Habicht-Benthin, formerly a social science lecturer at Kiel University. In 1970 he led the group of young rebels in the German Africa Society who ousted the conservative board of the Society. Under his leadership and through the Society's monthly *Afrika Heute* (*Africa Today*) the German Africa Society became the most articulate critic of Germany's policies towards Africa. *Afrika Heute* identified itself with the African cause by opposing Germany's collusion with South Africa and calling for the observance of UN sanctions against Rhodesia and compliance with UN resolutions on Namibia (where West Germany then still maintained a consulate). It also advocated support for the liberation movements and better relations with the Third World in general.
35. *Der Stern*, 28 November 1974, in a story about the German secret service entitled 'Die letzte Schlacht des alten Generals' ('The last battle of the old generals').
36. Wischnewski's statement was recorded in an official protocol of the meeting between the Board of the German Africa Society and the Foreign Ministry of 29 October 1974, published in the circular letter no. 1/75 of the German-Africa Society.
37. In a conversation with the authors.
38. See note 36 above.
39. Item 3:
    'Visit of experimental establishment; study of methods and instrumentation used for calibration of magnetic and acoustic sensors. Discussion of performance and merits of mines offered to S.A. Navy, in particular the acoustic directional mine (Italmeccanica, type ITM/ACS/300). Discussion of pressure mines.
    Italmeccanica: to obtain precise technical details (sensitivity, frequency, etc.) of acoustic directional mine, type ITM/ACS/300. Discussion of sweeping possibility and study possible adaptation of this principle to our own monitor systems.'
    Item 7:
    'Study of latest developments in French mines offered to S.A. Navy. Discussion of possibility of detecting mines that incorporate active sensors. Obtain information on sensitivity of active sensors of MCC 23, ED 21 and ED 28.'
40. The Documentation 'Conspiracy to arm apartheid continues' (*Progress Dritte Welt*, Bonn, 1977) lists the following South African military researchers as visitors to Federal Republic of

Germany: M. E. Beyers, of the Aeronautics Research Unit of the National Mechanical Engineering Research Institute, who has been involved in free flight tests in supersonic wind tunnels and visited laboratories at Göttingen (DFVLR) and at Porz; Dr H. G. Denkhaus, Director of the National Mechanical Engineering Research Institute, and Dr T. J. Hugo, Director of the National Institute for Defence Research, who are listed as frequent visitors of Siemens military research laboratories; H. N. Jungwirth, of the Nuclear Science Group of the National Physical Research Laboratory interested in design of cyclotron magnets, magnet materials and magnetic field measurements, and Prof. W. L. Rautenbach, who came to Germany as Project Leader of the Board of Control Feasibility Study for National Open Sector Cyclotron, who visited Karlsruhe Nuclear Research Centre; I. Rodger of the National Institute of Defence Research who has been a regular visitor of Messer-schmidt-Bölkow Blohm – the largest military and weapons manufacturer in the Federal Republic and one of the largest in the world. Others were: N. J. Smit of the National Institute of Defence Research; C. G. Van Niekerk, Head of the Aeronautics Research Unit; and J. D. N. van Wyk, Director of the National Electrical Engineering Research Institute, who has often travelled to Western Germany on behalf of the South African Armaments Board.

41. See article by Andrew Wilson, 'Secret visit by South Africa's arms chief', in *The Observer* (London) of 9 May 1976.
42. ibid.
43. Letter of 19 June 1974, file no. U/5/4/13, signed for him by A. W. D. Chilton.
44. The case of the *Otto Hahn* is not isolated. Other incidents have come to light of semi-clandestine shipments using vessels outside normal commercial channels. The *President Steyn*, a South African Navy ship, was observed during an official visit to Kiel in West Germany (27-29 September 1971). It was being loaded with packages during the night by West German troops, perhaps evading the normal customs controls. Another curious incident was an 'emergency' call at Durban by the *Deutschland*, flagship of the West German Navy, after an official visit to Simonstown on 21 March 1974. This was ostensibly to land an Ethiopian sailor with acute appendicitis; however, the illness appears to have been fictional, and it seems that something more concrete may have been landed instead.
45. The South African Scientific Liaison Office.

421

46. *Aviation Week and Space Technology*, 13 March 1972, 19 March 1973, 11 March 1974.
47. *Wehrtechnik*, April 1976.
48. We asked the Swedish Military Attaché in Bonn, Colonel Hjelmer Martenson, whether the military attachés of other non-NATO powers enjoyed similar privileges. The answer was an emphatic no.
49. 'Was Ihrer Delegation bei der Bundeswehr gezeigt werden könnte, wäre ein Mehrrechnerbetrieb, bei dem 2 Rechner auf einen gemeinsamen Datenbestand zurückgreifen und die Peripheriegeräte wahlweise auf den einen oder anderen Rechner schaltbar sind.'

Chapter Four
1. Documents of Rio Tinto Zinc, Australia, made available to the authors by Friends of the Earth, Australia.
2. C. S. McLean and T. K. Prentice, 'The South African Uranium Industry', paper submitted to the First International Conference on the Peaceful Uses of Atomic Energy, Geneva, 1956.
3. *South African Mining Survey* no. 69, September 1971.
4. R. S. Cooke, President of the Chamber of Mines of South Africa, 'Straining at the Leash: The Future of South Africa's Uranium Industry', *Southern Africa* (London), 15 August 1970.
5. McLean and Prentice, *op.cit.*
6. *The Star* (Johannesburg), 30 July 1965.
7. McLean and Prentice, *op.cit.*
8. R. B. Hagart, 'National Aspects of the Uranium Industry', conference paper (mimeo.), n.d.
9. For elaboration of this process, see Barbara Rogers, *White Wealth and Black Poverty: American Investments in Southern Africa* (Greenwood Press, Westport, Conn., 1976).
10. McLean and Prentice, *op.cit.*
11. Rogers, *op.cit.*
12. *Nuclear Energy Report*, 17 November 1975.
13. Mike Muller in *New Scientist* (London) cited in *Rand Daily Mail* (Johannesburg), 20 May 1974.
14. *Rand Daily Mail*, 16 December 1976.
15. ibid., 6 August 1976.
16. McLean and Prentice, *op.cit.*
17. *The Star* (Johannesburg), 14 October 1972.
18. *Nuclear Energy Report*, 17 November 1975.
19. *Southern Africa* (London), 15 August 1970.
20. *South African Digest* (London), 27 September 1974.
21. *The Star*, 2 January 1971.

22. ibid., 22 October 1970.
23. Dr Diederichs, quoted in *South African Digest*, 6 February 1976.
24. *Nuclear Energy Report*, 17 November 1975.
25. *Nuclear Active*, South African Atomic Energy Board, July 1973, pp. 5-8; *Financial Gazette* (Johannesburg), 16 January 1970.
26. *Nucleonics Week*, 21 October 1976.
27. ibid.; *Rand Daily Mail*, 16 October 1976.
28. *Die Burger*, 25 August 1973.
29. *Nucleonics Week*, 5 August 1976.
30. Richard West, 'Tony Benn: heir to Cecil Rhodes', *The Spectator* (London), 19 March 1977.
31. *Financial Gazette*, 9 November 1973.
32. *Southern Africa*, 15 August 1970; *Rand Daily Mail*, 16 October 1976.
33. *Wall Street Journal*, 26 January 1976.
34. *Nuclear Energy Report*, 17 November 1975; *Financial Gazette*, 28 May 1976.
35. In 1974 the official South African estimate of the total population was 850,000: African and Coloured, 750,000; Whites (Afrikaans- and English-speaking South Africans, and Germans) 100,000. The UN estimates the population at just over one million.
36. *Rand Daily Mail*, 9 January 1973.
37. RTZ Annual Report, 1974.
38. *The Star*, 30 June 1973.
39. Bernard C. Duval, 'The Changing Picture of Uranium Exploration', p. 11, International Symposium on Uranium Supply and Demand, 22-24 June 1977.
40. *Namib Times*, 12 September 1975.
41. RTZ Annual Report, 1976, p. 43, and *Rio Tinto Zinc in Namibia*, Christian Concern for Southern Africa, 1976, p. 7.
42. *Rand Daily Mail*, 26 May 1977.
43. *Financial Times*, 18 June 1976; *Rand Daily Mail, op.cit.*
44. SA Atomic Energy Act No. 90 of 1967 repealed the previous Act No. 35 of 1948. Uranium Enrichment Act No. 37 of 1974.
45. *The Star*, 25 June 1974; Resource and Energy Agency, Japanese Ministry of International Trade and Industry, *Kozan*, December 1973.
46. SWAPO Press Statement, 20 October 1976.
47. Africa Bureau, Fact Sheet 51, May-June 1977.
48. *Rand Daily Mail*, 2 March 1970; *Palaminer* (RTZ house journal), January 1973.
49. *Mining*, March 1972.

50. RTZ Press Release, 13 October 1976.
51. Chairman's statement, 24 May 1977. *New African Development*, September 1977, p. 932.
52. De Zoete and Bevan; Scott, Goff and Hancock; February 1976.
53. *Nuclear Active*, July 1973, p. 34.
54. *Nuclear Engineering International*, November 1976, p. 80; *Nuclear Active*, July 1974, pp. 20-21.
55. *Minerals Yearbook*, 1970, p. 1155, US Department of Commerce, citing State Department communication of 5 March 1970. Panmure Gordon and Co., November 1975.
56. *Mining Journal*, 2 April 1976.
57. RTZ Annual Report, 1975, p. 34; 1975, p. 33.
58. *Daily Mail*, 21 July 1976.
59. 'The Pattern of Uranium Production in South Africa', p. 12, International Symposium on Uranium Supply and Demand, 16 June 1976. *E/MJ*, December 1976, p. 22.
60. *The Star*, 22 November 1975.
61. *Financial Mail*, 15 April 1977.
62. *Nucleonics Week*, 18 February 1974.
63. *Financial Mail*, 15 April 1977.
64. Economist Intelligence Unit, *Economic Review of Southern Africa*, no. 2, 1977, Angola and Mozambique; *Financial Times*, 25 June 1976; *The Star*, 22 January 1977.
65. Quoted in *Namibia News*, London, May/June 1973, p. 7.
66. Interview with Dr Kienlin of Urangesellschaft (mimeo.), June 1972.
67. Letter from the Federal Ministry for European Cooperation to Georg Stingl, 24 March 1971, quoted in *Die Lage Namibias und das Uranschonfungs Projekt bei Swakopmund*, Information Zentrum Dritte Welt, Freiburg 1971.
68. Deutsche Bundestag, 7 Wahl periode, Drucksache 7/3706, Fragen A94 and A95.
69. 'The Nuclear Metals', *Mining Annual Review*, May 1966.
70. See for example Arthur Findlay, letter to *New Statesman* (London), 9 June 1972.
71. Quoted in Roger Murray, 'South West African Uranium: A new dilemma for Labour', *Financial Times* (London), 3 May 1974.
72. Quoted in *X-Ray on Current Affairs in Southern Africa* (Africa Bureau, London), vol. 1, no. 3, September 1970, p. 1.
73. *Atom*, UK Atomic Energy Authority, no. 205, November 1973.
74. *Atom*, no. 195, January 1973.
75. *Nucleonics Week*, 24 December 1970.
76. Statement circulated as Annex II of letter to the United

Notes

Nations Secretary-General from the UK Chargé d'Affaires; United Nations Document A/9918, 4 December 1974.

77. Lord Caradon, UK Permanent Representative to the United Nations; UN Official Records, General Assembly, Session 21, 1448th meeting, 19 October 1966.
78. David Fishlock in *Financial Times*, quoted in *The Star* (weekly), 30 November 1974.
79. Nelson F. Sievering, Jr., US Energy Research and Development Agency; United States Congress, House of Representatives, Committee on International Relations, Subcommittee on International Resources, Food, and Energy, *Resource Development in South Africa and U.S. Policy*, hearings of 25 May and 8 and 9 June, 1976 (US Government Printing Office, Washington DC, 1976) p. 58.
80. For example, Dr Anton Rupert, quoted in *Namib Times*, 1 April 1977.
81. *The Star*, 8 December 1973.
82. *State of South Africa Year-Book, 1971* (Da Gama Publishers, Johannesburg, 1971).
83. *Nucleonics Week*, 25 October 1976.
84. *Financial Mail* (Johannesburg), 3 September 1976. We are also indebted to Friends of the Earth, Australia, for copies of the original documents.
85. Westinghouse suit, cited in *Nucleonics Week*, 21 October 1976.
86. Peter Rodgers in *Sunday Times*, 20 June 1976.
87. RTZ documents handed to the US Department of Justice by Friends of the Earth (Australia), cited in *Financial Mail* (Johannesburg), 3 September 1976.
88. See *Rand Daily Mail*, 16 December 1976.
89. ibid.
90. See, for instance, copies of *Government Mining Engineer* (South Africa) at the time.
91. *Minerals: a Report for the Republic of South Africa* (South African Department of Mines Quarterly Information Circular).
92. Dr Roux, cited in *Rand Daily Mail*, 16 October 1976.
93. *Financial Mail*, 21 January 1966.
94. *Southern Africa* (London), 28 November 1966.
95. *The Guardian* (London), 14 July 1977.
96. Leonard Beaton, *Must the Bomb Spread?* (Penguin Books, London, 1966).
97. *Safeguarding the Atom: A Soviet-American Exchange* (UNA-USA, New York, 1972).
98. *Financial Mail*, 3 September 1976.
99. *Southern Africa* (London), 15 August 1970; *Financial Gazette*, 9

May 1969.

100. Quoted in *South African Scope* (Pretoria), November 1974, p. 2.
101. *Forbes* (New York), 15 January 1975.
102. ibid.; also *Not Man Apart* (US Friends of the Earth), October 1976.
103. *Today's News* (South African Embassy, London), 24 July 1970.
104. Speech at the National Development and Management Foundation, Johannesburg, quoted in *Today's News*, 2 September 1970.
105. A. J. A. Roux, 'South Africa in a Nuclear World', *South Africa International*, South Africa Foundation, Johannesburg, vol. 4, no. 3, January 1974, p. 157.
106. Robert Alvarez, statement before the US House Science and Technology Subcommittee on Nuclear and Fossil Energy Research and Development, 4 March 1977.
107. *The Star* (weekly), 13 July 1974.
108. *Financial Mail*, 21 May 1971.
109. Speech at a seminar in New York organised by Safto/Senbank, 22 June 1977; quoted in *The South Africa Foundation News*, vol. 3, no. 7, July 1977. Shortly before his departure from the Treasury, Simon was reported by *The Times* (18 October 1976) as getting very close to the South African Government. '. . . Mr Simon seems to be getting on extremely well with these particular (South African) ministers right now . . . and he notes that after resigning in January he plans to go to South Africa to lecture at universities there.'

Chapter Five
1. IAEA Board of Governments, provisional record of the 501st meeting, 16 June 1977 (Doc. GOV/1849).
2. D. S. Greenberg, 'South Africa (1)', *Science* (American Association for the Advancement of Science), 10 July 1970, p. 162.
3. ibid.
4. Quoted, ibid.
5. *Financial Mail* (Johannesburg), 8 October 1971.
6. Greenberg, *loc.cit.*
7. Quoted in *The Star* (Johannesburg), 9 June 1966.
8. *Südafrika* (Zurich), no. 1 of 1964.
9. *Rand Daily Mail* (Johannesburg), 13 March 1971.
10. *Washington Post*, 16 February 1977.
11. *Rand Daily Mail*, 13 March 1971.
12 *Financial Times* (London), 18 February 1971.
13. *Rand Daily Mail*, 13 March 1971.
14. A. J. A. Roux in *Nuclear Active*, South African Atomic Energy

Board, January 1972, p. 17.
15. *The Star*, 19 March 1965.
16. *The Times* (London), 27 July 1965.
17. *South African Scope*, February 1965.
18. Letter from Mr William A. Anders, Chairman of the United States Nuclear Regulatory Commission, to Senator John Glenn, 6 June 1975, Attachment 1.
19. *Cape Times* (Cape Town), 27 February 1965.
20. *The Times*, 27 July 1965.
21. *Natal Mercury* (Durban), 8 December 1970.
22. *Cape Times*, 2 March 1965; *The Times*, 27 July 1965.
23. John J. Flaherty, US Atomic Energy Commission; United States Congress, House of Representatives, Committee on Foreign Affairs, Subcommittee on Africa, *US Business Involvement in Southern Africa*, Part 2 (US Government Printing Office, Washington DC, 1972), pp. 40-76.
24. Responses by the Department of State to additional questions submitted in writing by Congressman Diggs; House Subcommittee on International Resources, Food, and Energy, *Resource Development in South Africa and U.S. Policy* (US Government Printing Office, Washington DC, 1976), p. 236. (Hereafter referred to as House hearings on Resource Development.)
25. ibid.; *The Times*, 27 July 1965.
26. House hearings on Resource Development, pp. 236-37.
27. *Cape Times*, 8 May 1965; *The Star*, 19 March 1965.
28. *Südafrika von Woche zu Woche* (South African Consulate, Cologne), no. 98, 11 October 1963 and no. 113, 21 March 1964.
29. *Nuclear Active*, January 1974, p. 5.
30. *The Star* (weekly), 3 March 1973.
31. ibid., 16 February, and 9 and 16 March 1974; *Financial Gazette* (Johannesburg), 19 March 1965.
32. *New York Times* and *Washington Post*, 1 June 1976.
33. *The Star*, 7 August 1976.
34. *Sunday Times* (Johannesburg), 26 September 1976. The structure of the French consortium led by Framatome is as follows:
   40% Framatome, whose capital in turn belongs to:
   — 51% Creusot-Loire which belongs to the *French-Belgian* industrial group Schneider-Empain.
   — 30% to the *French government-owned* nuclear commissariat CEA.
   — 15% to the American company Westinghouse, the world's largest constructor of nuclear reactors. These

15% go to Creusot-Loire in 1982.

40% Spie-Batignolle, a construction company 91%-owned by Schneider, SA (société anonyme).

20% Alsthom, a subsidiary of France's largest electric concern, CGE.

Framatome will supply the nuclear reactors, Alsthom the turbo-generators and Spie-Batignolle the installations. 30% of the orders must go to South African suppliers.

35. *Energy Developments*, vol. XIX, 15 July 1976, p. 21.
36. *Financial Gazette*, 13 August 1976.
37. *Sunday Times*, 26 September 1976.
38. Nelson F. Sievering, Jr., US Energy Research and Development Agency; House hearings on Resource Development, p. 60.
39. ibid.; Jean Bernard, 'Note complémentaire sur l'accord franco-sud-africain', *Esprit* (Paris), no. 463, December 1976, p. 807.
40. Myron Kratzer, State Department; House hearings on Resource Development, p. 18; *Nucleonics Week*, 20 May 1976.
41. ibid., 3 June 1976.
42. *Le Monde* (Paris), 1-2 June 1976.
43. *Financial Gazette*, 23 April 1976.
44. *Nucleonics Week*, 3 June 1976.
45. *Financial Gazette*, 23 April 1976.
46. Myron Kratzer, State Department, statement before the Subcommittee on African Affairs, Senate Foreign Relations Committee, 27 May 1976, p. 4.
47. *Washington Post*, 30 May 1976.
48. *Sunday Times*, 15 August 1976.
49. *Nucleonics Week*, 3 June 1976.
50. ibid.
51. I am indebted to a South African source close to ESCOM for this. She cannot be named because of the stringent penalties provided by the Atomic Energy Act for such disclosures.
52. *The Star* (weekly), 9 March 1974.
53. *The Times*, 1 July 1974; *Die Transvaler* (Johannesburg), quoted in *The Afro-American* (Washington DC), 6-10 August 1974; *The Observer* (London), 30 June 1974; *Manchester Guardian Weekly* (London), 6 July 1974.
54. *The Star* (weekly), 6 July 1974.
55. *African Development* (London), September 1974, p. 45.
56. Reuter cable from Cape Town, 11 April 1975; *Financial Gazette*, 11 April 1975.
57. *The Guardian* (London), 31 May 1974.
58. *The Star*, 13 November 1976.

59. *Financial Gazette*, 27 October 1972.
60. *Sunday Times*, 15 August 1976.
61. ibid.; Reuters, 7 March 1975.
62. See 'Southern Africa: Era of prolonged restraint', Supplement on World Banking XXXVIII, *Financial Times* (London), 2 May 1977.
63. *Sunday Times*, 15 August 1976.
64. 'Going Fission', *Financial Mail*, 13 August 1976.
65. ibid.; *Sunday Times*, 15 August 1976.
66. *Financial Mail*, 13 August 1976.
67. See Barry Commoner, *The Poverty of Power: Energy and the Economic Crisis* (Jonathan Cape, London, 1974).
68. *Financial Gazette*, 15 February 1974.
69. *Sunday Times*, 26 September 1976.
70. *Financial Mail*, 13 August 1976.
71. *Nucleonics Week*, 21 February 1974.
72. *Sunday Times*, 26 September 1976.
73. *Financial Gazette*, 15 February 1974.
74. Dr A. J. A. Roux, *South Africa in a Nuclear World* (pamphlet prepared by the South African Department of Information and distributed by the Information Service of South Africa [ISSA], New York).
75. *The Star*, 31 July 1976.
76. *Today's News*, South African Embassy, London, 24 July 1970.
77. *Financial Times* (London), 22 October 1970 and 18 February 1971.
78. *The Star* (Johannesburg), 18 November 1971.
79. *Financial Times*, 23 July 1970.
80. *Wall Street Journal*, 23 October 1970.
81. See for example A.J.A. Roux, *South Africa in a Nuclear World, op. cit.*
82. Interview with *The Star* (weekly), 8 June 1974.
83. *Wall Street Journal*, 23 October 1970.
84. *Today's News*, 25 November 1971.
85. *Nuclear Industry*, quoted in *The Guardian* (London), 24 October 1970.
86. *Financial Times*, 3 November 1972.
87. David Fishlock, 'Pretoria pursues the atom', *Financial Times*, 18 February 1971.
88. Dr Roux, interview in *Sunday Times* (Johannesburg), cited in *Financial Times*, 3 November 1972.
89. Fishlock, *loc.cit.*
90. Dr Roux, statement of July 1970, cited in *Science* (American Association for the Advancement of Science), vol. 188, p.

1090.
91. ibid.
92. Statement in the House of Assembly, 7 April 1975: *House of Assembly Debates (Hansard)*, 7 April 1975, col. 3602.
93. *Southern Africa* (London), 1 August 1970.
94. David Fishlock, 'South Africa's way with uranium', *Financial Review* (Australia), 10 June 1975.
95. ibid; Fishlock, 'South Africa's nuclear hopes may be receding', *Financial Times*, 3 November 1972.
96. *South African Scope*, February 1965; United States Congress, House of Representatives, Committee on Foreign Affairs, Subcommittee on Africa, *U.S. Business Involvement in Southern Africa*, Part II (U.S. Government Printing Office, Washington DC, 1971), p. 53.
97. *Nuclear Engineering International*, April 1974, p. 255.
98. *Der Spiegel* (Hamburg), no. 18 of 1975, p. 161.
99. Fishlock in *Financial Review*, 10 June 1975.
100. Dr Roux, quoted in Fishlock, 'S. Africa to build big nuclear fuel plant', *Financial Times*, 23 April 1975.
101. ibid; 'Enriching U and SA', *Financial Mail* (Johannesburg), 28 June 1974.
102. Fishlock in *Financial Times*, 23 April 1975.
103. Dr A.J.A. Roux, *South Africa in a Nuclear World, op.cit.*, p. 6.
104. ibid.
105. ibid., p. 5.
106. *Hansard*, 7 April 1975, col. 3601.
107. *Financial Mail*, 11 April 1975.
108. ibid.
109. *Wall Street Journal*, 26 January 1976.
110. *Rand Daily Mail*, 6 October 1975; Reuters, 16 April 1975.
111. *Financial Mail*, 28 June 1974.
112. ibid.
113. Quoted in *Rand Daily Mail*, 23 April 1975.
114. ibid.
115. *International Herald Tribune*, 13 October 1975; *The Economist* (London), 6 December 1975.
116. *International Herald Tribune* and *The Guardian* (London), 4 May 1977.
117. *Pretoria News*, 29 December 1977.
118. Interview with Dr Roux in *Financial Mail*, 17 December 1976; and Dr Koornhof, quoted by *Agence France Presse*, 12 June 1975. See also David Fishlock, 'SA steps up its forays into the nuclear fuel cycle', *Nuclear Energy Report*, 17 November 1975.
119. Fishlock in *Nuclear Energy Report*.

Notes

120. *Hansard*, 11 June 1975, col. 7967.
121. ibid., col. 7968.
122. *Rand Daily Mail*, 24 October 1970.
123. *South African Digest*, 19 December 1969. For allegations about British collaboration with South Africa on its own hex process, see articles by Fishlock cited above.
124. *Rand Daily Mail*, 2 March 1970.
125. Cited in *Financial Gazette* (Johannesburg), 28 May 1976.
126. *Sunday Times* (Johannesburg), 15 September 1974, *Financial Times*, 23 April 1975; *Rand Daily Mail*, 23 April 1975.
127. *Nuclear Active*, South African AEB, July 1973, p. 35.
128. ibid.
129. *The Star* (weekly), 25 August 1973.
130. Fishlock in *Nuclear Energy Report*.
131. *Financial Times*, 11 April 1975; 'Top Companies', Special Survey by the *Financial Mail*, 27 April 1973.
132. Fishlock in *Nuclear Energy Report*.
133. Dr Roux, cited in *Wall Street Journal*, 26 January 1976.
134. Reginald E. Worall, uranium adviser to the Chamber of Mines of South Africa, quoted in *Nucleonics Week*, 24 June 1976.
135. ibid; *Financial Mail*, 28 June 1974.
136. *The Star* (weekly), 10 March 1973.
137. *Hansard*, 18 September 1974, col. 1505.
138. Cited by Fishlock in *Financial Times*, 23 April 1975.
139. Roux, *South Africa in a Nuclear World*, p. 7.
140. *Hansard*, 11 June 1975, col. 7966.
141. *Die Transvaler* (Johannesburg), 25 August 1973.
142. *Financial Gazette*, 1 February 1974.
143. Peter Rodgers in *Sunday Times* (London), 20 June 1976.
144. 1 gigawatt = 1,000 megawatts, the equivalent of one very large power reactor.
145. I am indebted to Robert Alvarez of the Environmental Policy Center, Washington DC, for this data.
146. See nuclear trade journals, especially *Nucleonics Week* and *Nuclear Engineering International*.
147. *Nucleonics Week*, 17 October 1974.
148. United States Congress, Office of Technology Assessment, *Nuclear Proliferation and Safeguards* (Office of Technology Assessment, Washington DC, prepublication draft, April 1977), p. II-9.
149. Interview with *Financial Mail*, 17 December 1976.
150. *Financial Times*, 18 February 1971.
151. ibid.
152. *The Star* (weekly), 18 June 1974.

153. Roux, *South Africa in a Nuclear World*, p. 7.
154. *Financial Times*, 4 April 1975.
155. *Hansard*, 11 June 1975, col. 7968.
156. *Financial Times*, 23 December 1977.
157. *The Guardian*, 24 December 1977.
158. *Financial Mail*, 17 February 1978.
159. Albert Wohlstetter et al., *Moving toward Life in a Nuclear Armed Crowd?*, Final Report, ACDA/PAB-263 (Pan Heuristics, Los Angeles, 1976), pp. 200-1.
160. Office of Technology Assessment, *op.cit.*, p. II-43.
161. Statement to the Senate Subcommittee on Africa, quoted in Peter Younghusband, 'Nuclear Power Politics', *European Community*, August-September 1976, p. 23.
162. Jim Hoagland, 'S. Africa, with U.S. Aid, Near A-Bomb', *Washington Post*, 16 February 1977.
163. *Der Spiegel*, 1977 (exact date not available).
164. Personal communications from Peter Katjavivi of SWAPO, and other members of a delegation from the United Nations Council for Namibia, meeting with IAEA officials on 29 April 1977.
165. Jim Hoagland in *International Herald Tribune*, 18 February 1977. It is curious that when French Foreign Minister Guiringaud was asked by President Kaunda of Zambia about the reports of South Africa having nuclear weapons, during his Africa tour of mid-1977, Guiringaud categorically denied that South Africa had any nuclear weapons capability — as Kaunda later told a sympathetic western journalist.
166. ibid.
167. *The Guardian* (London), 12 August 1977.
167. *Ta Kung Pao* (Hong Kong), 30 October 1969. This publication is generally regarded as an outlet for important statements by Peking.
169. Reported in *Daily Times* (Malawi), 24 August 1977.
170. Office of Technology Assessment, *op.cit.*, p. G-2.
171. *The Guardian*, 13 June 1977.
172. Robert Gillette in *Los Angeles Times*, 14 September 1977.
173. *Newsweek*, quoted in *Sunday Times* (Johannesburg), 19 November 1967.
174. *Daily Telegraph* (London), 22 July 1970; *Financial Times* (London), 21 July 1970.
175. Office of Technology Assessment, *op.cit.*, p. II-52
176. Frank Barnaby in *Africa* (London), no. 69, May 1977, p. 92.
177. Office of Technology Assessment, *op.cit.*
178. *Washington Post*, 15 May 1976.

Notes

179. Cited in *New York Times*, 1 June 1976.
180. Personal communication from Patrick Keatley; Jean Bernard, 'Note complémentaire sur l'accord franco-sud-africain', *Esprit* (Paris), no. 463, December 1976. pp. 806-7.
181. Office of Technology Assessment, *op.cit.*, p. II-47.
182. *New York Times*, 28 September 1976.
183. Dr Albert Wohlstetter, statement to the International Relations Committee, U.S. Senate, reported in *The Star* (Johannesburg), 17 June 1976.
184. Personal communication from Patrick Keatley.
185. United States Congress, Senate, Committee on Foreign Relations, *Nonproliferation Issues* (US Government Printing Office, Washington DC, 1977), p. 30.
186. Personal communication from Robert Alvarez, Environmental Policy Center.
187. Bernard, 'Note complémentaire', p. 807.
188. Office of Technology Assessment, *op.cit.*, p. II-47.
189. For criticism of the IAEA safeguards system, see US Senate, *Nonproliferation Issues*, op.cit.; Office of Technology Assessment, *op.cit.*; and Wohlstetter et al., *op.cit.*
190. United States Congress, House of Representatives, Committee on International Affairs, Subcommittee on International Resources, Food, and Energy, *Resource Development in South Africa and US Policy* (US Government Printing Office, Washington DC, 1976), p. 83.
191. Pearce Wright, Science Correspondent of the London *Times*, 'South Africa joins the world's nuclear club', *Sydney Morning Herald*, 21 August 1973.
192. US House of Representatives, *Resource Development*, p. 56.
193. For example, Dr Koornhof, cited in *Financial Gazette* (Johannesburg), 12 December 1975.
194. John F. Burns, 'South Africa's mysterious A-plant', *New York Times* and *International Herald Tribune*, 4 May 1977.
195. 'Memorandum' circulated by the Federal Government to journalists, 17 January 1977, para. 1.b.
196. The 1977 *Yearbook* of the Stockholm Peace Research Institute (SIPRI) gives a figure but there is no indication of the source.
197. Dr A.J.A. Roux, *South Africa in a Nuclear World*, *op.cit.*, p. 6.
198. *Washington Post*, 16 February 1977.
199. *Der Spiegel* (Hamburg), 20 December 1976.
200. A.J.A. Roux in *Nuclear Active* (South Africa AEB), July 1971, p. 4.
201. *Rand Daily Mail* (Johannesburg), 23 August 1973; *Financial Times*, 22 August 1973.

433

202. Quoted in *Washington Post*, 16 February 1977.
203. *Newsweek*, 21 February 1977.
204. Nelson F. Sievering, Jr., United States Energy Research and Development Administration, at Congressional hearings: US House of Representatives, *Resource Development*, p. 61.
205. Office of Technology Assessment, *Nuclear Proliferation*, p. II-47.
206. International Atomic Energy Agency, restricted document, *Safeguards: (a) Special Safeguards Implementation Report*, GOV/1842, 8 June 1977, p. 5.
207. ibid., pp. 35-36.
208. ibid., pp. 8, 23.
209. ibid., p. 10.
210. ibid., pp. 20-21.
211. ibid., p. 22, Table 2.
212. ibid., pp. 29-30.
213. ibid., pp. 8, 11.
214. Responses by the Department of State to questions in writing by Congressman Diggs: U.S. House of Representatives, *Resource Development*, p. 236.
215. AEB annual report for 1969, cited in *Japan Times*, 8 August 1970.
216. *The Star* (Johannesburg), 17 January 1970.
217. *Daily Times* (Malawi), 24 August 1977.
218. Confirmed in an interview by the authors with a source close to the C.I.A.
219. See *US Business Involvement in Southern Africa* (3 vols), hearings of the House Subcommittee on Africa.
220. Personal communication from Martin Walker.
221. Annual Report of the AEB, cited in *Cape Times*, 7 June 1967.
222. Personal communication from Martin Walker.
223. *South African Panorama*, January 1974.
224. Roux, *South Africa in a Nuclear World*, op.cit., p. 8.
225. Dr Carel de Wet, Minister of Mines, press statement quoted in *Southern Africa* (London), 1 August 1970.
226. *Southern Africa*, 5 March 1960.
227. ibid., 5 January 1962.
228. *Cape Times*, 19 February 1965.
229. *South African Digest* (London), 13 August 1965.
230. *New York News*, 28 February 1965.
231. *Sunday Express* (London), 22 December 1968.
232. *The Times* (London), 12 July 1974.
233. *Washington Post*, 12 July 1974.
234. *The Times*, 12 July 1974.
235. *Newsweek*, 17 May 1976.

236. *Bild-Zeitung* (Hamburg), 8 June 1976.
237. *Daily Times* (Malawi), 24 August 1977.
238. *The Guardian*, 25 October 1977.
239. *Africa* no. 74, October 1977, p. 53.
240. United Nations, General Assembly, Document A/C.1/PV.1571, 20 May 1968, p. 57.
241. *The Star*, 15 June 1968.
242. UN General Assembly, A/C.1/PV. 1571.
243. *Cape Times*, 6 June 1968.
244. UN General Assembly, A/PV.214, 5 October 1973.
245. SIPRI, *The Near-Nuclear Countries and the Nuclear Non-Proliferation Treaty* (Humanities Press, New York, 1972).
246. ibid., p. 38.
247. *The Sun* (Baltimore), 10 December 1972.
248. Quoted by John F. Burns in *New York Times* and *International Herald Tribune*, 4 May 1977.
249. Sverre Lodgaard, 'International Nuclear Commerce: Structures, Trends and Proliferation Potentials', *Bulletin of Peace Proposals*, Oslo, vol. 8, no. 1, 1977, p. 23.
250. *The Guardian*, 28 September 1977.

Chapter Six

1. United States Congress, Office of Technology Assessment, *Nuclear Proliferation and Safeguards* (Office of Technology Assessment, Washington DC, prepublication draft, April 1977), p. II-35.
2. *Der Spiegel*, cited in *Windhoek Advertiser*, 16 March 1977.
3. *The Guardian*, 4 June 1977.
4. William Epstein, 'Why States Go – and Don't Go – Nuclear', *Annals of the American Academy of Political and Social Science*, vol. 430, March 1977, p. 17.
5. Editorial comment in *Die Volksblad*, 6 May 1969.
6. *Die Beeld*, Pretoria, 26 July 1970.
7. 'S'il y avait une intervention internationale, de quelque pays que cela vienne, contre les Blancs d'Afrique du Sud, il ne faut pas que ces pays oublient que nous possedons la bombe atomique.' Quoted in *Paris Match*, 13 May 1977.
8. Quoted in *Der Spiegel*, 6 October 1975.
9. *Richmond Times-Dispatch*, 12 October 1975.
10. *Washington Star*, 13 November 1975; *Washington Post*, 28 May 1976; *The Times*, 7 August 1976.
11. Mason Willrich, Testimony before the US Senate Committee on Foreign Relations, Subcommittee on African Affairs, on

Proposed US Nuclear Power Exports to South Africa, 26 May 1976, p. 3.

12. Responses by the Department of State to questions submitted in writing by Congressman Diggs: US House of Representatives, *Resource Development*, p. 240.
13. Epstein, *op.cit.*, p. 28.
14. *The Star* (weekly), 24 April 1976.
15. Quoted in *Sunday Times Business News* (London), 19 December 1976.
16. Bowen Northrup, 'How Westinghouse lost out', *Wall Street Journal*, 2 July 1975.
17. Quoted in *The Observer*, 29 May 1977.
18. Interview cited in *The Guardian*, 30 May 1977.
19. International Institute for Strategic Studies, *Strategic Survey 1976* (IISS, London, 1977), quoted in *The Guardian*, 30 April 1977.
20. Willrich, *loc.cit.*, p. 6.
21. Epstein, *op.cit.*, p. 16.

Chapter Seven
1. Stockholm International Peace Research Institute (SIPRI), *Yearbook 1977*, p. 15.
2. *Southern Africa* (London), 20 June 1966; *Africa Confidential* (London), 3 November 1972.
3. Lewis H. Gann, 'The Military Outlook: Southern Africa', *Military Review*, July 1972; *Africa Report*, January 1966, p. 31.
4. *Air Enthusiast* and *Flight International*, quoted in *The Star* (weekly), 15 September 1973.
5. *The Star*, January 1977, quoted in *Apartheid Non* (Paris), no. 14, 17 March – 15 April 1977, p. 3.
6. *Air Enthusiast* and *Flight International*, *op.cit.*
7. International Institute of Strategic Studies, *The Military Balance 1976-1977* (IISS, London, 1970), pp. 246-47. See also 'World Armaments and Disarmament', *SIPRI Yearbook 1976* (MIT Press, Cambridge, Mass., 1976; and Amqvist and Wiksell International, Stockholm, 1976), pp. 274-75.
8. *Southern Africa* (London), 27 May 1968.
9. SIPRI, *Southern Africa: The Escalation of a Conflict* (Praeger, New York, 1976), pp. 130-32.
10. IISS, *Military Balance*, pp. 246-47.
11. SIPRI, *World Armaments*, pp. 274-75.
12. IISS, *op.cit.*, p. 45.
13. *The Observer* (London), 23 September 1973.
14. *Die Burger*, 28 June 1971. See also Carl T. Rowan, 'France

South Africa in Nuclear Deals,' *Chicago Daily News*, 18 July 1967.

15. *Die Transvaler* (Johannesburg), quoted in *The Afro-American* (Washington DC), 6-10 August 1974.
16. ibid.
17. *The Star*, cited in *Apartheid Non*, *op.cit.*
18. *The Times* (London), 28 October 1963.
19. SIPRI, *Southern Africa*, pp. 142-43.
20. Quoted in Abdul Minty, *South Africa's Defence Strategy* (The Anti-Apartheid Movement, London, 1973).
21. *Africa Today* (New York), August/September 1965.
22. SIPRI, *World Armaments*, pp. 182-83.
23. *The Star* (weekly), 23 May 1970.
24. IISS, *op.cit.*, p. 45.
25. *Rhodesia Herald* (Salisbury), 25 April 1968.
26. Quoted in Minty, *op.cit.*, p. 7. See also SIPRI, *Southern Africa*, pp. 142-43.
27. See Barbara Rogers, *Divide and Rule: South Africa's Bantustans* (International Defence and Aid Fund, London, 1976).
28. Personal communication from Patrick Keatley.
29. SIPRI, *Southern Africa*, pp. 142-43.
30. *Der Spiegel*, 27 October 1976.
31. *Frankfurter Allgemeine Zeitung*, 10 and 29 April 1964, quoted in *Memorandum on co-operation between the West German Federal Republic and the Republic of South Africa in the military and atomic fields* (Afro-Asian Solidarity Committee in the GDR, Berlin, September 1964). See also SIPRI, *Southern Africa*, pp. 142-43.
32. SIPRI, *Southern Africa*, pp. 142-43.
33. ibid.
34. *Frankfurter Allgemeine Zeitung*, 10 and 29 April 1964.
35. ibid.
36. SIPRI, *Southern Africa*, pp. 142-43.
37. Martin Walker, in *The Guardian* (London), 10 September 1974.
38. IISS, *op.cit.*, p. 45.
39. *New York Times*, 12 December 1971.
40. Denis Archer (ed.), *Jane's Pocket Book of Naval Armament* (Macdonald and Jane's, London, 1976), p. 33.
41. SIPRI, *World Armaments*, pp. 274-75.
42. ibid., pp. 246-47.
43. *Baltimore Sun*, 26 August 1976; *Christian Science Monitor*, 12 and 14 April and 17 August 1976; *Washington Post*, 9 August 1976.
44. *Daily News* (Dar es Salaam), 6 December 1976.
45. Frank C. Barnaby, 'How States Can "Go Nuclear"', *Annals of the American Academy of Political and Social Science*, vol. 430,

March 1977, pp. 40-41.
46. West German operations in Zaire, almost at the heart of Africa, are of some interest. An article appearing in *Penthouse* magazine by Tad Szulc has alleged that the CIA and its West German counterpart, the *Bundesnachrichtendienst*, have helped to set up a West German cruise-missile and rocket base in Zaire, and that 'four or five cruise-missile prototypes' had been developed and flown over the leased enclave. ('V-3', *Penthouse*, March 1978, pp. 19-96.) Szulc claimed that the operation, although ostensibly carried on by a private concern called OTRAG (Orbital Transport und Raketen AG), is in fact financed and carried out by the most important companies in West Germany's military-industrial complex, with support from their counterparts in France, and in collaboration with the Federal Government itself.

This and similar reports have been challenged in some detail by a number of other publications, notably with a lead story in *The Observer* saying that they amounted to 'a storm of propaganda' directed by the Soviet Union. (Andrew Wilson and Colin Legum, 'Dr. Strangelove and a rocket hoax', *The Observer* [London], 5 February 1978.)

Chapter Eight
1. *Washington Post*, 16 February 1977.
2. *US Business Involvement in South Africa*, vol. II, Hearings before the House Subcommittee on Africa (US Government Printing Office, Washington DC, 1973), p. 69.
3. *South African Mining Survey* No. 69, September 1971.
4. ibid.
5. *State of South Africa Year-Book, 1971* (Da Gama Publishers, Johannesburg, 1971); R.B. Hagart, *National Aspects of the Uranium Industry*, conference paper (mimeo.), n.d.
6. *The Star* (Johannesburg), 30 July 1965.
7. United States Congress, House of Representatives, Committee on International Relations, Subcommittee on International Resources, Food, and Energy, *Resource Development in South Africa and US Policy*, hearings of 25 May, 8 and 9 June 1976 (US Government Printing Office, Washington DC, 1976), p. 58. (Hereafter referred to as House hearings on Resource Development.).
8. *The Star*, 8 May 1965; *Rand Daily Mail*, 29 May 1965.
9. Statement before the Subcommittee on Africa, House Foreign Affairs Committee, 27 May 1976, p. 2.

10. TIAS 3885 (1957), 5129 (1962), 6312 (1967) and 7845 (1974).
11. *Africa Today*, August/September 1965.
12. Jim Hoagland in *Washington Post*, 16 February 1977.
13. Letter to Senator John Glenn from William A. Anders, Chairman of the Nuclear Regulatory Commission, 6 June 1975.
14. ibid., attachment 2.
15. ibid., attachment 3.
16. House hearings on Resource Development, p. 59.
17. Statement by G. Wayne Kerr, Chief, Agreements and Exports Branch, NRC, before the Subcommittee on African Affairs, House Foreign Affairs Committee, 27 May 1976, p. 6.
18. Statement by Myron Kratzer, State Department, before the Subcommittee, p. 4.
19. ibid.
20. Letter from B.D. Wilson of General Electric to the NRC, 1 June 1976.
21. Statement by Kerr, p. 29.
22. *Washington Post*, 15 May 1976.
23. Statement by Stephan Minikes, Senior Vice-President, Research and Communications.
24. Wilson, *loc.cit.*
25. *Le Monde*, 1 January 1976.
26. Statement by Kratzer, p. 4.
27. Statement by Minikes, pp. 3-4.
28. Nelson F. Sievering, Jr., Energy Research and Development Administration, in House hearings on Resource Development, p. 60.
29. The challenge is in the form of a suit on behalf of the Congressional Black Caucus and others, by the Natural Resources Defence Council, which earlier challenged American exports of nuclear materials to India.
30. *Washington Post*, 16 February 1977.
31. Letter to Senator Glenn, *op.cit.*, attachment 4.
32. Statement of G. Wayne Kerr, Chief of Agreements and Export Branch, US Nuclear Regulatory Commission, to the Subcommittee on African Affairs, Senate Foreign Relations Committee, 27 May 1976; attachment 1.
33. Letter from Senators Ribicoff, Percy and Glenn to James T. Lynn, Office of Management and Budget, 15 July 1975; letter from Lynn to Sen. Ribicoff, 15 September 1975.
34. *Congressional Record*, 22 September 1975, pp. H 10855-56.
35. *Africa Today*, August/September 1965.
36. Personal communication from John Fialka of the *Washington*

Star, based on ERDA documents obtained under the Freedom of Information Act.

37. Personal communications from South Africa students in the United States.
38. *Rand Daily Mail*, 13 March 1971.
39. *South African Scope*, September-October 1972, p. 9.
40. D.S. Greenburg, 'South Africa (1)', *Science* (Washington DC), 10 July 1970, p. 162.
41. *New York Times*, 1959, n.d.; interview with a former US intelligence official.
42. Annual report of the South African Atomic Energy Board, cited in *Cape Times*, 7 June 1967.
43. *Washington Post*, 28 May 1976.
44. ibid., and 16 February 1977.
45. Greenberg in *Science*.
46. House hearings on Resource Development, p. 16.
47. ibid., p. 19.
48. ibid., p. 18.
49. Rowland Evans and Robert Novak, column in *Washington Post*, 18 March 1977.
50. Proinsias Mac Aonghusa in *Sunday Press* (Dublin), 19 September 1976.
51. *The Guardian* (London), 24 May 1976.
52. *Baltimore Sun*, 21 May 1976.
53. *The Guardian*, 27 May 1977.
54. Statement before the Senate Foreign Relations Committee, 27 May 1976; from the uncorrected transcript.
55. Cited by Prof. John Marcum in House hearings on Resource Development, p. 9.
56. See for example the interview with Prime Minister Vorster in *Newsweek*, 17 May 1976.
57. Mr P.W. Botha in House of Assembly, quoted in *South Africa News* (Information Service of South Africa, New York), 10 May 1976.
58. *Washington Post*, 8 March 1977.
59. *Ta Kung Pao*, Hong Kong (a recognised outlet for intelligence leaks from mainland China), 30 October 1969; also *The Bonn-Pretoria Alliance*, Memorandum of the Afro-Asian Solidarity Committee of the German Democratic Republic (Berlin), 1967, Chapter 3.
60. *Nuclear Fuel*, 25 July 1977.
61. Interview on ABC TV's 'Issues and Answers', quoted in *New York Times*, 31 October 1977.
62. Interview on CBS TV's 'Face the Nation', quoted ibid.

63. *The Economist* (London), 22 October 1977.
64. R.W. Johnson, *How Long Will South Africa Survive?* (Macmillan, London, 1977), pp. 73-110, 214-67.
65. International Monetary Fund, confidential document: *Monthly Report of Gold Prices in World Markets*, DM/77/44, 23 May 1977, p. 3.
66. Johnson, *op.cit.*, pp. 234-36.
67. ibid., p. 4.
68. Quoted ibid., p. 4.
69. ibid.
70. ibid., p. 5.
71. For analysis of these loans, see James Morrell and David Gisselquist, *How the IMF Slipped $464 Million to South Africa* (Center for International Policy, Washington DC, 'Special Report', January 1978).
72. International Monetary Fund, Minutes of Executive Board Meeting 76/6, 21 January, 1976, pp. 13-15.
73. Registration statement to the Department of Justice, 6 February 1976.
74. Statement of 6 August 1975.
75. IMF, Minutes of Executive Board Meeting 76/156, 10 November 1976, pp. 12-13.
76. ibid., p. 15.
77. Johnson, *op.cit.*, p. 108N.
78. See *Washington Post*, 4 June 1976; *New York Times*, 2 June 1976; *The Observer*, 16 May 1976.
79. *The Guardian*, 29 April 1977.
80. United States Congress, Senate, Committee on Foreign Relations, *Nonproliferation Issues*, hearings on 19 March, 16 and 28 April, 18 and 22 July, 21 and 24 October, 1975; 23 and 24 February, 15 March, 22 September and 8 November 1976 (US Government Printing Office, Washington DC, 1977), pp. 123-59. (Hereafter referred to as Senate hearings on Nonproliferation Issues.).
81. *The Star* (weekly), 24 April 1976.
82. House hearings on Resource Development, p. 67.
83. ACDA official interviewed by the authors.
84. State Department official interviewed by the authors.
85. ACDA official.
86. *The Star* (weekly), 26 July 1975.
78. *Africa News* (Durham, N.C.), 31 July 1975.
88. A.L. Bethel, Vice-President, Westinghouse Corporation, in Senate hearings on Nonproliferation Issues, p. 135; Robert Alvarez, Environmental Policy Center, in House hearings on

Resource Development, p. 73. See also Richard J. Barber Associates, Inc., *LDC Nuclear Power Prospects, 1975-1990: Commercial, Economic, and Security Implications. Report prepared for the US Energy Research and Development Administration*, p. C-14.

89. *The Guardian*, 21 February 1977.
90. Letter to Senator Ribicoff from James Lynn, 15 September 1975.
91. ibid.
92. ibid.
93. ibid.
94. ibid.
95. For details, see *Washington Intrigue*, vol. 1, no. 1.
96. *Washington Post*, 28 May 1976.
97. See Barbara Rogers, *White Wealth and Black Poverty: American Investments in Southern Africa* (Greenwood Press, Westport, Conn., 1976).
98. Johnson, *op.cit.* describes some of the methods available.
99. Press statement by Rep. Les Aspin, Washington DC, cited in *Washington Post*, 14 April 1975, and elsewhere.
100. *New York Times*, 24 April 1975.
101. *Nucleonics Week*, 29 July 1976.
102. Interview with Jacob Scherr, Attorney, NRDC, Washington DC.
103. *Petition of the Natural Resources Defense Council, Inc., the Sierra Club and the Union of Concerned Scientists for leave to intervene before the US Nuclear Regulatory Commission in the Matter of General Electricity Company and GE Technical Services Operations, Inc.*, Docket No. 50-563, 28 May 1976.
104. Press statement by Senator Dick Clark, Washington DC, 18 May 1976.
105. *Petition of Congressman Charles Diggs et al. to intervene before the US Nuclear Regulatory Commission in the Matter of General Electric Company and GE Technical Services Operations, Inc.*, Docket Nos 50-563 and 70-2280, 28 May 1976.
106. Interview on ABC TV's 'Issues and Answers', quoted in *New York Times*, 31 October 1977.
107. For an analysis of South African propaganda in the US, see Barbara Rogers, 'Propaganda in America', in *The Great White Hoax* (The Africa Bureau, London, 1977), pp. 64-84.
108. *Rand Daily Mail*, 29 June 1974.
109. *The Star*, 1 February 1975.
110. *The Star* (weekly), 11 May 1974.
111. *Sunday Times* (Johannesburg), 13 April 1975.
112. Paragraph omitted for legal reasons.

113. Paragraph omitted for legal reasons.
114. Registration statement by the South Africa Foundation to the US Department of Justice, December 1975; Response to Item 11.
115. *Herald-Despatch*, 13 June 1974.
116. General S.L.A. Marshall, *South Africa: The Strategic View* (New York: American-African Affairs Association, 1967).
117. ibid.
118. *Africa News*, 14 November 1974.
119. See Harold Orlans, *Contracting for Atoms* (Brookings Institution, Washington DC, 1967).
120. Robert Alvarez, in House hearings on Resource Development, pp. 85-86.
121. *The Guardian*, 20 October 1976.
122. *Science*, 25 July 1975.
123. *Not Man Apart* (Sierra Club), December 1975; *Nucleonics Week*, 11, 18 and 25 September 1975.
124. Robert Alvarez, statement before the House Science and Technology Subcommittee on Nuclear and Fossil Energy Research and Development, 4 March 1977.
125. Barber Report, p. C-6.
126. ibid., pp. C-10 – C-22.
127. Senate hearings on Nonproliferation Issues, p. 184.
128. Robert Gillette in *Science*, 25 July 1975.
129. ibid.
130. *The Star*, 2 January 1971.
131. *Nucleonics Week*, 5 August 1976.
132. *Die Burger* (Cape Town), 25 August 1973; *Financial Gazette* (Johannesburg), 31 August 1973.
133. *Science*, 25 July 1975.
134. *Not Man Apart*, March 1976.
135. *Nucleonics Week*, 18 March 1976.
136. Peter Younghusband, 'Nuclear Power Politics', *European Community*, August-September 1976, p. 25.
137. *Wall Street Journal*, 15 March 1976.
138. News release, Office of Technology Assessment, US Congress, 4 April 1977. See also *Trends in Nuclear Proliferation* (The Hudson Institute).
139. Alvarez, statement of 4 March 1977, pp. 2, 6.
140. See Barbara Rogers, *White Wealth*, especially pp. 131-32.
141. See *Nucleonics Week*, and summary by Alvarez in House hearings on Resource Development, p. 72.
142. Personal communication by Michael Klare.
143. Anti-Apartheid Bewegung, *Western Nuclear Shield for Apartheid*,

(AAB, Bonn, 21 December 1977), p. 9.
144. *Nuclear Fuel*, 8 August 1977.
145. *Science*, 25 July 1975.
146. Michael T. Klare, *U.S. Arms Deliveries to South Africa: The Italian Connection*, TNI Special Report (The Transnational Institute/ IPS, Washington DC, February 1977).
147. Bruce Oudes in *Baltimore Sun*, 22 February 1976.
148. Dr Theodore B. Taylor, statement to the Senate Government Operations Committee, 29 January 1976.
149. South African *House of Assembly Debates*, May 1968, cited in Abdul Minty, *South Africa's Defence Strategy* (Anti-Apartheid Movement, London, 1973), p. 6.

Chapter Nine
1. H. E. Schimmelbusch and H. Hardung-Hardung, 'The Manufacture of Fuel Elements for the FRG', *Atomwirtschaft*, December 1957, pp. 421-22.
2. *Fourth Atomic Programme of the Federal Republic of Germany for 1973-1976* (Federal Ministry for Research and Technology, Bonn, 1974), pp. 49-50.
3. *Kerntechnik*, December 1975, p. 521.
4. *Fourth Atomic Programme*, p. 49.
5. *The Bulletin of the Press and Information Office of the Government of the Federal Republic of Germany*, 4 August 1976, p. 2.
6. ibid.
7. *Agreement between the United Kingdom of Great Britain and Northern Ireland, the Federal Republic of Germany and the Kingdom of the Netherlands on Collaboration in the Development and Exploitation of the Gas Centrifuge Process for Producing Enriched Uranium, Almelo, 4 March 1970*. Cmnd. 4315. The Agreement was signed the day before the NPT went into effect.
8. Centec press release, 2 August 1971; *Atom*, October 1970, pp. 226-27.
9. Thomas Tuohy (former Chief Executive of Urenco and Centec), paper presented at the International Conference on Uranium Enrichment, Reston, Va., 23-26 April 1974.
10. L. N. Chamberlain, J. A. B. Gresley, J. E. E. Meilof, G. E. Steele, F. Stockschlaeder, 'Operational Experience on Centrifuge Enrichment Plants', paper presented at the International Conference on Uranium Isotope Separation, London, 5-7 March 1975.
11. *Fourth Atomic Programme*, p. 50.
12. Annual Report of the DSM Co. (Nederlands Staatsmijnen), 1974, p. 17.

13. Peter Boskma, Wim Smit, Gerard de Vries, *Uranium Enrichment: History, Technology, Market* (Twente Technical University, Enschede, 17 June 1975), p. 123.
14. ibid., pp. 123-24.
15. David Fishlock and Robert Mautner, 'Behind the Gas Centrifuge Argument', *Financial Times* (London), 14 November 1969.
16. *Proceedings of the Second Chamber 1967-68, Report on the Debate Concerning Uranium Enrichment*, 3 April 1968, p. 1719. The foreign policy aspects of the Dutch gas centrifuge programme are discussed in Wim Klinkenberg's *The Gas Centrifuge 1937-70: Hitler's Bomb for Strauss?* (Van Gennep, and Baarn: In Den Toren, Amsterdam, 1971).
17. See Barbara Rogers, *The Expansion of Foreign Oil Companies in Southern Africa* (National Council of Churches, New York, 1976).
18. 'Gas Centrifugation Behind VMF Move', *Algemeen Handelsblad* (Amsterdam), 6 January 1969.
19. *Nucleonics Week*, 1 November 1976.
20. Editorial, *NRC/Handelsblad*, 7 March 1973.
21. *Hengelo's Dagblad*, 25 January 1973.
22. ibid.
23. 'Brazil Wants Almelo Uranium', *De Volkskrant* (Amsterdam), 1 December 1976.
24. *The Guardian*, 4 and 6 March, 1978.
25. Letter from Professor Dr J. Blok to Prof. Dr J. Verkuyl, 1 November 1975.
26. *Sechaba* (London), 4th quarter 1976, p. 49.
27. Statement of chief Federal Government spokesman Bölling, Bonn, 6 October 1975.
28. The sources cannot be named.
29. Stephen Salaff, *The Anglo-Dutch-West German Consortium for the Enrichment of Uranium*, unpublished working paper, p. 46.
30. ibid., pp. 50-51.
31. Netherlands Government, Memorandum of Reply, *Proceedings of the Second Chamber 1970-71*, p. 4.
32. 'German Centrifuge Plant', *Nuclear News*, January 1977, pp. 71-72; 'West Germany Ready with Uranium Plan', *De Volkskrant*, 3 February 1977.

Chapter Ten   "Israel Joins ...."   p. 311 - 312
1. In an interview with Raphael Basham of *Yediot Aharanot*, 16 August 1973.
2. A. Hodes, 'The Implications of Israel's Nuclear Capability', in

Wiener Library Bulletin, vol. XXII, Autumn 1968.

3. *The Times* (London), 12 December 1960.

4. *Israel-Arab States, Atom-Armed or Atom-Free*, (AMIKAM, Tel Aviv, 1963, in Hebrew), quoted by S. Flapan in his essay 'Israel's attitude towards the NPT' in *Nuclear Proliferation Problems* (SIPRI, Stockholm, 1974): main source for the historical background of Israel's nuclear programme, as summed up in this chapter.

5. S. Flapan, *op.cit.*, p. 274.

6. *New Outlook . . .*, vol. 5, no. 9, November-December 1962.

7. S. Flapan, *op.cit.*, p. 281.

8. *New York Times*, 5 June 1965.

9. Quoted by S. Flapan, *op.cit.*

10. *The Jewish Observer* (London), 2 July 1965.

11. Alon, Y., *Jewish Observer* (London), 24 December 1965.

12. *New York Times*, 7 January 1966.

13. *op.cit.*, 13 February 1966.

14. *op.cit.*, 28 February 1966.

15. *Yediot Ahronot* (Tel Aviv), 4 April 1966, quoted by S. Flapan, *op.cit.*

16. *New York Times*, 19 May 1966.

17. G. H. Quester, 'Israel and the Nuclear Non-Proliferation Treaty', in *Bulletin of the Atomic Scientists*, June 1969, p. 7.

18. L. Beaton, *New Middle East*, April 1969, p. 9.

19. J. B. Bell in *Middle East Journal*, Autumn 1972, p. 382.

20. S. Flapan, *op.cit.*

21. *SIPRI Yearbook 1977*.

22. 5 May 1977.

23. 12 September 1977.

24. 29 May 1968.

25. It should be recalled that until 1945 the South African Nationalist Party had pursued a blatantly anti-Semitic policy (from 1948 relations between the South African Jewish community and the Nationalist Government were more cordial). Israel, on its part, had in 1956 launched a diplomatic offensive to win Black Africa and in 1961 voted with the Third World countries for the UN Resolution condemning *apartheid* as 'being reprehensible and repugnant to the dignity and rights of peoples and individuals'. (Quoted by Richard P. Stevens in his analysis 'The Paradoxical Triangle (Zionism, South Africa and Apartheid)', in *Israel-South Africa — co-operation of Imperialistic Outposts*, (Progress Dritte Welt (PDW), Bonn, 1976).

26. The wedge between African states and Israel, which also drew

Israel closer to South Africa, was the Israeli position on a number of African questions. For example, Israel voted against Algerian independence in the UN in 1965, and against the UN programme to hold general elections in Cameroon in 1959. Israel also voted against censuring France's explosion of a nuclear device in the Sahara. In 1960, Israel abstained in the voting for the independence of Tanganyika, Rwanda and Burundi, only one year after voting against the Liberian proposal to grant self-rule to African colonies. Israel's support of secessionist movements in Africa also earned her some enmity. She supported Biafra in Nigeria. On the Congo crisis, she took an ambivalent stand. Africans also came to hear about the visit of Thomas Tshombe (brother of Moise Tshombe, leader of the Katanga movement) to Israel, during which he claimed to have discussed with Israeli officials the recognition of an independent Katanga. Israel also abstained on seating the Congolese delegation at the UN. She actively supported secessionist activities in southern Sudan.

A critical aspect of Israeli activities in Africa has been Israel's collaboration with the Portuguese effort to maintain its colonial rule in Guinea-Bissau, Mozambique and Angola. Louis Cabral (brother of Amilcar), of the political bureau of the African Party for the Independence of Guinea and Cape Verde (PAIGC) said in a statement: 'regarding the role played by the US imperialists, Zionism and NATO in support of Portugal, I would like to remind you . . . that most of the arms of the Portuguese are Israeli'. (Quoted by H. S. Rogers, in 'Imperialism in Africa', *Black Scholar*, January 1972.)

The dilemma of Israeli policy was that it could not at the same time fight the Arab boycott against Israel and support the African boycott against South Africa, the latter being the crucial question for Africans. Contrary to widespread opinion that Africa broke relations with Israel under the pressure of the oil-boycott, this process had started much earlier (see Table below). It was initiated at the OAU's 1973 Summit conference. The following table shows the breaking off of diplomatic relations with Israel by African countries:

Guinea	6.6.1967	Madagascar	20.10.1973
Uganda	30.3.1972	Central African	
Chad	28.11.1972	Republic	21.10.1973
Congo	5.12.1972	Ethiopia	23.10.1973
Niger	10.12.1972	Nigeria	25.10.1973
Mali	5.1.1973	Gambia	26.10.1973
Burundi	16.5.1973	Zambia	26.10.1973

Togo	21.9.1973	Ghana	28.10.1973
Zaire	4.10.1973	Senegal	28.10.1973
Benin	6.10.1973	Gabon	30.10.1973
Rwanda	9.10.1973	Sierra Leone	30.10.1973
Upper Volta	11.10.1973	Kenya	1.11.1973
Cameroun	15.10.1973	Liberia	2.11.1973
Equatorial Guinea	15.10.1973	Ivory Coast	8.11.1973
Tanzania	18.10.1973	Botswana	13.11.1973

27. For a detailed analysis of the economic and military co-operation between Israel and South Africa see Birgit Sommer, 'Military-economic collusion', in *Israel-South Africa: Co-operation of Imperialist Outposts* (PDW, Bonn, 1976).

28. Source: *Israel-South Africa: Co-operation of Imperialist Outposts, op.cit.* See also Appendices.

29. *Jerusalem Post Weekly*, 13 April 1976.

30. *Israel-South Africa: Co-operation of Imperialist Outposts, op.cit.*

31. ibid.

32. ibid.

33. ibid.

34. For a detailed analysis of Israel's strategic thinking, see F. Jabber, *Israel and Nuclear Weapons* (Chatto and Windus, London, 1971).

35. Quoted by Colin Legum in *Africa Contemporary Record 1970-71* (Rex Collings, London), p. B 216.

Chapter Eleven

1. The OAU Delegation in Teheran in 1977.

2. *SIPRI Yearbook 1977*, p. 321.

3. See John K. Cooley, 'More fingers on nuclear trigger? Despite denial, Iran believed ready to move with own plans for atomic weapons', *Christian Science Monitor*, Boston, 25 June 1974.

4. *SIPRI Yearbook 1977*, p. 228. The total volume of Africa's arms trade in 1975 stood at $9280 m.

5. ibid., p. 316.

6. 16 June 1976.

7. 4 November 1975.

8. No. 253 of 1977.

9. 20 July 1976.

10. 23 June 1974.

11. Agence France Press report published by *International Herald Tribune* (Paris), 24 June 1974, under the title 'A-weapons for Iran Soon: Shah'.

12. *Africa Contemporary Record 1970-71*, ed. Colin Legum (Rex Collings, London, 1972).

Notes

13. *African Development*, September 1974, p. 45.
14. *New York Times*, 28 June 1974.
15. *International Herald Tribune*, 3 July 1975: 'Brazil adopts a "bifocal" view toward US'.
16. *International Herald Tribune*, 2 June 1975: 'W. German Firm to provide Nuclear capability to Brazil'.
17. Argentina concluded bilateral agreements on the peaceful uses of energy with Canada (30 January 1976) and with the USA (25 July 1969).
18. *Der Spiegel*, 31 January 1977: 'Bombengeschäft – Geschäft mit der Bombe?'.
19. On 26 April 1977, in a letter addressed to the Foreign Minister Hans-Dietrich Genscher, 98 scientific workers of the Hahn-Meitner Institute for Nuclear Research in West Berlin protested against the agreement with Brazil on the grounds that 'it provides the Brazilian military régime with a nuclear capability which the régime had openly admitted it wished to acquire.' 'Wissenschaftler halten Atomvertrag mit Brasilien für gefährlich', *Frankfurter Rundschau*, 27 April 1977.
20. *Der Spiegel*, 31 January 1977.

Index

NOTES 1. References to countries are in bold type as follows: **(B)** Brazil, **(C)** Canada, **(F)** France, **(I)** Israel, **(N)** Netherlands, **(S)** Switzerland, **(UK)** United Kingdom, **(US)** United States, **(WG)** West Germany. 2. The Notes on pages 412-49 have not been indexed.

Names

450

Index

Subjects and Places

Index

Index

461

Enrichment Corporation (SA)
122-3; **hexafluoride ('hex')** 35, 75,
113-4, 153, 157, 184-5, 203, 248,
346; **hijacked** 322-24, 327; **supply**
also Namibia, Rössing
Urenco (Centec) 271, 299-310 *and*
see 'Troika'
Uruguay 127
US Nuclear Inc. (US) 250
US Steel Corporation (US) 115, 250
Utah Mining (US) 115

Vacuumschmelz (WG) 96
Van de Graaff accelerator 163, 313, 332
Valindaba (SA) 7, 10, 160, 173, 179-80, 191, 203
Venezuela 81
Vereinigte Stahlwerke (WG) 33
Vernon Craggs (US) 251
Vietnam 226-7, 275
VMF (N) 305-6
Volksblad, Die 222
Volkskrant, Die 306

Waffen und Luftrüstung AG (WG) 235
Wall Street Journal 176
Warsaw Pact 344 *see also* Soviet Union
Washington Post 77, 195, 198, 203, 329-30

Wehrmacht 16
Wehrtechnik 100, 311, 317-18
Weizman Institute (I) 311
West Germany *see* Germany
WEU *see* Western European Union
Western Allies and South Africa 170, 336
Western defence 14-16, 80, 336-7, 345
Western European Union (WEU) 22, 29
Western Knapp Engineering (US) 127,296
Westinghouse (US) 247, 270, 276, 291-2, 295-8, 338
'Witleveen Facility' 265
Witwatersrand, University of 208, 252
World Armaments: the nuclear threat 393
World Bank 266, 268
World Nuclear Fuel Market (US) 290
World War Two 11-12, 24, 30-5, 51, 91, 109,180, 341-2

Zaire xv, 117, 241, 278-82
Zambia 90-3, 117, 234, 328
Zeit, Die 29
Zippe 37
Zululand 233